T0314270

Resilient Power Electronic Systems

Resilient Power Electronic Systems

Shahriyar Kaboli
Department of Electrical Engineering, Sharif University of Technology, Tehran, Iran

Saeed Peyghami
Department of AAU Energy, Aalborg University, Aalborg, Denmark

Frede Blaabjerg
Department of AAU Energy, Aalborg University, Aalborg, Denmark

Registered Offices
John Wiley & Sons, Inc., 111 River Street, Hoboken, NJ 07030, USA
John Wiley & Sons Ltd, The Atrium, Southern Gate, Chichester, West Sussex, PO19 8SQ, UK

Editorial Office
The Atrium, Southern Gate, Chichester, West Sussex, PO19 8SQ, UK

For details of our global editorial offices, customer services, and more information about Wiley products visit us at www.wiley.com.

Wiley also publishes its books in a variety of electronic formats and by print-on-demand. Some content that appears in standard print versions of this book may not be available in other formats.

Library of Congress Cataloging-in-Publication Data

Names: Kaboli, Shahriyar, 1975– author. | Peyghami, Saeed, author. | Blaabjerg, Frede, author.
Title: Resilient power electronic systems / Shahriyar Kaboli, Saeed Peyghami, Frede Blaabjerg.
Description: Hoboken, NJ : Wiley, 2022. | Includes bibliographical references and index.
Identifiers: LCCN 2021062788 (print) | LCCN 2021062789 (ebook) | ISBN 9781119772187 (cloth) | ISBN 9781119772194 (adobe pdf) | ISBN 9781119772200 (epub)
Subjects: LCSH: Power electronics. | Electric current converters.
Classification: LCC TK7881.15 .K335 2022 (print) | LCC TK7881.15 (ebook) | DDC 621.31/7–dc23/eng/20220124
LC record available at https://lccn.loc.gov/2021062788
LC ebook record available at https://lccn.loc.gov/2021062789

Cover Design: Wiley
Cover Images: © IM_photo/Shutterstock; Sean McAuliffe/Unsplash; Graph drawn by the authors

Set in 9.5/12.5pt STIXTwoText by Straive, Pondicherry, India
Printed and bound by CPI Group (UK) Ltd, Croydon, CR0 4YY

C9781119772187_290722

Contents

Preface

The application of power electronics in industries has grown considerably in recent years due to a global paradigm shift to carbon-free energy technologies such as renewable energy resources, micro- and smart-grids, and e-transportation, which strongly depend on power electronics. Therefore, modern electronic-based power systems are almost young in comparison with conventional power systems. Regarding the wide usage of the above-mentioned systems in industries, an estimation of their effective operative life and reliability is considered to be crucial. Furthermore, penetration of power electronics converters is increasing in power systems and the classical reliability assessment tools and concepts in power systems need to be modified taking into account the reliability of power converters. One of these concepts is resilience. Resilience is the property that enables a system to continue operating properly in the event of the failure of some of its components. As the application of power electronic converters becomes more and more crucial in the coming years, the necessity of having non-stop operation in converters is undergoing rapid growth. In some industries, any stop in operation of power converters leads to a great penalty. On the other hand, a power electronic converter is faced with many internal and external faults. These faults can interrupt the continuous operation of the power converter. Therefore, implementing smart techniques for non-stop operation of power converters are high in importance. In other words, power electronic systems must be resilient in mission critical applications.

This book deals with resilience and effective operative life concepts in the field of power electronics. Resilience is almost the only method for achieving a desired reliability in a converter that operates with non-zero fault possibility. However, resilience is a bit different from fault tolerance. In some applications, resilience means achieving "zero" probability of failure with regards to the mission profile. In this book, advanced methods for resilient power electronic converters are presented. Furthermore, the fault mechanism is explained to determine the reason for failure in power converters. Finally, various methods are presented to improve the resilience of the power converters. The following aspects are covered and discussed in this book:

- Analysis of failure mechanisms in power electronic converters including internal and external faults are described.
- Fault prognosis and diagnosis concepts are described and the certain methods for fault detection in power electronics are presented.
- Advanced techniques for reducing the noise effect on the fault detection and prognosis are explained.
- Classic methods of resilience are reviewed and a comprehensive perspective is provided.
- Advanced methods of resilience are presented regarding the mission of the power electronic converter.

- Reconfiguring of a power electronic converter as a design consideration is presented.

The book will have the following specific objectives:

- Enhancement of the knowledge on the subject of resilience in power electronics converters.
- Explore the challenges and concerns of resilience and fault tolerance in power electronics.
- Introduce some new concepts such as the tradeoff between the resilience and efficiency in power electronics.
- Show the basic principles and provide guidelines for design of a resilient power converter.
- The book will lead to the advancement of the current state-of-the-art advantages of resilient power converters. Moreover, the book will generate relative practical case studies and experimental results for most of the chapters.

The book has been prepared in four parts. This division helps the readers to follow the contents easily. The first part, *Resilient Power Electronic Systems*, presents a general view of the resilience concept in power electronic systems and is contained in four chapters. Chapter 1 provides an introduction for entering the concept of resilience in power electronics. In this chapter, the resilience concept is described in its general form and mission critical systems are introduced. In this chapter, the reader is also introduced to the techniques that are used in industries and nature for reaching a resilient performance. The contents of this chapter are used in the following chapters, with a focus on the appearance of these concepts in the field of power electronics. In Chapter 2, we introduce the concept of resilience in power electronic converters with an introduction to these devices and a recognition of their main functions as well as their importance. Some typical industrial examples are presented and the elements of power electronics are introduced. We used this introduction to describe the reasons for faulty conditions in power electronic converters, described in the next part of the book. Finally, mission critical power electronic converters are introduced. In Chapter 3, the resilience concept is described in power electronic systems. The possible faulty conditions are explained in a power electronic system and the conditions for supporting the load during the fault are presented. Internal and external faults are explained and their effects on the converter resilience are presented. The requirements for resilience of a power electronic system are described. The main part of this chapter deals with the difference between fault tolerance and resilience. In Chapter 4, a survey is presented about the state of the resilience in power electronic converters. In the second part of the book, *Useful Life of the Power Electronic Systems*, the failure mechanisms of the power electronic systems are presented. In Chapter 5, the concept of useful life and the methods for useful life modeling are described. These definitions are used to group the faults. This chapter provides a quantitative view to the reader about evaluation of the system useful life and can be used in the next chapters for achieving the resilient characteristics. In Chapter 6, internal faults of the power electronic systems are reviewed at the converter level, where the main important issues at the design and montage stages of the converters are presented. In Chapter 7, the random faults and wear-out failures of the power electronic systems are discussed. Various types and reasons for wear-out failures are presented and packaging of the power electronic modules is explained. Thermal and mechanical shocks, which are two important factors of wear-out failures, are described. In Chapter 8, the external faults that lead to unavailability of the power electronic converters are described. It is shown that these faults act as stressors and affect the lifetime of the converter components. On the other hand, the external faults interact with the protection system of the converter and lead the converter to be out of service. The right decision in the external fault period is explained. In Chapter 9, the availability of electric power converters is described. One of the most important factors for this undesired state is the influence of noise.

In this chapter, electromagnetic interference and certain methods for reducing its undesired effects on electric power converters are presented. Implementation of the methods for resilience achievement needs to have enough information about the condition of the converter. The third part of the book, *Health Estimation of the Power Electronic Systems*, presents the methods for system health monitoring. In Chapter 10, commonly used methods for condition monitoring the converters are presented. In Chapter 11, the methods of fault prognosis in power electronic systems are described. It is important to have an expectation about the useful life of a system before its construction or even its remaining useful life during its operation. The fault prognosis in the power electronic systems are presented in both converter-level and element-level categories. In Chapter 12, fault diagnosis in the power electronic converters is described. Two goals of the fault diagnosis, fault isolation and fault root cause analysis, are explained. Some of the methods for fault diagnosis in power electronic systems are presented. In the last part of the book, *Methods of Resilience in Power Electronic Systems*, guidelines for achieving resilience are presented. These methods are used in both design and operation processes of the converter. In Chapter 13, methods for reducing the stresses on the power electronic systems are described at both system and component levels. Algorithms for derating a faulty power supply are described. In Chapter 14, resilient operation of power electronic converters against external faults such as a load short circuit is studied. The subject of this chapter is the converters that are not damaged but cannot operate normally. In this chapter, the availability of electric power converters as a most important parameter in the topic of resilience is described. In Chapter 15, some of the methods and techniques for inherently resilient operation of the power electronic converters are reviewed. In these cases, the failure factor is applied to the converter but its effect is not sensed by the converter. The main requirement of resilient operation is a short recovery time and a small drop in the system performance index. One of these methods is the application of fault-tolerant structures for the power converters. Applications of active replacement methods and usage of highly reliable elements are described.

This book is a good guide for the researchers, senior undergraduate and graduate students, and professional engineers related to this field to investigate these topics. The book shows them what challenges they will face when they tend to operate resiliently, and provide some suggestions on how to solve the related problems. Although the fundamentals of resilience in power electronics will be discussed, this book focuses on the advanced methods of resilience enhancement and will provide the analysis and test results of nearly every technique described in the book. The book is useful for researchers, scientists, professional engineers, and graduate students studying power electronics and renewable energy as their major in postgraduate levels, and is especially useful for researchers and engineers majoring in power electronics. The prerequisite is basic courses in power electronics and control theory. In previously published books, an in-depth and comprehensive presentation of resilience in power electronics has not yet been provided. It is clear that this book will be distinct from existing ones and an excellent and important addition to an existing library. The book references are mainly previously published papers by the authors, who are specialists in the book's subject and benefit from experiences of many years working in the field of power electronics. All of the references are the most cited papers in this field. We hope to provide the readers with an exciting and knowledgeable book in the field of resilient power electronic systems.

About the Companion Website

This book is accompanied by a companion website.

www.wiley.com/go/kaboli/resilientpower

This website includes:

- Presentation Slides (PPTs)

Part I

Resilient Power Electronic Systems

1

Resilient Systems

1.1 Introduction

One of the most tragic accidents is that of a commercial passenger aircraft crash. Picturing the death of tens of children, women, and men is extremely horrifying and dreadful. Even though commercial passenger aircrafts are constructed with the utmost reliability with very low accident statistics, the occurrence of even one accident is not acceptable. A survey of aircraft crashes shows that most accidents occur near origin or destination airports. This fact leads to the necessity for a very important capability of the aircrafts: resilience. If the airplane is resilient it can withstand faults until a safe landing point is found. This safe point is usually the nearest airport. Figure 1.1 shows a conceptual diagram about this situation. The good path is for the aircraft with a resilient capability as this aircraft is able to protect its passengers. However, a non-resilient aircraft flies through the bad path and can lead to a tragedy.

In this example, three concepts are seen:

- A system with a critical mission
- A faulty condition
- Resilient operation of the system

The aircraft here is a mission critical system (with the safety of hundreds of lives) and the resilience capability helps to pass the mission after a faulty condition. It is the key phrase of this book: *Passing the faulty condition without failing in the mission.* This means that the topic of this book is not about reliable systems! The topic is about the reliable systems that have the capability to survive a faulty situation. The above-mentioned aircraft may be very reliable, which means that it has very rare faults, but this does not mean that it is necessarily resilient. A reliable and non-resilient system may not necessarily finish its mission in a faulty condition as even its faults have a low probability of occurrence.

As this chapter deals with resilience in its general form, we continue with an explanation about the definition of systems and their characteristics. To enter the subject of resilience, these concepts are now described.

1.2 Definition of a System

Our world consists of systems. A system is an integrated collection of various parts to meet a goal. The goal of the system usually appears in the output(s) of the system. The systems also accept one or more inputs that provide the required information and energy for the system [1]. The system

Resilient Power Electronic Systems, First Edition. Shahriyar Kaboli, Saeed Peyghami, and Frede Blaabjerg.
© 2022 John Wiley & Sons Ltd. Published 2022 by John Wiley & Sons Ltd.
Companion website: www.wiley.com/go/kaboli/resilientpower

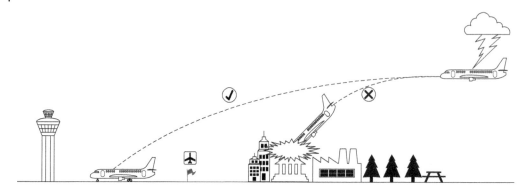

Figure 1.1   A basic diagram about the resilience concept for a mission critical system.

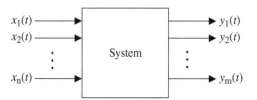

Figure 1.2   A system block diagram with some inputs and outputs.

inputs and outputs determine the system boundaries, as shown in Figure 1.2. In this figure, the shown system has n inputs, i.e. $x_1(t)$ to $x_n(t)$, and m outputs, i.e. $y_1(t)$ to $y_m(t)$. The inputs and outputs may be constant or time variant.

1.2.1   Elements of the System

The systems may be very simple and consist of a few parts. For example, a heater is a simple system consisting of a resistor. The input voltage is dissipated in the resistor and generates an amount of heat as the system output. On the other hand, there are more complex systems with many parts. A complex system may consist of some smaller systems where each acts as a part of the complex system [1]. Figure 1.3 shows a complex system consisting of some smaller systems. Each of these smaller systems may include other smaller systems. For example, a computer is a complex system consisting of several smaller systems: the mother board, memories, hard disk, graphic peripherals, monitor, and power supply. Figure 1.4 shows a Laptop as a system that consists of some smaller systems, such as memory, power supply, etc. Decomposition of a system into the smaller sections (systems) leads to the elements. An element is defined as a part that is not decomposable to smaller sections [1]. For example, in the power supply of a computer, transistors of the voltage regulators are considered as elements. It is obvious that the system decomposition can be continued to much lower layers of the system. A transistor can be considered as a system consisting of layers of semiconductors. Each semiconductor layer can be decomposed to atoms if it is considered as a system. Therefore, an important question is: what is the stop point of this system? The answer comes from another question: what is the required system study level? We can stop this process if the defined elements satisfy the accuracy of the study [1].

1.2.2   System Performance Criteria

The systems are designed based on the required mission profile(s). A high-quality system tries to close its output(s) to the planned mission target as much as possible [2]. System performance

Figure 1.3 A system consists of some subsystems.

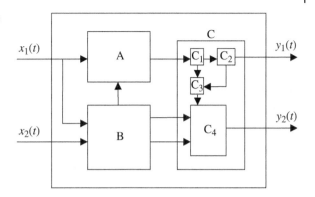

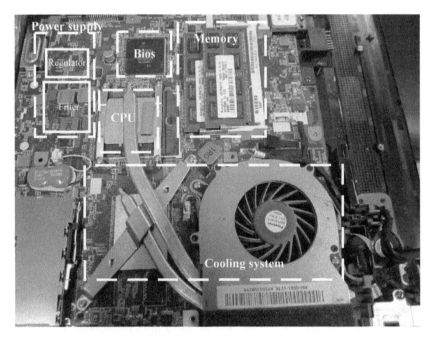

Figure 1.4 Various subsystems in a laptop.

criteria are quantitative indices used to characterize the performance of the system relative to its alternatives [2]. The difference between the pre-defined value of a system performance index and its respective actual value is the system error. A high-quality system keeps its errors close to zero. There are three different categories of system performance indices, as listed in the following:

- The indices about the system output(s)
- The indices about the system input(s)
- The indices about the relation between the system input(s) and output(s)

Figure 1.5 shows two different cars at the same level. Table 1.1 summarizes the performance indices of these cars in order to compare them [3]. The power is the performance index of the cars. The mileage is the index about the relation between the cars' input, the fuel volume, and the cars' output, the passed distance.

(a)   (b)

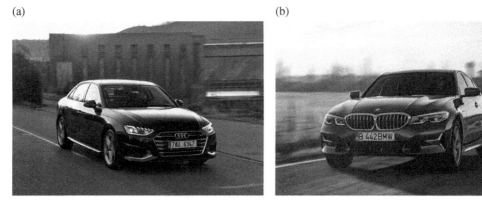

Figure 1.5 Two different cars at the same level. *Sources:* Jan Kliment/Adobe Stock; Gabriel/Adobe Stock.

Table 1.1 A comparison between the characteristics of two cars.

	Audi A4	BMW 3 Series
Power (bhp)	188	255
Mileage (km l^{-1})	17.84	16.13
Engine (cc)	1984	1998

1.2.3 System Useful Life and Reliability

The system useful life is an estimate of the time it is likely to remain in service. It is important to have an expectation about the useful life of a system before its construction or even its remaining useful life during its operation [4]. The useful life prediction is a tool for this goal. The useful life can be defined based on the performance degradation of the system. In this definition, the useful life is defined as the time interval to the point where the system performance falls below a threshold, as shown in Figure 1.6. In many cases, the useful life is defined as a probability. The useful life is the probability of performing adequately to achieve the desired aim of the system. There are useful life prediction techniques that depend on the knowledge about design. As more details of the design are known, more accurate methods become available. These methods use part failure rate models, which predict the failure rates of parts based on various part parameters, such as technology, complexity, package type, quality level, and stress levels. Predictive methods attempt to predict the useful life of a part based on some model typically developed through empirical studies and/or testing [5, 6]. An attempt is made to identify critical variables such as materials, application environmental and mechanical stresses, application performance requirements, duty cycles, and manufacturing techniques. Typically, a base failure rate for the component is assigned, which is then multiplied by factors for each critical variable identified. Some predictive models assume a constant failure rate over the lifetime of a product. This ignores higher failure rates typically seen at the beginning and end of the component life, infant mortality, and wear-out, respectively [6]. Predictive methods can provide a relatively accurate reliability estimate in cases where good studies have been done to analyze field failures. Reliability is the probability of performing adequately to achieve the desired aim of the system. This can be mentioned as a time-dependent equation. The reliability concept has more importance in specific applications, such as in space and military

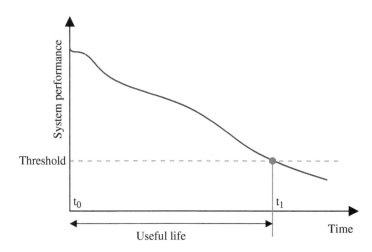

Figure 1.6 Definition of the useful life concept base on the system performance degradation.

equipment, where on a mission equipment can hardly be replaced or be performed by another system instead of the failed part. In order to improve the system reliability, different research has been carried out and several methods have been introduced. Fault occurrence is a relatively random phenomenom. Randomness means a lack of pattern or predictability in events. Therefore, essential methods of reliability prediction are based on probability analysis. In statistics, a random variable is an assignment that has a numerical value for each possible outcome of an event space. This association facilitates the identification and the calculation of probabilities of the events. Random variables can appear in random sequences. A random process is a sequence of random variables describing a process whose outcomes do not follow a deterministic pattern, but follow an evolution described by probability distributions [7]. These and other constructs are extremely useful in probability theory and the various applications of randomness. It is usual for some kinds of merits such as efficiency to be widely accepted by many users. However, reliability is less applicable than these labels.

In order to improve system reliability, different research projects have been done and several methods have been introduced. Many of these methods need information about the failure rate of the system. Each system contains a number of components. One method to enhance reliability is improvement of component reliability. This goal may be achieved by component-specific derating or by improvement of component specifications. At this level, the failure rate of the components should be modeled properly. The other method is to use a redundant or fault-tolerant system in which, after a partial fault, the rest of the system can work adequately to achieve the goal of the whole system. At this level, a systematic analysis of reliability is necessary. For example, a systematic analysis indicates that one problem of using a redundant system is load balancing. The other method that may be useful is a comparison of various possible topologies or different operating conditions to choose a proper state to achieve the goal of design. Because some systems are not available to be repaired or maintained, maintenance can rarely be used in these systems. Also, this method is not meaningful for each system element. For example, only a fan can be maintained to work properly and do its duty for cooling a car engine while for an injector, maintenance does not have a practical meaning. Rather than a theoretical reliability evaluation with standards, there are some accelerated or aging tests that manufacturers use to evaluate the reliability of their products.

In this method, at each test, one or some parameters of environmental conditions are stressed more than a typical state in order to reduce the test time less than the real state. There are some determined relations between these accelerated test results and typical condition results that are used for finding failure rates and reliability evaluations.

1.3 Resilience Concept

In a system, resilience is defined as the capability to recover from an abnormal state to another state where this new state guarantees the continuance of the operation of the system. According to this definition, there are some key points in the resilience concept:

- Normal state
- Abnormal state
- New state

In an ideal condition, the new state is the previous state of the system before failure. However, as shown in Figure 1.1, the new state is defined based on the mission profile of the system. It may be just a safe point in a damaged aircraft. The bomber B-17 shown in Figure 1.7 is a very famous example of this situation. The B-17 Flying Fortress became symbolic in the United States of America's air power and many of them came back to the origin points in a huge damaged state, as shown in Figure 1.8. The *All American* was a World War II Boeing B-17 Flying Fortress bomber aircraft that was able to return safely to its base after having its rear fuselage nearly cut off by an in-flight collision with a German BF-109 over enemy-held territory. The bomber's flight is said to have yielded one of the most famous photographs of World War II, as shown in Figure 1.9 [8].

In the fields of engineering and construction, resilience is the ability to absorb or avoid damage without suffering complete failure and is an objective of design, maintenance, and restoration for buildings and infrastructure, as well as communities. In this regard, resilience has a close meaning to availability. The concept of availability was originally developed for repairable systems that are required to operate continuously and are at any random point in time either operating or down because of failure and are being worked upon so as to restore their operation in a minimum time. Availability means the probability that a system is operational at a given time, i.e. the amount of

Figure 1.7 The famous Bomber B-17. *Source:* Carlo Borella/Unsplash.com.

(a)   (b)

Figure 1.8 Two damaged B-17 bombers that came back to the origin airport, (a) during flying, (b): after landing. *Sources:* Aviation C2017/Alamy Images; US Air Force/Wikipedia Commons/Public Domain.

Figure 1.9 The *All American* Flying Fortress bomber aircraft. *Source:* US Air Force/Wikipedia Commons/ Public Domain.

time a device is actually operating as the percentage of total time it should be operating. High-availability systems may report availability in terms of minutes or hours of downtime per year. Availability features allow the system to stay operational even when faults occur. A highly available system would disable the malfunctioning portion and continue operating at a reduced capacity. In contrast, a less capable system might crash and become totally non-operational. Availability is typically given as a percentage of the time a system is expected to be available. Availability of a system is typically measured as a factor of its reliability—as reliability increases, so does availability. A resilient control system is one that maintains state awareness and an accepted level of operational normalcy in response to disturbances, including threats of an unexpected and malicious nature. In computer networking, resilience is the ability to provide and maintain an acceptable level of service in the face of faults and challenges to normal operation.

Figure 1.10 shows the time diagram of a resilient system. When a fault occurs in a system, the system performance index falls down. Both the resilient system and fault-tolerant systems try to keep the index during the fault and then try to recover the system performance index. The key differences between a resilient system and a fault-tolerant system are the value of the system performance index drop and the time of adaptation and recovery phases. In a mission critical system, the values of this index drop and time interval are very important. For example, in a power supply, the recovery time of the output voltage after a short-circuit fault determines whether the system is resilient or just fault tolerant.

Resilience has two important aspects: (i) system performance and (ii) transition time. Figure 1.10 shows the phases of resilience during a fault. The vertical axis shows the level of resilience, while time describes the transition time as:

- Phase 1: Resilient State
  The resilience begins before the fault. Reducing the impact of the fault helps the system to reduce its quality drop and the transition time.
- Phase 2: Post-event Degraded State
  In this phase, the system performance falls and resilience determines its depth. The system capabilities for resilient operation such as redundancy and reconfiguration reduce the system degradation.
- Phase 3: Restorative State
  In the restorative state, the key factors are fast response and short recovery time.
- Phase 4: Post-restoration State
  After the restorative actions, the system enters into a post-restoration state. It may take a longer time depending on the impact of the fault.

Thus, a resilient system should consider the resilience level and transition time between each phase. The resilience of a system could be improved by reducing the degradation depth and reducing the transition time.

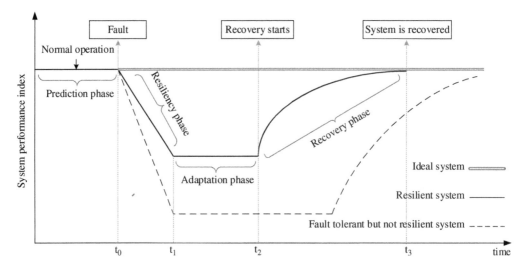

Figure 1.10 Time diagram of a comparison between a resilient system and a fault tolerant system.

Like a fault-tolerant system, there are some solutions for a resilient system. For example, redundancy is a solution. However, redundancy brings a number of penalties: an increase in weight, size, power consumption, cost, as well as time to design, verification, and test. Therefore, a number of choices have to be considered to determine which components should be redundant. These are critical operations of the component, probability of the component failure and its cost.

The focus of some methods in the design stage is on the element's stress reduction in the design process. Reliability of a system increases if it operates at a set point with low stress. It is assumed that the system is under a design process or operates without fault. In the field of power electronics, methods for reducing an electric field are described at both the system and printed circuit board levels. Low temperature operating conditions for an electric power converter are described and tools for this goal are presented. Series connection for voltage sharing and parallel connection for current sharing is explained. Novel control methods of power converters for reducing the complexity and reliable operation are presented. Control of an inrush current as a typical transient problem in electric power converters is presented. Methods for preventing the high stress condition on the components in faulty cases are described. Techniques for reducing mechanical and environmental stress are expressed. Mechanical dampers for preventing high amplitude vibration and insulating colors against humidity are presented.

Another technique for resilience is derating. Within the topic of derating, uninterrupted operation of a faulty system with catastrophic damages in some of its parts is investigated. A faulty system can continue to work with degraded specifications. This technique can be used for both a faulty system because of its uninterrupted operation and a normal system because of its extensive lifetime. Here, the question is: what is the level of derating? If this derating level meets the required performance index of the system, the system is resilient. If not, the system may be fault tolerant but not within the acceptable margins of the system performance indices.

Resilience always has a conflict with the protection system. Protection systems act when a fault occurs in a converter. Their performance is very important; isolation of the converter is not always the best choice because this strategy has a bad effect on the availability of the converter. Protective device coordination is the process of determining the "best fit" timing of interruption when abnormal electrical conditions occur. It is about a solution too close to resilience. The goal is to minimize an outage to the greatest extent possible.

Resilience is a time-dependent factor: the level of resilience during the fault determines the recovery time of the system. In a power electronic converter, resilience can be considered at various levels. In the component level, application of more resilient components leads to resilience of the system. This means that a diode-based rectifier has lower performance indices than a modern synchronous rectifier, but it shows more resilience than the synchronous rectifier.

However, the topic of this book relates to the system performance after the fault occurrence when the protection system acts, when the concept of derating, availability, and fault tolerance are presented, and when the system recovery is important. In stress reduction, we tried to postpone the failure time by reducing the stress on the converter. However, the converter must operate and it is logically affected by stress. It is true that we reduce the stress, but the stress still exists and causes failure in the long term. Suppose that we are faced with a fault in the converter. What should we do? One of the methods for preventing catastrophic damage in faulty converters is to use protection systems. However, the protection system usually causes the system to shut down. The main question is what is the recovery time and the degrading level of the system during the fault? A system can be very reliable but not resilient. Here, reliability means that there are low probability faults in the system and none of them leads to permanent damage. However, these are not enough reasons for resilience. The time of recovery and the level of system performance degrading are the main factors. Protection techniques prevent catastrophic failures in electric power converters. However, as described previously, protection systems cannot prevent every fault. There are two main approaches to a faulty converter. Isolating the faulty converter is one of the approaches and is used when the converter mission can be interrupted for repair and maintenance. If the mission of the converter is important at the time of failure, users should use another method. Derating is a method for allowing the faulty converter to continue its mission. The derating concept is a commonly used technique for faulty systems. Under this topic, uninterrupted operation of a faulty power conversion system with catastrophic damages in some of its parts is investigated. It is shown that a faulty electric power converter can continue to work with degraded specifications. This algorithm is named "derating for accessibility." This technique can be used for both a faulty system because of its uninterrupted operation and a normal system because of its extensive lifetime. Although the derating technique allows the system to operate after faulty conditions, the system degradation may be as high as the unacceptable performance. On the other hand, the system recovery time is also important, which is not considered in the derating methods.

1.3.1 Mission Profile

In the field of reliability, the mission profile has been one of the key concepts since the start of the scientific examination of the subject of reliability. Mission profiling is the process of creating an overview of all the influences that a product is subjected to throughout its life cycle, which typically comprises transport, storage and use environments [9]. A mission profile takes the following into account: field of application, geographical area, user impact, and use profile. A mission profile contains both environmental impacts, such as thermal, mechanical, humidity, water, dust, chemical, electromagnetic impacts, and functional influences related to the pattern of use. Mission profiles are an important part of the product development process and form the basis of product requirements. Together with acceleration models, mission profiles also form the basis for accelerated life testing and other relevant tests. A mission profile is a simplified representation of relevant conditions that the system in focus will be exposed to in its intended application. In photovoltaic systems, the mission profile can be defined as a dataset of annual power generation, energy estimation, and environmental and graphical results; from these, designers are able to extract the minimum, average, and maximum current, voltage, and temperature values in which the photovoltaic system will operate and build a relevant operation map. Figures 1.11 and 1.12 shows two mission profiles for a human and a solar plant, respectively. A mission profile directly affects the useful life of the system. In the example of a human mission profile, it is accepted that a human without a good sleep program ages faster than one with a good and enough sleep program.

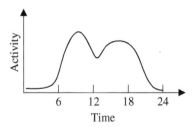

Figure 1.11 Mission profile of a human during one day.

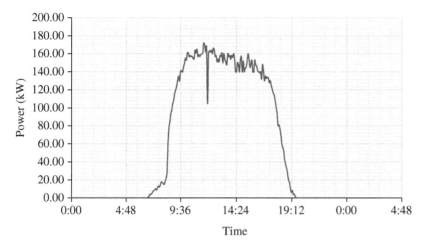

Figure 1.12 Mission profile of a solar plant during one day.

1.3.2 Mission Critical

The topic of this book is resilience in power electronics. Resilience is often supposed to be fault tolerance. However, there is an important difference between a fault-tolerant system and a resilient system. This difference is mission critical considerations, which become more important in applications involving critical missions. These systems are named mission critical systems. A mission critical system is a system that is essential to the survival of a mission [10]. As an example, databases and process control software are considered mission critical to a company that runs on mainframes or workstations. Emergency call centers, computerized hospital patient records, data storage centers, stock exchanges, and other operations dependent on computer and communication systems have to be protected against breakdowns due to the system's mission critical functions. In each of these cases, the failure of a mission critical service can cause severe disruption of services, heavy financial losses, and even danger to people. Many essential functions depend on electricity. When a mission critical system fails or is interrupted, system operations are significantly impacted. Mission essential equipment and mission critical applications are also known as mission critical systems. Mission critical systems are well known in business and organizations. Examples of mission critical systems are: an online banking system, railway/aircraft operating control systems, electric power systems, and many other computer systems that will adversely affect business and society when they fail. As an example, an aircraft is highly dependent on its navigation system. Another example is a nuclear plant. A nuclear reactor controls and contains the sustained nuclear chain reaction. It is usually used for generating electricity. Nuclear reactors have been one of the most concerning systems for public safety worldwide because the malfunction of a nuclear reactor can cause serious disasters. Figure 1.13 shows various applications with mission critical systems. An aircraft is a very visible example. Aircraft reliability programs are essential to ensure safety and dependability. To achieve the highest level of aircraft dispatch reliability at the lowest cost, it is essential to use operational data, pilot reports, and assess the effectiveness of maintenance and training programs.

This is the most important part of this section. The difference between the fault-tolerant systems and resilient systems comes from attention to the concept of mission critical. In other words, the

(a)

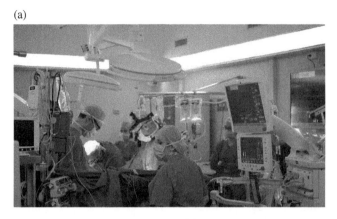

(b)

(c)

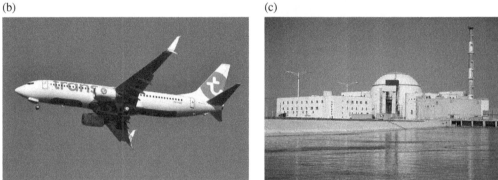

Figure 1.13 Some applications with mission critical. (a) surgery room, (b) air craft, (c) nuclear plant. *Sources:* Pfree2014/Wikipedia Commons/CC BY-SA 4.0; papazachariasa/Pixabay; (c) ITAR-TASS News Agency/Alamy Images.

time needed for survival of the mission of the system determines its level of resilience. As an example, consider a power supply in an aircraft. This converter can be fault tolerant in the ground base with a long time for recovery. However, this converter must supply the loads in the aircraft without being absent for a long time. The reason is obvious: if the power supply of the aircraft fails for a long time, it will be damaged! As a famous historic example, the Japanese men's gymnastics team in the 1976 Montreal Olympics is presented in reference [11]. In this Olympics, the Japanese men's gymnastics team won five consecutive gold medals in the team event for the first time in Olympic history. The team members were Hiroshi Kajiyama, Eizo Kenmotsu, Hisato Igarashi, Sawao Kato, Shun Fujimoto, and Mitsuo Tsukahara. Soon after Fujimoto stepped onto the mat for the preliminary floor exercise disaster struck. Landing awkwardly during a tumbling run, the Japanese gymnast shattered his kneecap—an excruciatingly painful injury that looked to have dashed his (and perhaps Japan's) Olympic dreams. The heroic gymnast decided against telling his teammates about his injury and continued on to the pommel horse and ring exercises—two events where clean dismounts are vital and where a painful impact to his broken knee could not be avoided. With his team trailing the Soviet Union by little more than a point, Fujimoto stepped up to the rings, where in order to play his part in Japan's victory, he would need to land a dismount from 2.4 m above the ground. Fujimoto's performances that day proved key to Japan's quest for gold, as shown in Figure 1.14.

Figure 1.14 The Japanese men's gymnastics team in 1976 Montreal Olympics; Shun Fujimoto with a broken knee is seen. *Source:* Don Morley/Hulton Archive/Getty Images.

1.4 Faulty Condition in a System

A faulty condition is a state of the system where the system mission is interrupted or the system quality is reduced to an unacceptable level. Faults are the factors generating the faulty condition in a system. In the general form, there are two types of faults: intrinsic (internal) faults and extrinsic (external) faults. These two categories of the faults occur together. External faults cause an increase in the stress on the system elements.

1.4.1 Internal Faults

Intrinsic faults are triggered by internal failure causes in the system. Failures of the elements and subsystems are the reason for system intrinsic faults. Figure 1.15 shows an example of an internal fault in an aircraft.

1.4.2 External Faults

On the other hand, extrinsic faults are forced from outside of the system. They are applied to the system via inputs and outputs of the system. On 15 January 2009, US Airways Flight 1549, an

Airbus A320 on a flight, struck a flock of birds shortly after take-off, losing all engine power. Unable to reach any airport for an emergency landing due to their low altitude, pilots glided the plane to a ditching in the Hudson River off Midtown Manhattan. All 155 people on board were rescued by nearby boats, with a few serious injuries, as shown in Figure 1.16. This water landing of a powerless jetliner with no deaths became known as the Miracle on the Hudson [12]. In this example, the birds were the reason for the external fault.

Figure 1.15 Internal fault in the engine of an aircraft. *Source:* aapsky/Adobe Stock.

Figure 1.16 Miracle on the Hudson. *Source:* Greg L/Wikimedia Commons.

1.4.3   Malfunctioning

The internal and external faults may act as the factors that force the system to work in an abnormal state. In this state, the system works but its outputs are not acceptable. No system element nor subsystem is damaged in this condition but they are forced to work abnormally. False data are the main reason for this malfunctioning. Air France Flight 447 was a scheduled international passenger flight from Rio de Janeiro, Brazil, to Paris, France. On 1 June 2009, the Airbus A330 serving the flight stalled and did not recover, eventually crashing into the Atlantic Ocean, as shown in Figure 1.17. Temporary inconsistency between the measured speeds was likely to have been the result of an obstruction of the pitot tubes by ice crystals [13].

1.5   System Health Awareness

One of the most important requirements of a system for resilient operation is the system health awareness. Thus, the sign of the faults can be detected at an earlier time and make the system ready for reacting. As an example, in biology, the nervous system is a highly complex part of an animal that coordinates its actions and sensory information by transmitting signals to and from different parts of its body. The nervous system detects environmental changes that impact the body and then works in tandem with the endocrine system to respond to such events. Neurons have special structures that allow them to send signals rapidly and precisely to other cells. They send these signals in the form of electrochemical impulses traveling along thin fibers called axons, which can be directly transmitted to neighboring cells through electrical synapses or cause chemicals called neurotransmitters to be released at chemical synapses. A cell that receives a synaptic signal from a neuron may be excited, inhibited, or otherwise modulated. The connections between neurons can form

Figure 1.17   Air France Flight 447. *Source:* Alamy Images/Abaca Press.

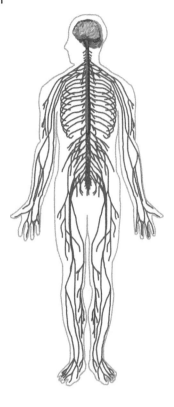

Figure 1.18 The human nervous system.

neural pathways, neural circuits, and larger networks that generate an organism's perception of the world and determine its behavior. (Figure 1.18).

1.5.1 System Condition Monitoring

Condition monitoring is the process of monitoring a parameter of condition in the systems to identify a significant change that is indicative of a developing fault. The use of conditional monitoring allows maintenance to be scheduled or other actions to be taken to prevent failure and avoid its consequences. Condition monitoring allows information to be obtained about the growth of failure in the systems. Our decision about this failure is related to the data that are obtained from the condition monitoring system. For example, monitoring of a long-term high-output current beyond the nominal specification of a power electronic converter usually causes the protection system to operate and take it out of service. There are two main approaches for condition monitoring: sensor-less and sensor-based methods. In sensor-based methods, a physical sensor is used to measure and monitor the desired parameter or variable. In the sensor-less method, the job is done based on calculations. Sensor-based methods are historically the first method for condition monitoring. They have some features and advantages that allows them to be still applicable in power converters. Faster access to data derived from online measurements provides a more efficient response. Although the benefits and even the necessity of using online sensors are recognized, several factors need to be considered prior to their installation in industrial applications. Typical industrial environments often involve fluids, temperatures, pressures, and flows that could be harmful to the delicate electronic sensor components. Sensor-less methodology is a solution for eliminating the physical sensors required for the control process.

It is historically applied to servomotor systems for controlling goals. However, sensor-less methods are also used for normal condition monitoring. Monitoring the state of mechanical machine tool components is of growing importance for increasing machine tool availability and reducing inspection efforts and costs. Figure 1.19 shows some examples of condition monitoring systems.

1.5.2 Fault Prognosis

Prognosis is the capability to use the observations to predict the future states of a machine or forecast a fault before its occurrence. Engineering systems, such as aircraft, industrial processes, manufacturing systems, transportation systems, electrical and electronic systems, etc. are becoming more complex and are subjected to failures that impact adversely their reliability, availability, safety, and maintainability. Such critical assets are required to be available when needed, and

(a)

(b)

Figure 1.19 Some examples of the condition monitoring systems. (a): in a plant, (b): in a car. *Sources:* RGtimeline/Adobe Stock; Pexels/Pixabay.

maintained on the basis of their current condition rather than on the basis of scheduled or break-down maintenance practices. Moreover, online, real-time fault diagnosis and prognosis can assist the operator to avoid catastrophic events. Recent advances in condition-based maintenance and prognostics have prompted the development of new algorithms for fault, or incipient failure, diagnosis, and failure prognosis aimed at improving the performance of critical systems. Potential benefits to industry include reduced maintenance costs, improved equipment uptime and safety. As an example, the Cooper test shown in Figure 1.20 was one of the most commonly used fitness tests to measure the fitness levels of both amateur and professional football referees. Another example is the blood test for illness prognosis as shown in Figure 1.21.

Figure 1.20 The Cooper test.
Source: mainathlet/Pixabay.

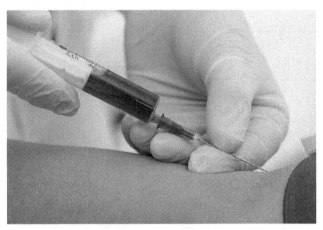

Figure 1.21 Blood test as a prognosis of the human illness. *Source:* Photographee.eu/Adobe Stock.

1.5.3 Fault Diagnosis

Fault diagnosis is determining which fault has occurred. Process fault diagnosis involves interpreting the current status of the plant from given sensor readings and process knowledge. Fault detection and diagnosis techniques are based upon the use of process models. Data from the plant is fed into these algorithms and the outputs are compared with the corresponding plant outputs. If there are discrepancies, it is an indication that at least one fault has occurred. The location of the fault is determined using a representative model. Figure 1.22 shows the diagnosis of a fault in a car brake disc.

1.6 Methods for Resilience

According to the above-mentioned explanations, any method that prevents a fault occurrence or fault influence (after an occurrence) can be defined as a method of resilience. Some of these methods help the systems to prevent the fault occurrence. Prevention cannot be ideally realized. This means that the methods reduce the failure rate of the system to such a low level that the system does not fail between two consecutive maintenance processes. However, in a very important mission critical system, even one fault is important. On the other hand, some faults are outside the control of the method used for fault prevention. Here, the other methods of resilience enter the problem. They act after the fault occurrence and keep the system performance at an acceptable level until the mission is passed.

1.6.1 Protection

The system must operate and it is logically affected by stress. It is true that we reduce the stress but the stress still exists and may cause failure in the long term. Then we are faced with a fault in the system. Protection systems act when a fault occurs in the system. Their performance is very important; isolation of the system is not always the best choice because this strategy has a

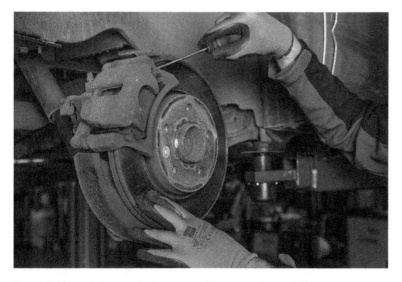

Figure 1.22 Inspection of a brake disc. *Source:* geraldoswald62/Pixabay.

bad effect on availability of the system. Protection methods save the system against non-catastrophic faults. However, this method saves the system but it also takes the converter out of service. If the protection system is very fast, it protects the system but conflict with noise is possible. Therefore, protection systems should be fast enough to protect the system but not so fast as to interrupt the operation. The protection system operates against both the external faults and internal faults. Examples of protection against external faults are the chest in a male body, shown in Figure 1.23, and the car guard, shown in Figure 1.24. An example of the protection against internal faults is the fire extinguisher of aircraft engines, shown in Figure 1.25.

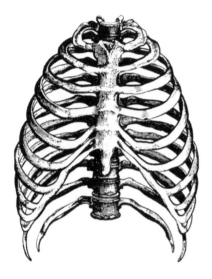

Figure 1.23 Chest as a protective part of the body. *Source:* StarGladeVintage/Pixabay.

Figure 1.24 Protective guard of a car. *Source:* Kruglovsasha/Wikimedia Commons.

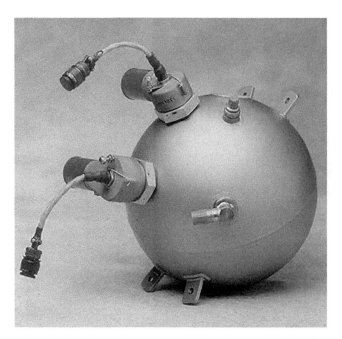

Figure 1.25 Fire extinguisher of aircraft engines. *Source:* AviationNuggets.

1.6.2 Derating

Protection techniques prevent catastrophic failures in electric power converters. However, as described before, protection systems cannot prevent any fault. There are two main approaches to a faulty converter. Isolating the faulty converter is one of the approaches and is used when the converter mission can be interrupted for repair and maintenance. If the mission of the converter is important at the time of failure, users should use another method. Derating is a method for allowing the faulty converter to continue its mission. Although the derating method is mainly a method for faulty systems, it can also be used for extending life in normal cases. Component failure rates generally decrease as applied stress levels decrease. Thus, derating or operating components at levels below their ratings (for current, voltage, power dissipation, temperature, etc.) will increase reliability. Derating increases the margin of safety between part design limits and applied stresses, thereby providing extra protection for the part. By applying derating in an electrical or electronic component, its degradation rate is reduced and reliability and life expectancy are improved. As an example, aging is the process of becoming older. The term refers especially to humans and many other animals. In humans, aging represents the accumulation of changes in a human being over time and can encompass physical, psychological, and social changes. Reaction time, for example, may slow with age, as shown in Figure 1.26, while memories and general knowledge typically increase.

Another example is Alzheimer's disease, which is a neurodegenerative disease that usually starts slowly and progressively worsens. The most common early symptom is difficulty in remembering recent events. As the disease advances, symptoms can include problems with language, disorientation, mood swings, loss of motivation, self-neglect, and behavioral issues.

The heart attack is another example. A heart attack occurs when a coronary artery becomes completely blocked and a large portion of the muscle stops receiving blood. A serious heart attack

Figure 1.26 Aging as a derating process in a man. *Source:* FreeToUseSounds/Pixabay.

can cause significant damage. However, a heart attack may occur only partially and in this case the affected coronary artery is only partially blocked.

In industries, the Boeing 777 is a good example. The Boeing 777 is an American wide-body airliner, which is shown in Figure 1.27. It is the world's largest twin-engine jet aircraft and can fly with only one engine for three hours. These regulations allowed twin-engine airliners to make ocean crossings at up to three hours' distance from emergency diversionary airports.

1.6.3 Stress Reduction

Reliability of a system increases if it operates at a set point with low stress. The failure mechanism of a system is started immediately after the first operation. At the beginning, the system operates normally but under stress of failure factors:

- Power losses in the system cause the temperature to rise in various parts of the system.
- Applied voltage causes an electric field to be applied to insulators.
- Mechanical forces lead to vibrations.
- Environmental factors.

These factors act from the beginning of system application. In the long term, they cause the system to age and lead to its failure. The time interval for changing a stress factor to a failure factor is directly related to the amount of stress. Higher stress leads to a shorter time to failure and vice

Figure 1.27 Boeing 777: the largest twinjet aircraft that fly with one engine for three hours. *Source:* Aero Icarus/Wikipedia Commons/CC BY-SA 2.0.

Figure 1.28 An electric soft starter used for reducing the stress on the electric motors during power on. *Source:* Demjas/Wikipedia Commons/CC BY-SA 4.0.

versa. Thus, the first method of reliability improvement is reducing the stress factors on the system. As an example, a motor soft starter shown in Figure 1.28 is a device used with AC electrical motors to temporarily reduce the load and torque in the powertrain and electric current surge of the motor during start-up. This reduces the mechanical stress on the motor and shaft, as well as the electrodynamic stresses on the attached power cables and electrical distribution network, extending the lifespan of the system. It can consist of mechanical or electrical devices, or a combination of both. Mechanical soft starters include clutches and several types of couplings using a fluid, magnetic forces, or steel shot to transmit torque, similar to other forms of torque limiter. Electrical soft starters can be any control system that reduces the torque by temporarily reducing the voltage or current input, or a device that temporarily alters how the motor is connected in the electric circuit. A soft starter continuously controls the motor's voltage supply during the start-up phase. In this way, the motor is adjusted to the machine's load behavior. Mechanical operating equipment is accelerated smoothly. This lengthens service life, improves operating behavior, and smooths work flows.

1.6.4   Immunity Against False Data

Two approaches exist in this section: reducing the data rate in the case of an unknown data channel and shielding against false data. As an example, in electrical engineering, electromagnetic shielding is the practice of reducing the electromagnetic field in a space by blocking the field with barriers made of conductive or magnetic materials. Shielding is typically applied to enclosures to isolate electrical devices from their surroundings and to cables to isolate wires from the environment through which the cable runs.

1.7   Inherently Resilient Systems

Some of the systems reach to resilience by providing the no-fault state. Figure 1.29 shows two inherently resilient aircrafts. Lockheed aircrafts U-2 and SR-71 have an unattainable flying altitude and speed, respectively. Therefore, they are inherently resilient against any attack as the external fault.

Mongooses shown in Figure 1.30 are one of at least four known mammalian taxa with mutations in the nicotinic acetylcholine receptor that protect against snake venom [14]. Their modified receptors prevent the snake venom α-neurotoxin from binding.

Achieving the inherently resilient characteristic may be valid for a certain time interval. The kidney operation is an example. Most humans are born with two kidneys, shown in Figure 1.31, as the functional components of what is called the renal system, which also includes two ureters, a bladder, and a urethra. The kidneys have many functions, including regulating blood pressure, producing red blood cells, activating vitamin D, and producing some glucose. Most evidently, however, the kidneys filter body fluids via the bloodstream to regulate and optimize their amount, composition, pH, and osmotic pressure. Life is incompatible with a lack of kidney function. However, unlike the case with most other organs, we are born with an overabundant kidney capacity. Indeed, a single kidney with only 75% of its functional capacity can sustain life very well. If only one kidney is present, that kidney can adjust to filter as much as two kidneys would normally. In such a situation, the nephrons compensate individually by increasing in size—a process known as hypertrophy—to handle the extra load. This happens with no adverse effects, even over years.

(a)   (b)

Figure 1.29   Two inherently resilient aircrafts, (a) U-2, (b) SR-71. *Sources:* (a) US Air Force/Wikipedia Commons/Public Domain; (b) USAF/Judson Brohmer/Wikipedia Commons/Public Domain.

Figure 1.30 Mongoose as an inherently resilient animal against external fault of the snake venom. *Source:* Dan Dennis/Unsplash.com/.

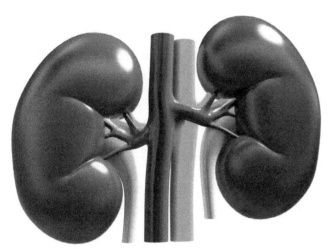

Figure 1.31 Most humans are born with two kidneys that they can operate redundantly for resilience. *Source:* abhijith3747/Adobe Stock.

1.8 Summary and Conclusions

In this chapter, a brief introduction of the concept of resilience in systems was presented. The necessity of resilience in systems and its requirements were described. The results of this chapter are summarized as follows:

1) The resilience concept is defined as the capability to recover from an abnormal state to another state where this new state guarantees the continuance operation of the system.
2) Resilient operation is defined in systems with a critical mission. A mission critical system is a system that is essential to the survival of a mission.

3) The requirements for resilient operation are system health monitoring, fault-tolerant characteristics, an intelligent protection system, and operation under low stress.
4) Some of the systems are inherently resilient. In these systems, the fault origins are removed by the advanced characteristics of the system.

References

1 Hitchins, D.K. (2007). *Systems Engineering: A 21st Century Systems Methodology*. Wiley.

2 Wasson, C.S. (2015). *System Engineering Analysis, Design, and Development: Concepts, Principles, and Practices*, 2e. Wiley.

3 www.drivek.co.uk/car-comparison-technical-details-features-dimensions, 2021.

4 Si, X., Wang, W., Hu, C. et al. (2012). Remaining useful life estimation based on a nonlinear diffusion degradation process. *IEEE Transactions on Reliability* 61 (1): 50–67.

5 Kapur, K.C. and Pecht, M. (2014). *Reliability Engineering*. Wiley.

6 Chung, H.S., Wang, H., Blaabjerg, F., and Pecht, M. (2015). *Reliability of Power Electronic Converter Systems*. IET Press.

7 Rohatgi, V.K., Md, A.K., and Saleh, E. (2015). *An Introduction to Probability and Statistics*, 3e. Wiley.

8 Boeing B-17F. National Museum of the U.S. Air Force. Archived from the original on August 8, 2016. Retrieved 8 August 2016.

9 Ma, K., Liserre, M., Blaabjerg, F., and Kerekes, T. (2015). Thermal loading and lifetime estimation for power device considering mission profiles in wind power converter. *IEEE Transactions on Power Electronics* 30 (2): 590–602.

10 Banerjee, A., Venkatasubramanian, K.K., Mukherjee, T., and Gupta, S.K.S. (2012). Ensuring safety, security, and sustainability of mission-critical cyber–physical systems. *Proceedings of the IEEE* 100 (1): 283–299.

11 https://olympics.com/en/news/fujimoto-bravery-helps-japan-make-it-five-golds-in-a-row-gymnastics, 1976.

12 Clark, A. (2009). Plane crashes in Hudson River in New York. via www.theguardian.com.

13 Clark, N. (2017). Report on air france crash points to pilot training issues. *The New York Times*.

14 Barchan, D., Kachalsky, S., Neumann, D. et al. (1992). How the mongoose can fight the snake: the binding site of the mongoose acetylcholine receptor. *Proceedings of the National Academy of Sciences* 89 (16): 7717–7721.

2

Mission Critical Power Electronic Systems

2.1 Power Electronic Converters

Power electronics is the application of solid-state electronics to convert one form of electrical power to another form. Changing the voltage levels, increasing or decreasing the frequency of the load voltage, and converting one form of electrical power to another form are the main applications of power electronics. This task is done using power electronic converters. A power electronic converter consists of theoretically lossless elements as well as solid-state switches. Figure 2.1 shows a power electronic converter used for converting the AC form of electrical power to the DC form. This conversion is named rectification. The converter uses some solid-state diodes as the switching elements. The shown inductor, L, and capacitor, C, form a low-pass filter to suppress the ripple of the DC output voltage. Both of these elements are theoretically loss less. Thus, the main advantage of the power electronic converters is their high-efficiency operation. In the real world, the most powerful electronic converters have an amount of power loss but this loss is still negligible in comparison to the converted nominal power. There are special linear power electronic converters whose power loss is considerable, but they are a small portion of the converters in the field of power electronics. Application of solid-state devices in power electronic converters leads to other advantages, such as less weight in comparison to other solutions. Figure 2.2 shows a comparison between two new and old solutions of power conversion. Figure 2.2a is an old rotary power converter in train applications with 15 kW nominal power. Figure 2.2b shows the new solution of this power conversion for the same application. This is a 40 kW converter with the same dimensions and less weight.

2.1.1 Elements of Power Electronics

A power electronic converter is a circuit where all of its elements act as a lossless operation. Thus, a power electronic converter contains all of the electronic elements but none of them operates as a resistor with power losses. Therefore, power electronic converters consist of passive elements such as capacitors, inductors, and active elements such as solid-state switches that act as lossless resistors. Figure 2.3 shows a power electronic converter that is used to adjust the DC voltage of the load. This converter consists of some active elements as the switching device and some inductors and capacitors as the passive elements.

Resilient Power Electronic Systems, First Edition. Shahriyar Kaboli, Saeed Peyghami, and Frede Blaabjerg.
© 2022 John Wiley & Sons Ltd. Published 2022 by John Wiley & Sons Ltd.
Companion website: www.wiley.com/go/kaboli/resilientpower

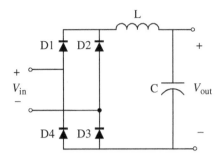

(a)

Figure 2.1 A rectifier circuit diagram as a power electronic converter.

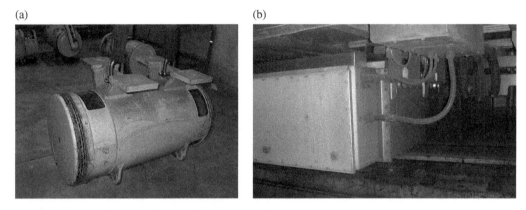

(a)   (b)

Figure 2.2 Application of power electronic converters for reducing the weight and increasing the power density: (a) old traction rotary converter, (b) new power electronic static converter.

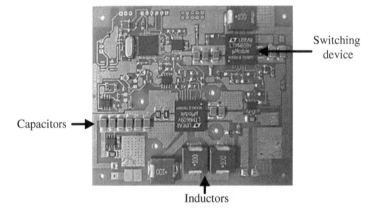

Figure 2.3 Passive and active elements of a power electronic converter.

2.1.1.1 Passive Elements

The passive elements have constant characteristics. It is in the opposing active elements that their characteristics vary considerably. There are three passive elements: resistor, capacitor and magnetic device such as an inductor, and transformer.

- *Resistor*

  The main applications of resistors in power electronic converters are dissipating the stray energy in avoiding energy accumulation, measuring devices such as series resistor used for current

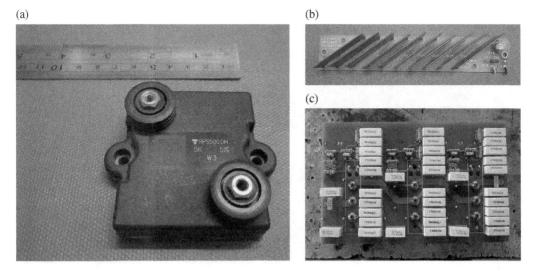

Figure 2.4 Some types of resistors used in power electronics: (a) a resistor for current measurement, (b) series resistors in a high-voltage divider, (c) resistors used for dissipating the stray energy in a power electronic converter.

measurement and voltage divider, and limiting fault currents. Figure 2.4 shows three types of the resistors used in power electronic converters.

- *Capacitor*
  As one of the energy storage elements, capacitors have many applications in power electronics. Capacitors are connected in parallel with the power circuits of most electronic devices and larger systems to shunt away and conceal current fluctuations from the primary power source and to provide a clean power supply for the load, as shown in Figure 2.5a. In electric power distribution, capacitors are used for power-factor correction, as shown in Figure 2.5b. Capacitors are used to suppress any instantaneous voltage spike across the solid-state devices. Figure 2.5c shows a high-voltage switch with capacitors used in the voltage limiter circuit.
- *Magnetic elements*
  Inductors and transformers are two magnetic elements used in the world of power electronics. The inductors are used for smoothing the current and storing the magnetic energy. Figure 2.6a shows four inductors used in a low-power DC motor drive. The transformers are applied to change the voltage level, providing isolation and impedance matching. Figure 2.6b shows the transformers in a high-voltage power supply. Ferromagnetic cores are usually used in these elements, as shown in Figure 2.6c.

2.1.1.2 Active Elements
Solid-state devices are important tools in the design of power electronic converters. Power electronic devices may be used as switches or as a variable resistor. In the switching operation state, an ideal switch is either open in the off state with zero current or closed in the on state with zero voltage. As the voltage or current of the switches is zero in these two states, they have no power dissipation. The real semiconductor switches approximately show this ideal property and so most power electronic applications rely on switching devices on and off, which makes systems very efficient. The losses that a power electronic device generates should be as low as possible because of the importance of efficiency. There are various types of solid-state switches. A diode is a device that is either turn on or turn off depending on the polarity of its current and voltage. Power devices

(a)

(b)

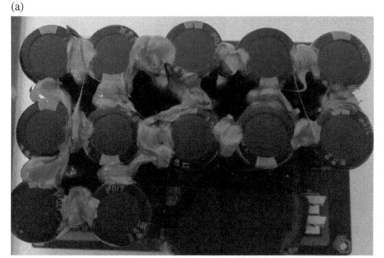

(c)

Figure 2.5 Various applications of the capacitors in power electronics: (a) capacitor bank of the DC power supply output, (b) capacitors used for power factor correction, (c) capacitors in the voltage limiting circuit of a high-voltage switch.

such as thyristors have the ability to control the start of conduction. Therefore, they are named semi-controlled devices. Other devices provide full switching control only. They are named fully-controlled devices. Another difference between solid-state switches is their difference from a switching time viewpoint. Devices are different in switching speed. Some diodes and thyristors are operated at a low switching frequency. There are devices that operate at higher switching speeds.

- *Uncontrolled switches*
  A diode is a unipolar and uncontrolled switching device. This means that the diode turns on and turns off according to its voltage and current in the circuit and it is not under the control of the user. A power diode is a type of diode that is commonly used in power electronic circuits. Just like a regular diode, a power diode has two terminals and conducts current in one direction. A power diode varies in construction from a standard diode to enable this higher current rating. Figure 2.7 shows two types of diodes.

(a)
(b)
(c)

Figure 2.6 Applications of magnetic elements in power electronics: (a) inductors, (b) transformers, (c) ferromagnetic cores.

(a)
(b)

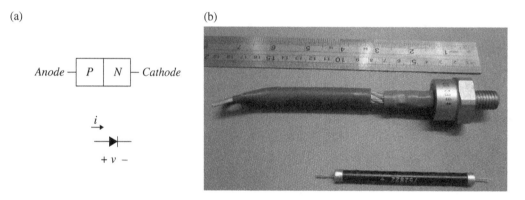

Figure 2.7 The diode as a switch in power electronics: (a) typical semiconductor layers in a diode and its symbol, (b) two diodes, one with a high nominal current and the other with a high nominal voltage.

- *Semi-controlled switches*

  A thyristor is a semi-controlled switching device. This switch turns on when a gate signal is applied while the anode–cathode voltage is positive. In the on-state, a thyristor operates as a unidirectional switch like a diode. The device cannot be turn off via the gate signal. It turns off when the current falls to zero. Other devices in the family of thyristors are the GTO (gate turn-off) thyristor and Triac. GTO, of course, is a fully controlled switch but in the thyristor family. Figure 2.8a shows a high-current thyristor. Figure 2.8b shows the thyristors in a series connection used in a high-voltage switch.
- *Fully controlled switches*

  In these switches, both turn-on time and turn-off time are controllable. Therefore, these switches are suitable for high-frequency applications.

(a)

(b)

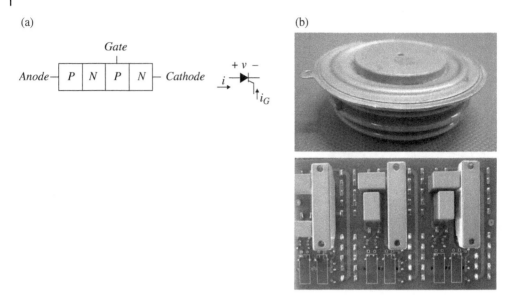

Figure 2.8 The thyristor and its application: (a): Typical thyristor semiconductor layers and its symbol, (b): A high current thyristor and a high voltage connected stack of thyristors.

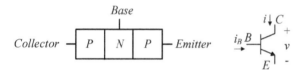

Figure 2.9 Typical semiconductor layers in a BJT and its symbol.

Bipolar Junction Transistor

The bipolar junction transistor (BJT) is a fully controlled switching device. It can be turned on and turned off by the user command. The BJT is not widely used in high-power switching converters because it is slower than the other new switching devices. The symbol of a BJT and its typical structure are shown in Figure 2.9.

MOS

The metal oxide semiconductor field effect transistor (MOSFET) is a commonly used semiconductor in digital and analog circuits and is also a useful power device. It is more widely used than the BJT as it requires a minimal current for load-current control. The level of conductivity can be increased from the "normally off" state when the MOSFET is set to the enhancement mode. Voltage transmitted via the gate can minimize conductivity from the "normally on" state. The main advantages of the power MOSFET is its higher switching speed in comparison to the BJT as well as a simpler gate drive. Figure 2.10a shows the structure of a MOSFET and its symbol. Figure 2.10b shows a high-frequency converter using MOSFETs.

IGBT

Insulated-gate bipolar transistor (IGBT) modules have been developed to be used as switching elements for the power converters of variable-speed drives for motors, uninterruptable power supplies, and others. An IGBT is a semiconductor device that combines the high-speed switching performance of a power MOSFET with the high-voltage/high-current handling capabilities of a

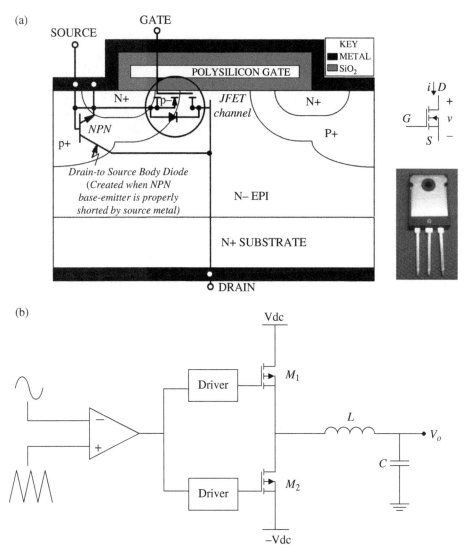

Figure 2.10 The MOSFET and its application: (a) typical semiconductor layers in a MOSFET. *Source:* Used with permission from SCILLC dba onsemi, (b) a high-frequency class-D amplifier using MOSFETs.

bipolar transistor. This device has the best characteristics of MOSFETs and BJTs. The IGBT has a high gate impedance like MOSFET. Thus, the gate drive circuit is simple. In addition, this device has a low on-state voltage drop like BJT. Figure 2.11a shows the circuit diagram of a three-phase inverter as a power electronic converter based on IGBTs. Figure 2.11b shows the inside view of the IGBT module, which consists of six IGBTs and respective antiparallel diodes.

2.1.2 Types of Power Electronic Converters

Any circuit with a power conditioning function is considered as a power electronic converter. In this book, we consider the converters with theoretically lossless features. Converters in this category are named switching power electronic converters. However, there is another important

(a)   (b)

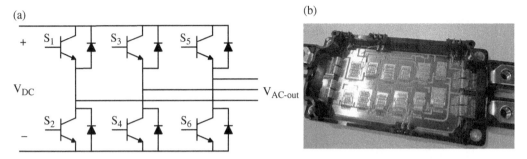

Figure 2.11 Three-phase inverter based on IGBT: (a) circuit diagram, (b) integrated module.

power converter, but with limited applications as it has considerable power losses. This category is named linear power electronic converters.

2.1.2.1 Linear Power Electronic Converters

A linear power electronic converter is a circuit that does not contain any switching components. It has some outstanding characteristics compared to switching converters, such as very low noise and ripple, simplicity, robustness, ease of design, and repair. Figure 2.12 shows a linear solid-state amplifier.

2.1.2.2 Switching Power Electronic Converters

In a switched-mode power converter, all of the solid-state devices operate as a switch. Switching occurs at a very high frequency and enables the use of transformers and filter capacitors that are much smaller, lighter, and less expensive than those found in linear power supplies operating at low frequency. As the solid-state elements operate in the switching state, switching power converters have a high efficiency. This is the main advantage of these types of electric power converters. Regarding the critical energy demand, operation with high efficiency is one of the most important requirements of electric power converters. Thus, switching power converters are used much more

Figure 2.12 A linear solid-state amplifier.

than linear types. In switching power converters, the solid-state elements operate in the switching state. Therefore, switching power converters have a high efficiency, which is the main advantage of these types of electric power converters. The voltage and current relation of an ideal switch is presented in the following, where it can be seen that the power loss is zero in the switching operation:

$$\begin{cases} i_D = 0 & Switch\,OFF \\ v_D = 0 & Switch\,ON \end{cases} \tag{2.1}$$

where i and v are the current and voltage of the switch, respectively.

Figure 2.13 shows a switching high-frequency power electronic converter.

2.1.3 Types of Power Conversion

In the following, we describe the switching type of electric power converters in detail. There are four different groups in this field:

- AC to DC converters, which are usually named as rectifiers
- DC to AC converters, which are usually named as inverters
- AC to AC converters, which are usually named as AC voltage controllers
- DC to DC converters, which are usually named as choppers

2.1.3.1 AC to DC Power Conversion

The main application of AC to DC converters is rectification. A rectifier converts alternating current to direct current. These converters are usually used for producing a constant DC voltage from an AC network. There are classic rectifiers and modern types. Rectifier circuits may be single-phase or multi-phase. Most low-power rectifiers for domestic equipment are single-phase, but three-phase rectification is very important for industrial applications. Figure 2.14 shows single phase and three phase classic rectifiers. Polyphase systems with more than three phases are easily accommodated into a bridge rectifier scheme. When polyphase AC is rectified, the phase-shifted

Figure 2.13 A high frequency switching power electronic converter.

(a)   (b)

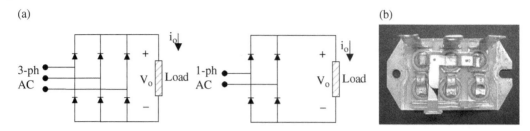

Figure 2.14 Rectifiers and their applications: (a): single- and three-phase classic rectifiers, (b) the inside view of the package of a three-phase rectifier.

pulses overlap each other to produce a DC output that is much smoother than that produced by the rectification of single-phase AC.

Rectifiers in Industry: High-Power Electrolysis

Electrolysis is commercially highly important as a stage in the separation of elements from natural sources using an electrolytic cell. Electrolysis is a method of using a direct electric current to drive a chemical reaction for separation of elements. The process needs a high level of DC current and this cannot be supplied by the sources with a limited lifetime, such as a battery. Thus, high-current rectifiers are used to produce the required DC voltage for the process via rectification of the AC network voltage.

2.1.3.2 DC to DC Power Conversion

DC to DC converters are important in portable electronic devices. Most DC to DC converters regulate the output voltage and can be categorized as the following:

- Isolated or not isolated
- Single or multi output

There are two types of DC/DC converters: linear and switched. A linear DC/DC converter uses a resistive voltage drop to create and regulate a given output voltage, while a switched-mode DC/DC converts by storing the input energy periodically and then releasing that energy to the output at a different voltage. The storage can be in either a magnetic field component like an inductor or a transformer, or in an electric field component such as a capacitor. Transformer-based converters provide isolation between the input and the output.

DC Voltage Controllers in Industry: Power Distribution Unit

The operating voltage of different electronic devices such as ICs can vary over a wide range, making it necessary to provide a voltage for each device. Many power electronic converters are used as adaptors of voltage levels between two or more subsystems. As an example, consider the power distribution of a satellite. The power subsystem consists of solar panels to convert solar energy into electrical power and batteries that store power and supply the satellite when it passes into the shadow. The satellite has other essential parts: the telemetry subsystem monitors the on-board equipment operations, transmits equipment operation data to the Earth control station, and receives the Earth control station's commands to perform equipment operation adjustments. The thermal control subsystem helps to protect electronic equipment from extreme temperatures due to intense sunlight or the lack of sun exposure on different sides of the satellite's body. The attitude and orbit control subsystem consists of sensors to measure vehicle orientation; control laws embedded in the flight software; and actuators to apply the torques and forces needed to re-orient the vehicle to a desired attitude, keep the satellite in the correct orbital position, and keep antennas positioning in the right directions. The power distribution unit receives input DC voltage from solar panels or batteries and converts it into some DC voltages with different amplitudes for the above-mentioned subsystems.

Power conditioning and voltage regulation are two of the other applications of a power electronic converter. This application is named as the power supply. A power supply may be implemented as a discrete, standalone device or as an integral device that is hardwired to its load. Examples of the latter case include the low-voltage DC power supplies that are part of desktop computers and consumer electronic devices.

A switched-mode supply of the same rating as a line-frequency supply will be smaller, is usually more efficient, but would be more complex. Figure 2.15 shows the circuit diagram of a classical voltage mode DC to DC converter and a commercial DC to DC switching converter.

2.1.3.3 AC to AC Power Conversion

Directly converting AC to AC electric power allows control of the voltage, frequency, and phase of the waveform applied to a load from a supplied AC system. This single power conversion benefits high efficiency, especially for extra high-power conversion, like power transmission between two countries. There are various different types of AC to AC converters, such as cycloconverters and matrix converters. Figure 2.16 shows a simplified circuit diagram of a single and three phase AC voltage controller.

(a)   (b)

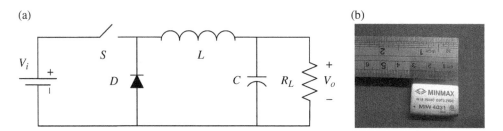

Figure 2.15 DC to DC converters and their applications: (a) circuit diagram of a step-down DC to DC converter, (b) a commercial step-down DC to DC converter.

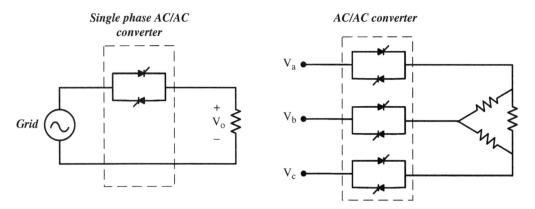

Figure 2.16 Single and three AC controller.

AC to AC Converters in Industry: Soft Starters

A soft starter is an effective and economical solution that can guarantee stable operation of a motor. A motor soft starter is used in series with AC electric motors to reduce current surge of the motor during startup. This reduces the mechanical stress on the motor and shaft, as well as the electrodynamic stresses on the attached power cables and electrical distribution network, extending the lifespan of the system. Figure 2.17 shows a high-power soft starter used in the iron industries.

2.1.3.4 DC to AC Power Conversion

The power inverter or DC to AC converter is a widely used device in our everyday life. It is a kind of high efficiency and convenient product. The equipment can not only adapt the capacity to the current low-carbon lifestyle, but can also adapt to the diversity of people's living characteristics, so the power inverter can be welcomed by all the people in the world [1]. DC to AC converters produce an AC output waveform from a DC source. Some applications of these types of power electronic converters include adjustable speed drives, uninterruptible power supplies, active filters, flexible AC transmission systems, voltage compensators, and photovoltaic generators. Topologies for these converters can be divided into two categories: voltage source inverters and current source inverters. Voltage source inverters use a constant-voltage source. Similarly, in current source inverters the controlled AC output is a current waveform. The DC to AC power conversion is commonly carried out by fully controllable semiconductor power switches, like IGBT and MOSFET. There are some modulation techniques for controlling the output voltage of the inverter. Voltage source inverters have practical uses in both single-phase and three-phase applications. Single- and three-phase types of inverter and its typical waveforms are shown in Figures 2.18–2.20, respectively.

Inverters in Industry: Motor Drives

Motor drives are the backbone of the modern global industry [2]. They play a key role in industries helping to make the world, and our way of living, more sustainable. Today, the most common usage of drives is for the control of fans, pumps, and compressors. A low-power motor drive is shown in Figure 2.21.

(a)

(b)

(c)

Figure 2.17 The Sirjan Iron refining factory in Iran: (a) high-power blowers, (b) high-power motor for the blower fan, (c) the soft starter used for speed control of the blower.

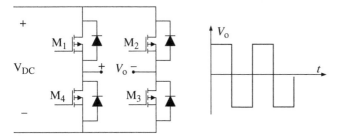

Figure 2.18 A single-phase high-frequency inverter and its output voltage waveform.

Figure 2.19 A three-phase inverter.

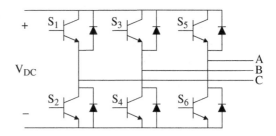

Figure 2.20 Output voltages of a three-phase inverter.

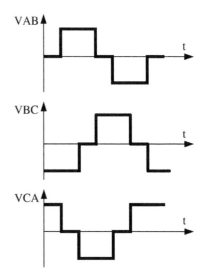

Motor drives go by various names, such as AC drives, adjustable speed drives, and variable speed drives (VSDs). By varying the frequency and voltage of the power supply to an electrical motor, drives can control its speed, making it possible to enhance process control, reduce energy usage, and generate energy efficiently or optimize the operation of various applications relying on electric motors. Drives also have the capacity to control ramp-up and ramp-down of a motor during start or stop, respectively. This decreases mechanical stress on motor control applications and improves ride quality in, for example, elevators. The major variable frequency design used in drives is pulse-width modulation (PWM). It requires switching the motor drive's inverter power devices–transistors or IGBT–on and off many times, to generate the proper voltage levels. Controlling and varying the width of the pulses is how PWM varies the output frequency and voltage.

Figure 2.21   A commercial motor drive.

A combination of power electronic converters in motor applications leads to a high-performance level in electric motor applications. A VSD is an equipment that regulates the speed and torque of an electric motor. Many industrial processes must operate at different speeds for different products. In starting a motor, a drive initially applies a low frequency and voltage, thus avoiding a high inrush current associated with direct online starting. However, motor cooling deteriorates and can result in overheating as the speed decreases such that prolonged low-speed motor operation with significant torque is not usually possible without separately motorized fan ventilation.

Inverters in Industry: Solar Energy

Solar inverter can convert the variable DC voltage generated by PV solar panels into AC with power frequency, which can be fed back to commercial transmission systems or to off-grid power grids [3]. All the power generated by the solar panels can be exported through the inverter, as shown in Figure 2.22. With inverters, DC batteries can be used to provide alternating current for electrical

(a)

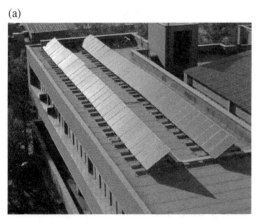

(b)

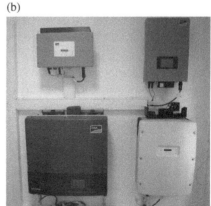

Figure 2.22   Application of a power electronic in renewable energy sources: (a) solar plant, (b) the inverter used for connecting the solar plant to the power grid.

appliances. Solar inverters have special functions in conjunction with photovoltaic arrays, such as maximum power point tracking and island effect protection.

2.1.4 Figure of Merits

An electric power converter is specified by its system performance indices. Many system performance indices, such as efficiency, are important in the design process of a system. A figure of merit is a quantity used to characterize the performance of a device, system or method, relative to its alternatives. The figure of merit of a power electronic converter represents an important quality as it pertains to its performance limits and is often used to drive the technology development in a specific direction. The converter figure of merits can be defined according to the input(s)/output(s) of the converter. The block diagram of a single-phase power electronic converter is shown in Figure 2.23.

To define the figure of merits, it is assumed that the converter input/output voltages/currents are periodic waveforms. A periodic waveform ϕ is defined as:

$$\phi(\omega t) = \phi(\omega t + 2\pi) \tag{2.2}$$

where ω is the fundamental angular frequency of the waveform ϕ.

2.1.4.1 Performance Indices at the Input of Power Electronic Converters

Electrical power quality is the study of how close to the ideal sinusoid the voltage and current waveforms are. In reality the voltage quality is of most concern as it is through the terminal voltage that the devices interact. Any deviation, whether momentary or sustained, in the voltage a device sees at its terminal is a power quality issue. A disturbance in the current that, flowing through the system and its impedance, results in poor voltage quality. Devices connected to the network will be subjected to this voltage waveform. The performance indices at the input of power electronic converters deals with standardization of the power quality at the power converter input. Total harmonic distortion (THD) is the well-known index used for evaluating the power quality. For the periodic waveform ϕ with a zero average value, the THD index is defined as

$$THD = \frac{\Phi_h}{\Phi_1} \tag{2.3}$$

where Φ_h and Φ_1 are the rms values of the harmonic content and fundamental part of the periodic waveform ϕ. The harmonic content of the periodic waveform ϕ is defined as

$$\phi(t) = \phi_1(t) + \phi_h(t) \tag{2.4}$$

where $\phi_1(t)$ and $\phi_h(t)$ are the harmonic content and fundamental part of the periodic waveform ϕ. Figure 2.24 shows two waveforms with poor THD.

Figure 2.23 A single-phase power electronic converter and its input and output variables.

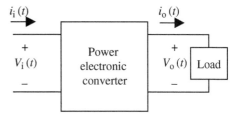

(a) (b)

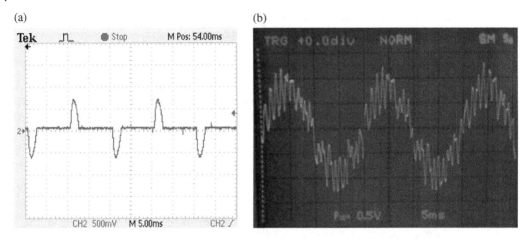

Figure 2.24 Two waveforms with poor THD: (a) low-frequency harmonics, (b) high-frequency harmonics.

2.1.4.2 Performance Indices at the Output of Power Electronic Converters

The quality of the converter output is evaluated by these indices. Some of the indices are defined with regards to the quality of the output waveforms. A non-ideal DC voltage waveform can be viewed as a composite of a constant DC component with an AC voltage as shown in Figure 2.25. The ripple component is often small in magnitude relative to the DC component. A ripple in electronics is the residual periodic variation of the DC voltage within a power supply that has been derived from an alternating current. This ripple is due to incomplete suppression of the alternating waveform after rectification. As a figure of merit, the ripple factor (RF) is defined to describe a non-ideal DC voltage. For the periodic waveform ϕ with a non-zero DC value, the RF index is defined as

$$RF = \frac{\Phi_{ac}}{\Phi_{dc}} \tag{2.5}$$

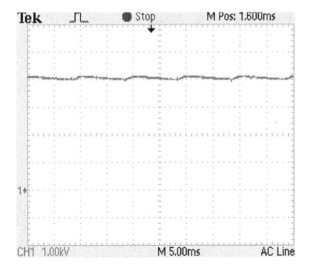

Figure 2.25 A real DC voltage with ripple voltage.

where Φ_{ac} and Φ_{dc} are the rms value of the AC content and average value of the periodic waveform ϕ, respectively. The AC content of the periodic waveform ϕ is defined as

$$\phi(t) = \phi_{dc}(t) + \phi_{ac}(t) \tag{2.6}$$

where $\phi_{ac}(t)$ and $\phi_{dc}(t)$ are the AC content and average value of the periodic waveform ϕ.

The ripple voltage usually originates as the output of the rectifiers. The amplitude of ripple voltage of a single-phase rectifier with a capacitive output filter is presented in the following. It can be seen that increasing the frequency leads to a decrease in the capacitance value of the converter:

$$V_{r_{PP}} \approx \frac{V_P}{R \cdot C \cdot f_0} \tag{2.7}$$

where f_o is the frequency of the voltage source.

Other indices can be defined to express the ability of the converter for continuous operation. The duty cycle index is a factor used to describe the operation mode of the converter. Figure 2.26 shows the label of a piece of equipment about its duty cycle.

2.1.4.3 Performance Indices for the Input/Output Relation

A huge power loss is the main drawback of linear converters. This power loss increases considerably when there is a great difference between the input and output voltages of the converter. Based on the presented efficiency relation, lower power losses leads to higher efficiency:

$$\eta = \frac{P_o}{P_o + P_{loss}} \tag{2.8}$$

where P_o is the output power and P_{loss} is the power loss.

2.1.4.4 Reliability Indices

Reliability is the probability of performing adequately to achieve the desired aim of the system [4]. This can be mentioned as a time-dependent equation. The reliability concept has more importance

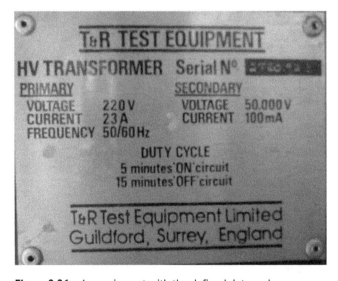

Figure 2.26 An equipment with the defined duty cycle.

in specific applications such as space equipment, where in the mission the equipment can hardly be replaced or be done by another system instead of the failed part. It is usual that some figure of merits such as efficiency are widely accepted by many users. However, reliability is less applicable than these labels. In general, the relation between reliability (*R*) and the failure rate can be deduced as

$$R(t) = e^{-\int_0^t \lambda(\tau)\,d\tau} \tag{2.9}$$

where λ is the system failure rate. Assuming a constant failure rate, this definition is changed to

$$R(t) = e^{-\lambda t} \tag{2.10}$$

Figure 2.27 shows the reliability curve for a power electronic converter.

2.2 Power Electronic Systems

The daily energy challenge and the issues caused by global warming leads to wide usage of renewable energy sources. Many of these energy sources are connected to the power grid by a power electronic interface. A combination of some energy sources, loads, and respective power electronic interfaces builds a power electronic system. Figure 2.28 shows a modern power system based on a high penetration ratio of the power electronic converters.

2.2.1 Power Electronic-Based Small Grids

Power electronic converters will serve as the fundamental components of modern power systems. A small grid can be as small as one converter and one load to greater grids with several converters and loads.

2.2.1.1 Microgrid
A microgrid is a decentralized group of electricity sources and loads that normally operates connected to and synchronous with the traditional wide-area synchronous grid, the macrogrid, but is able to disconnect from the interconnected grid and to function autonomously in island mode [6]. In this way, microgrids improve the security of supply within the microgrid cell, and can supply emergency power, changing between island and connected modes.

Figure 2.27 The reliability curve of a converter.

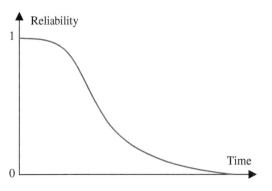

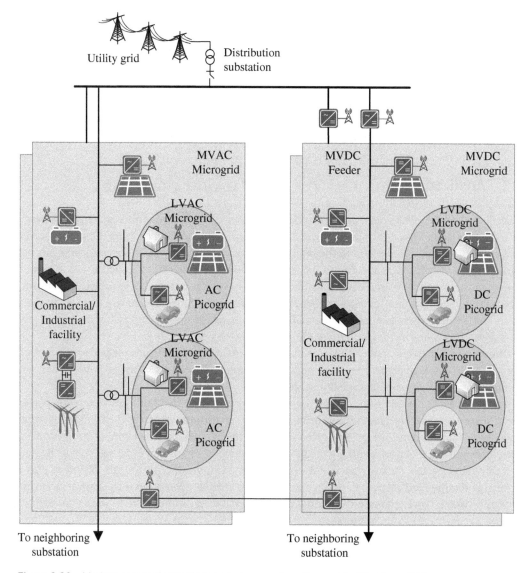

Figure 2.28 Modern power electronic based power system. *Source*: Modified from [10].

2.2.1.2 Picogrids

A picogrid is even smaller. Think of a cluster of homes connected to a single solar panel. The picogrid provides enough power to keep the lights on and charge cell phones. Picogrids fill in where people do not have access to a minigrid.

2.2.2 Power Electronic-Based Power Systems

Figure 2.1 shows a modern power system. As the penetration ratio of the renewable energy sources increases, the dependency of the power system to the power electronic converters increases. Therefore, the future power system is based on power electronics. In this power system, various power grids can be detected.

2.2.2.1 Local Grids

A local grid is a collection of electricity sources and loads that normally operates connected to the traditional wide area synchronous macrogrid, but is able to disconnect from the interconnected grid and to function autonomously in island mode. Microgrids and picogrids are examples of the local grids.

2.2.2.2 Large Networks

A combination of the traditional power grid based on the traditional big plants and power electronic-based local grids build a large network and supply all customers of the network.

2.3 Applications

Electric power converters deal with electrical energy, which is a commonly used type of energy in industries. Many industries need to use different types of electric motors as the prime mover and power electronic converters as the power conditioner for industrial processes. Demands for various types of converters are due to the fact that the most important characteristics of power converters vary with the type of application and the type of task they are expected to perform. There are some important industries where the electric power converters have a key role in their operation. Some of them are listed in the following. These are the most important industries in any country and consume a great amount of electricity.

Chemical industries: In a chemical factory, there are many applications for power converters. Power electronic converters are employed in a chemical factory as a power conditioner, for example for controlling the furnace temperature.

Food processes: In addition to similar applications of power electronic converters for power conditioning in a chemical factory and in a food factory, some types of power electronic converters have special applications in food industries. These are the converters with a special output voltage waveform and are used for sterilizing the foods. For example, a high-power pulsed voltage is used in a commercial accelerator for sterilizing vegetables and fruits without using any disinfectant.

Pulp and paper industries: In a paper factory, there are many motor-driven conveyers used for relocating wood. Power electronic converters are also used for controlling the process temperature and speed control of electric motors.

Metal forming industries: High-power presses and rolling systems are driven by electric motors. Precise speed control of these motors is a key factor in quality of the process and this task is performed with power electronic converters. Power electronic converters are also used for temperature control.

Petroleum process: Power electronics have been present in recent years in a wide number of applications within the oil and gas industry. Power electronic converters like VSDs and soft starters are suitable for extraction plants and chemical industries based on petroleum derivatives. Nowadays, under seas VSDs have an important role in petroleum extraction and are built based on power electronic converters.

Electronics: Advanced electronic industries are established based on the robotic process. A robot is driven and controlled using electric motors and power electronic converters, respectively.

Transportation: Nowadays, environmental pollution of the petroleum-based transportation vehicles has led to interest in electric transportation. An electric vehicle or electric train uses electric motors as the mover and power electronic converters as the controlling device of these motors. Power electronic converters are also used in modern industries for power conditioning on a very wide scale.

2.3.1 Power Range

The power range of the power electronic converters is typically from tens of watts to several hundred watts.

2.3.1.1 Sub-Watt Applications

Continuous reduction in transistor dimensions and constant advances in semiconductors, electronic devices, circuits, subsystems, systems, and memories have paved the way for the relentless progress of low-power electronics, which fuels digitalization in the modern era. Figure 2.29 shows a small power electronic converter in an integrated circuit.

2.3.1.2 Multi-Watt Applications

This category includes examples such as power adapters, personal battery chargers, and LED drivers. Figure 2.30 shows a high-voltage low-power DC to DC converter.

2.3.1.3 Kilo-Watt Applications

In industry, a common application is the VSD that is used to control an induction motor. The power range of VSDs start from a few hundred watts and end at tens of megawatts. Figure 2.31 shows typical applications of power electronics in this category.

2.3.1.4 Mega-Watt Applications

The increasing demand for establishing deep water enhanced oil recovery systems for subsea boosting, power generation, transmission from remote subsea marine current turbine farms, efficient power transmission, and operation of subsea dredgers and mining machines requires subsea-based power conversion, such as subsea pumps, compressors, and turbines that need to be operated at varying power and voltage levels. Even though technoeconomic studies indicate the absolute need for locating the multi-megawatt power electronic converters in the subsea, the recorded failure-in-time data of multi-megawatt capacity industrial standard power converters

Figure 2.29 A small power electronic converter in an integrated circuit.

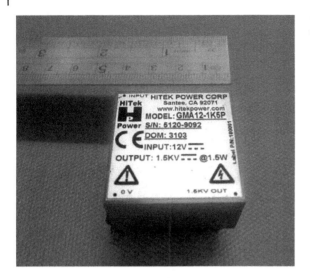

Figure 2.30 A high-voltage low-power DC to DC converter.

(a)   (b)

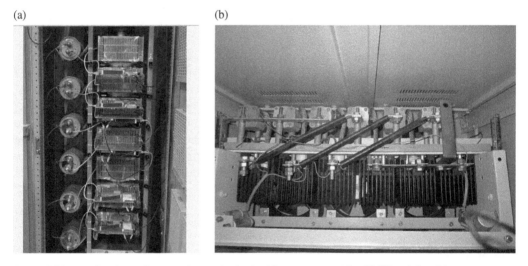

Figure 2.31 Applications of power electronics in the kilowatt range: (a) a 50 kW multilevel inverter, (b) a 20 kW controlled rectifier.

are not encouraging. Pulsed power is an example of a very high-power application, as shown in Figure 2.32.

2.3.1.5 Giga-Watt Applications

Long-term environmental concerns, population growth, and increased energy demand urge the development of clean- and green-energy-based power generation. High-power electronics plays a vital role in integrating the various renewable energy resources into the grid to address the current energy crisis [7]. Solar and wind are the fastest-developing sources of renewable energy. The grid should be able to accommodate such renewables without losing its reliability and robustness. The smart grid is an enhanced version of the conventional electricity grid, which enables

 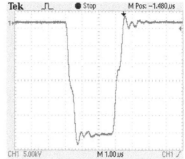

Figure 2.32 A high-voltage stacked switch with 6 MW nominal peak power and its output voltage.

 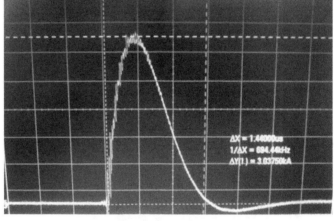

Figure 2.33 A high-voltage Marx generator with 1.2 GW nominal peak power and its load current.

energy security and reliability and the integration of various renewable energy resources. The pulsed power converters based on the gas-filled switches are also in this category, as shown in Figure 2.33.

2.3.2 Frequency Range

The power electronic converters operate from very low frequencies to very high frequencies.

2.3.2.1 Low Frequency Applications

Converters for high power drives operate at a low switching frequency in order to restrain the dynamic losses of the power semiconductor devices and solve problems such as EMI, differential-mode/common-mode voltages, and high cooling requirements. However, a lower switching frequency will lead to current distortion and reduction in the system control band; therefore, specific control strategies should be adopted. Figure 2.34 shows a water-cooled high power motor drive with a switching frequency of 800 Hz.

2.3.2.2 Medium Frequency Applications

The rectifier connected to the grid, conventional power supplies, and low-power motor drives operates at the medium frequency range. Figure 2.35 shows a high-voltage power supply that operates at medium frequency.

2.3.2.3 High Frequency Applications

Higher switching frequencies of a converter result in a reduction in size of the passive elements, such as capacitors, inductors, filter components, and transformers, which lead to a compact size,

Figure 2.34 A water-cooled high-power motor drive with a switching frequency of 800 Hz.

Figure 2.35 A high-voltage ozone generator that operates at 6 kHz.

Figure 2.36 A low-power high-frequency DC to DC converter that operates at 800 kHz.

weight, and the increased power density of the converters [8], as shown in Figure 2.36. There is a continuous demand to design a low-profile, high-power density, highly energy efficient, and fast dynamic response converter at a reasonable price for many portable appliances. The tremendous efforts in the improvement of switching devices such as GaN HEMTs/SiC MOSFETs, SiC diodes led to increased switching speeds of power semiconductor devices.

2.4 Mission Critical Applications

Resilient power electronic-based power systems are the leading providers of innovative energy solutions to electric utilities and their customers. They offer solutions that are reliable, cost effective, and versatile enough to meet the needs of today's ever-changing energy demands. The main goal of electricity provision is to supply consumer demand for whenever is needed. The growth in the number of electric vehicles, automation systems, and shifting to electric heaters increased the dependency of human activity on electricity. The designed new power system is expected to withstand disturbance and avoid significant power cuts in the system. The power system faces various disturbances including generator outage, components or line failure, animals or tree bark hitting the lines, etc. Power outage not only simply turns the bedside lamp off but also could lead to major disruption of industrial activities, discontinuance of health care center activities, communication network failure, and traffic control system interruptions [5]. The event of a major blackout may cause a huge financial loss depending on its severity. The electricity infrastructure is a critical lifeline system and of the utmost importance to our daily lives. Power system resilience characterizes the ability to resist, adapt to, and timely recover from disruptions. With an increasing awareness of such threats, the resilience of power systems has become a top priority for many countries. In this regard, resilience has a close meaning to availability. The concept of availability was originally developed for the systems that are required to operate continuously. Availability means the probability that a system is operational at a given time, i.e. the amount of time a device is actually operating as the percentage of total time it should be operating. High available systems may report availability in terms of minutes or hours of downtime per year. Availability features allow the system to stay operational even when faults do occur. A highly available system would disable the malfunctioning portion and continue operating at a reduced capacity. Availability is typically given as a percentage of a time period a system is expected to be in the operating state.

The meaning of failure in this chapter is the type of failure that leads to damage in a converter. Based on this view, four main reasons of failure are:

- Thermal shock
- Over voltage
- Mechanical forces
- Environmental effects

Some of the main reasons for failure in electric power converters are presented in this chapter. All these factors cause catastrophic damage in the systems. Over temperature, over voltage, mechanical forces, and environmental effects are the main factors of failure in power converters, which are described in detail in the following:

1) Over temperature is the most important failure factor in electric power converters. Most failure factors finally lead to an over temperature. Since the power conversion process is not ideal, heat generation is a common problem in all power converters. Of course, this is true for all electrical systems and not only for electric power converters. However, it is a serious problem in these converters because the amount of transferred energy is high and therefore their energy loss is considerable. Conduction loss in conductors and semiconductors, switching loss in switching devices, and core loss in the core of electric machines are the main mechanisms of energy loss in electric power converters. Transient phenomena like inrush currents with high power in a short time also cause over temperature damage. Damage due to over temperature is a long-term factor in comparison to damage due to electric breakdown.

2) Electrical breakdown or dielectric breakdown is a rapid reduction in the resistance of an electrical insulator when the voltage applied across it exceeds a specified voltage. Electric breakdown has a fast effect on the failure of converters. It acts to break down inside the body of the insulator or show a creepage on the surface of the insulator. Environmental factors have a direct effect on the creepage type of insulator breakdown. A breakdown in a solid insulator usually leads to permanent damage. Therefore, special attention to this failure factor is very important because low-power and low-voltage power converters use elements with solid insulators.

3) Mechanical factors affect over the long term in comparison to electrical factors. Vibration is an important mechanical factor in failure of rotary power converters. In addition, outside generated vibration causes mechanical damage of leads in electronic elements. In the mechanical world, a transient condition like mechanical shock occurs in a similar way to transient phenomena in the electrical world. Improper mounting of the elements can cause mechanical damage. Modal analysis is a tool for analyzing the mechanical behavior of structures and elements.

4) Operation in a harsh environment accelerates the failure due to the above-mentioned factors. Operation in a high ambient temperature accelerates the failure due to over temperature. Humidity and dust act by reducing the insulator resistance and cause electrical breakdown in lower voltages than the designed nominal values.

2.4.1 Cost Effects

The power sources that top businesses rely on should be resilient and always on; they should address both the causes and consequences of climate change [9]. Business continuity needs to be a top boardroom issue, and ensuring reliable energy must be part of every company's strategy. In manufacturing operations, even seemingly small power flickers can cause significant impact. Manufacturers are especially vulnerable to equipment damage during outages; electrical surges may occur when power is restored, resulting in costly damage to critical equipment that is difficult to repair. For high-volume manufacturers, a power outage can mean disruption that ripples through the entire supply chain. Shipping and receiving can come to a standstill, causing truck drivers and other means of transportation to be delayed or rescheduled. Data centers store and manage data for a wide range of companies. When they go down, a whole suite of businesses that rely on them for their mission-critical data storage are impacted.

2.4.2 Human Effects

Hospital power systems must run smoothly 24 hours a day, 7 days a week, 365 days a year–patient lives rely on the power always being on. That is why hospitals are required to have backup equipment, usually diesel generators, to ensure life-sustaining equipment stays on throughout an outage. However, these generators are often unreliable and require regular testing and maintenance to ensure they come on when they are needed most. Many high-voltage power electronic converters have human effects. Figure 2.37 shows an electromagnet used for producing radioactive drugs. As the lifetime of these drugs is very short, any interruption in the process leads to failure of the drug.

2.4.3 Qualification Effects

Some applications need to continue to keep the quality of the respective process or industry. Figure 2.38 shows a high-voltage power supply used for refining the outlet from a factory. It is seen that a short interruption in the power supply causes environmental issues.

Figure 2.37 An electromagnet used for producing radioactive drugs.

(a)   (b)   (c)

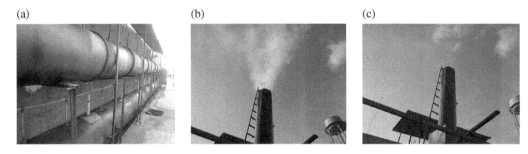

Figure 2.38 Application of a high-voltage power converter with a qualification effect: (a) the factory outlet and the positions of the high-voltage rods, (b) the factory outlet before the filter is powered on, (c) the factory outlet after the filter is powered on.

2.5 Summary and Conclusions

In this chapter, a brief introduction about the importance of power electronic converters was presented. Operation principles of these systems and their main failure factors were described. The outcomes of this chapter are summarized as follows:

1) Power electronic converters are electronic circuits used for power conditioning. They may be named as static power converters because they do not include any rotary part. Power conditioning is performed to achieve higher efficiency and better performance in an electrical energy conversion process. Power conditioning can be carried out with linear or switching electronic circuits. The main category of power electronic converters utilizes solid-state switches for various applications. These switches are the key parts of the converters for proper operation. As the power loss of the switch is very low, these types of power electronic converters have high efficiency. It is important today because loss minimization is a well-accepted method for reducing the electric energy demand. Four different types of switching power electronic converters are used in conversion of AC to DC, AC to AC, DC to AC, and DC to DC. All of these converters can be isolated by transformers or are not isolated. Some power electronic converters are faced with a power system from an input terminal (for example: rectifiers) or output terminals (for example: inverters). Thus, they are affected by problems occurring in the power system.

2) Over temperature is the most important failure factor in electric power converters. Most failure factors finally lead to over temperature. Damage due to over temperature is a long-term factor in comparison to damage due to an electric breakdown. An electrical breakdown or dielectric breakdown is a rapid reduction in the resistance of an electrical insulator when the voltage applied across it exceeds a specified voltage. Electric breakdown has a fast effect on the failure of converters. It acts to break down inside the body of the insulator or shows a creepage on the surface of the insulator. Environmental factors have a direct effect on the creepage type of insulator breakdown. Operation in a harsh environment accelerates the failure due to the above-mentioned factors. Operation in a high ambient temperature accelerates the failure due to over temperature. Humidity and dust act by reducing the insulator resistance and cause an electrical breakdown in lower voltages than the designed nominal values.

3) A resilient power electronic converter is a fault-tolerant converter with mission critical considerations. In fault tolerance, the system may be recovered successfully after the fault but a time interval is needed for recovery. This delay interval may not be accepted in mission critical applications.

References

1 Chen, Z., Guerrero, J.M., and Blaabjerg, F. (2009). A review of the state of the art of power electronics for wind turbines. *IEEE Transactions on Power Electronics* 24 (8): 1859–1875.

2 Emadi, A., Lee, Y.J., and Rajashekara, K. (2008). Power electronics and motor drives in electric, hybrid electric, and plug-in hybrid electric vehicles. *IEEE Transactions on Industrial Electronics* 55, 6: 2237–2245.

3 Carrasco, J.M., Franquelo, L.G., Bialasiewicz, J.T. et al. (2006). Power-electronic systems for the grid integration of renewable energy sources: A survey. *IEEE Transactions on Industrial Electronics* 53 (4): 1002–1016.

4 Yang, S., Bryant, A., Mawby, P. et al. (2011). An industry-based survey of reliability in power electronic converters. *IEEE Transactions on Industry Applications* 47 (3): 1441–1451.

5 Yang, S., Xiang, D., Bryant, A. et al. (2010). Condition monitoring for device reliability in power electronic converters: A review. *IEEE Transactions on Power Electronics* 25 (11): 2734–2752.

6 Flourentzou, N., Agelidis, V.G., and Demetriades, G.D. (2009). VSC-based HVDC power transmission systems: An overview. *IEEE Transactions on Power Electronics* 24 (3): 592–602.

7 Kouro, S., Malinowski, M., Gopakumar, K. et al. (2010). Recent advances and industrial applications of multilevel converters. *IEEE Transactions on Industrial Electronics* 57 (8): 2553–2580.

8 Hannan, M.A., Ker, P.J., Hossain Lipu, M.S. et al. State of the art of solid-state transformers: Advanced topologies, implementation issues, recent progress and improvements. *IEEE Access* 8: 19113–19132.

9 Taul, M.G., Wang, X., Davari, P., and Blaabjerg, F. (2020). Robust fault ride through of converter-based generation during severe faults with phase jumps. *IEEE Transactions on Industry Applications* 56 (1): 570–583.

10 Tafti, H.D., Konstantinou, G., Townsend, C.D. et al. (2020). Extended functionalities of photovoltaic systems with flexible power point tracking: recent advances. *IEEE Transactions on Power Electronics* 35 (9): 9342–9356.

3

Resilience in Power Electronics

3.1 Faulty Condition in Power Electronic Systems

Power electronic converters are faced with several faulty conditions during their operation. In a power electronic system, there are various types of connection between the source, the converter, and the load. The faulty conditions may be related to the converter or the load. Each of these faults leads to a level of system failure. Therefore, a system study is important to determine the requirements of resilience.

3.1.1 Single Source–Single Converter–Single Load

Many power electronic converters supply a single load via a single source, as shown in Figure 3.1a. Thus, the simplest configuration of the power electronic system is formed. For example, a house in a rural region is supplied with an inverter converting the direct current (DC) voltage of the solar panels to alternating current (AC) voltage. Of course, this is not a mission critical application. The mission critical isolated loads, communication transmitter/receivers in rural points, and traffic signals powered by a solar panel are several real case examples. In this case, any fault in the converter causes interruption of the load supply, as shown in Figure 3.1b. Therefore, the converter should be resilient against its internal faults. If a permanent fault occurs in the load, as shown in Figure 3.1c, resilient operation of the converter is meaningless because the load is permanently damaged. However, if the fault in the load is temporary, resilient operation of the converter is important because the converter can continue to supply the load after the fault is cleared. This is a very important and conceptual example of the resilience concept in power electronics. The converter may be highly reliable but not resilient. Reliable operation has various meanings. For example, the converter may have a fast protection system for isolating the converter from the load in a load short circuit fault. Figure 3.2a shows the switch, S, as an isolating protective switch. If a short circuit fault occurs in the load, the switch, S, opens the connection between the converter and the load, as shown in Figure 3.2b. If the load fault is permanent, resilient operation is not necessary because there is not a normal load to be supplied. However, if the fault is temporary, the load returns to its normal state and its supply can be continued. In this condition, if the protective switch, S, remains in the open state, the load supply is interrupted, as shown in Figure 3.2c. It is true that this protection system saves the converter from failures caused by the load short circuit current, but the load is not supplied after the fault is cleared. On the contrary, the resilient converter in Figure 3.3a is not rapidly isolated from the load in the load faulty conditions. It "resists" the load fault current for a

Resilient Power Electronic Systems, First Edition. Shahriyar Kaboli, Saeed Peyghami, and Frede Blaabjerg.
© 2022 John Wiley & Sons Ltd. Published 2022 by John Wiley & Sons Ltd.
Companion website: www.wiley.com/go/kaboli/resilientpower

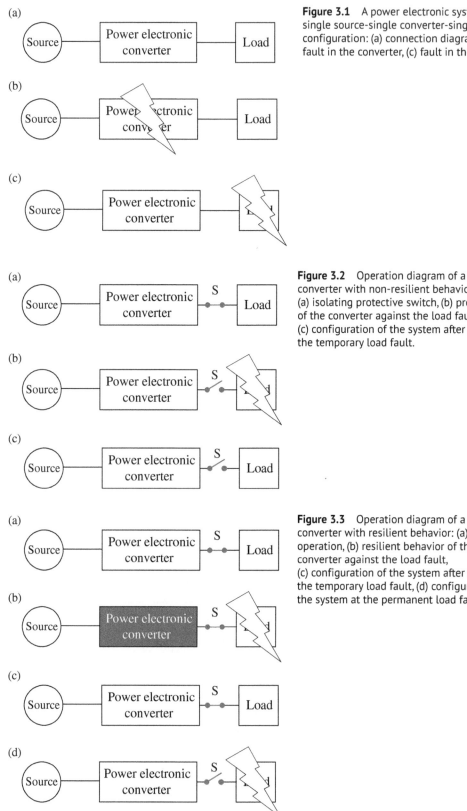

Figure 3.1 A power electronic system with single source-single converter-single load configuration: (a) connection diagram, (b) fault in the converter, (c) fault in the load.

Figure 3.2 Operation diagram of a converter with non-resilient behavior: (a) isolating protective switch, (b) protection of the converter against the load fault, (c) configuration of the system after removing the temporary load fault.

Figure 3.3 Operation diagram of a converter with resilient behavior: (a) normal operation, (b) resilient behavior of the converter against the load fault, (c) configuration of the system after removing the temporary load fault, (d) configuration of the system at the permanent load fault.

certain time interval, as shown in Figure 3.3b. If the fault is removed in this time interval, operation of the system is continued, as shown in Figure 3.3c. If the fault is permanent, the protection system operates to isolate the converter from the load as shown in Figure 3.3d.

3.1.2   Single Source–Single Converter–Multi Loads

Another configuration of the power electronic system is a power source that supplies several loads via one power electronic converter, as shown in Figure 3.4a. A satellite is an example of this configuration. In a satellite with a mission critical system, there are many subsystems whose operation is critical for the mission. An on board computer (OBC), camera, and actuators are examples of these subsystems. In this case, any fault in the converter leads to the interruption of the load supply, as shown in Figure 3.4b. Therefore, the converter should be resilient against its internal faults. Interesting situations occur during load faults, as shown in Figure 3.4c, d. In this case, the converter should be resilient, even against permanent faults in the load, because other loads could be supplied.

3.1.3   Single Source–Multi Individual Converter–Multi Load

In this configuration, several loads are supplied by individual converters, which are connected to a single power source, as shown in Figure 3.5. An example of this configuration is a power system with various required voltage levels from a single power source. In this case, the converters should be resilient against their internal faults. The permanent faults of the loads cancel the necessity for resilience in the converters. However, the converter should be resilient against the temporary faults of the loads, as explained in Section 3.1.1. In this case, it is important that no fault causes the failure of the source.

3.1.4   Multi Individual Source–Multi Individual Converter–Multi Load

The configuration shown in Figure 3.6 is the form of a system with multi power sources that are connected to the loads via individual converters. This is a configuration similar to the one described in Section 3.1.1. The only additional important issue is attention to the cross interference between the converters and loads during the faulty cases. In a comparison between this case and the case described in Section 3.1.2, it is noted that the faults of each load have no direct impact on the other loads in this case.

3.1.5   Multi Source–Multi Converter–Multi Load

In this case, shown in Figure 3.7a, each load can be supplied from one of the sources and its respective converter during a fault in the other converter, as shown in Figure 3.7b, c. In the case of a load fault, the other load can be supplied by both sources and the converters. In this case, the "system" is resilient, which means that the elements of the system may be non-resilient but the whole operation of the system is resilient. The effect of permanent or temporary faults and their relationship with a protection system should be considered.

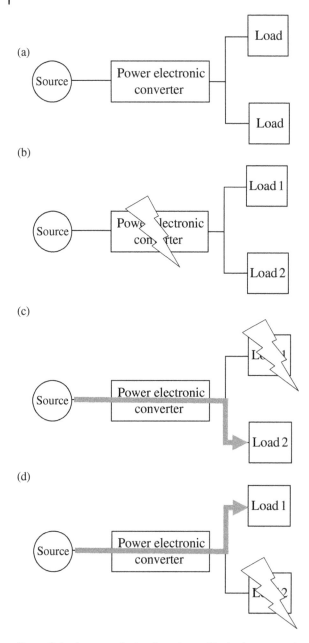

Figure 3.4 A power electronic system with single source–single converter–multi-load configuration: (a) connection diagram, (b) fault in the converter, (c) and (d) fault in the load.

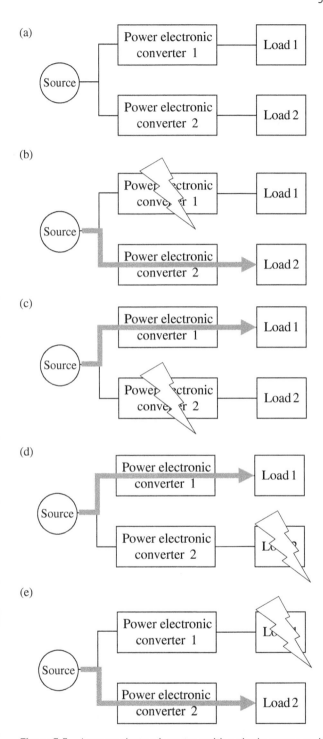

Figure 3.5 A power electronic system with a single-source multi-individual converter–multi-load configuration: (a) connection diagram, (b) and (c) fault in the converter, (d) and (e) fault in the load.

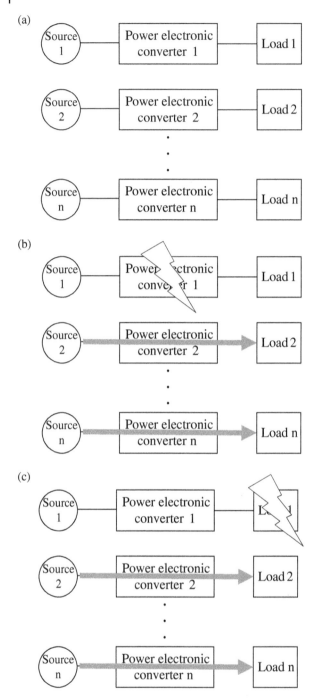

Figure 3.6 A power electronic system with a multi-individual source–multi-individual converter–multi-load configuration: (a) connection diagram, (b) fault in the converter, (c) fault in the load.

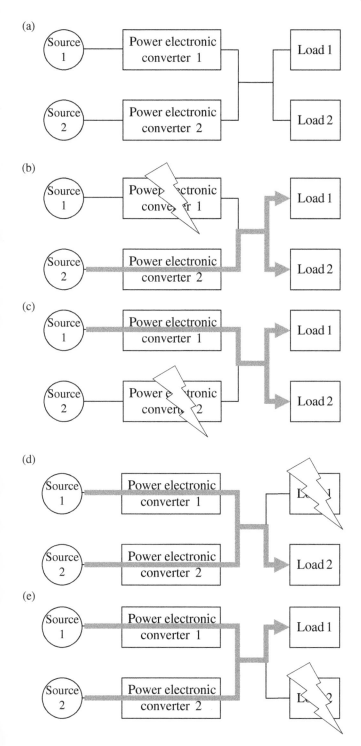

Figure 3.7 A power electronic system with a multi-source–multi-converter–multi-load configuration: (a) connection diagram, (b) and (c) fault in the converter, (d) and (e) fault in the load.

3.1.6 Single Source–Multi Converter–Multi Load

This configuration, shown in Figure 3.8, is similar to the previous case. The source should be capable to supply both the loads.

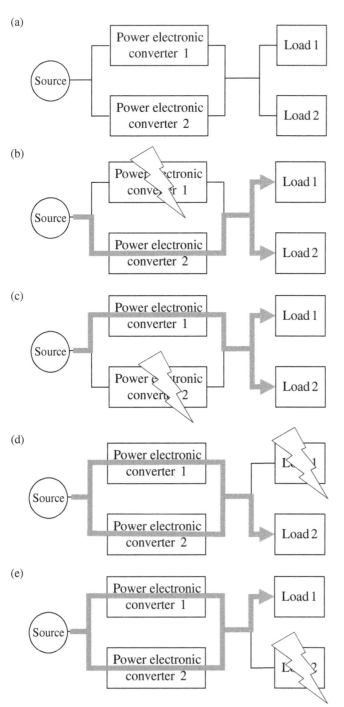

Figure 3.8 A power electronic system with single-source–multi-converter–multi-load configuration: (a) connection diagram, (b) and (c) fault in the converter, (d) and (e) fault in the load.

3.2 Fault Types

Power electronic converters are faced with several faulty conditions during their operation. Failure may be the type of failure that leads to damage in a converter or the faults that cause the system to malfunction. Based on this view, the main reasons for failure or fault in a power electronic system are:

- Thermal shock
- Overvoltage
- Mechanical forces
- Environmental effects
- Noise effect

These faults may be generated as internal or external faults. The meaning of an external fault is a fault at the converter input/output terminals that is propagated inside the converter.

3.2.1 Internal Faults

These faults occur in the elements of the converter. In the following, several internal faults in power electronic converters are presented.

3.2.1.1 Thermal Shock

Over temperature is the most important factor of failure in all of systems. Thermal damage is a very important factor in the failure of electric power converters. There are two scenarios for thermal damage: over temperature and thermal shock. Thermal shock occurs when a thermal gradient causes different parts of an object to expand by different amounts. At some point, this stress can exceed the strength of the material, causing a crack to form. Figure 3.9 shows some failed parts due to temperature problems.

3.2.1.2 Over Voltage

Electrical breakdown or dielectric breakdown refers to a rapid reduction in the resistance of an electrical insulator when the voltage applied across it exceeds the breakdown voltage. This results in a portion of the insulator becoming electrically conductive. Under sufficient electrical

(a)

(b)

(c)

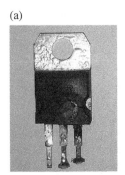  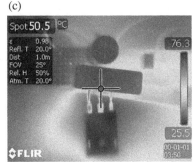

Figure 3.9 Some failed parts of the converters due to the temperature problems: (a) damaged switch with over temperature, (b) damaged transformer winding because of the over temperature, (c) infra-red (IR) photo of the temperature distribution of a converter.

(a)   (b)

Figure 3.10 Two failed parts due to the electric breakdown: (a) capacitor, (b) transformer winding.

stress, electrical breakdown can occur within solids, liquids, gases, or non-ideal vacuum. Electrical breakdown is often associated with the failure of solid or liquid insulating materials used inside high voltage transformers or capacitors in the electricity distribution grid. Dielectric breakdown within a solid insulator can permanently change its appearance and properties. Figure 3.10 shows two failed parts because of the overvoltage.

3.2.1.3 Mechanical Forces

One of the most likely failure modes is the power electronic package flexing, which can cause the solder joints on the package to crack, causing intermittent or complete failure of the connection. It is necessary to keep the package as compact as possible. The smaller the package, the smaller is the overall flex of the board. Figure 3.11 shows the effect of mechanical forces on the power electronic components.

3.2.2 External Faults

An external fault is the fault at the converter input/output terminals, which is propagated inside the converter. In the following, several external faults in power electronic converters are discussed.

3.2.2.1 Electrical Faults

The most likely failure modes of the power electronic converters begin when one part of the system fails. This external part may be outside the converter but its effect is sensed inside the converter. Thus, an "external" fault is changed to an "internal" fault. When this happens, nearby nodes must then take up the slack for the failed component. This in turn overloads these nodes, causing them to fail as well, prompting additional nodes to fail one after another in what is also known as a vicious circle. Cascading failure is common in power grids when one of the elements fails and shifts its load to nearby elements inside the converter. Those nearby elements are then pushed beyond their capacity so they become overloaded and shift their load onto other elements. Cascading failure is a common effect seen in high voltage systems, where a single point of failure on a fully loaded or slightly overloaded system results in a sudden spike across all nodes of the system. This surge current can induce the already overloaded nodes into failure, setting off more overloads and thereby taking down the entire system in a very short time. This failure process cascades through the elements of the system like a ripple on a pond and continues until substantially

(a)

(b)

(c)

(d)

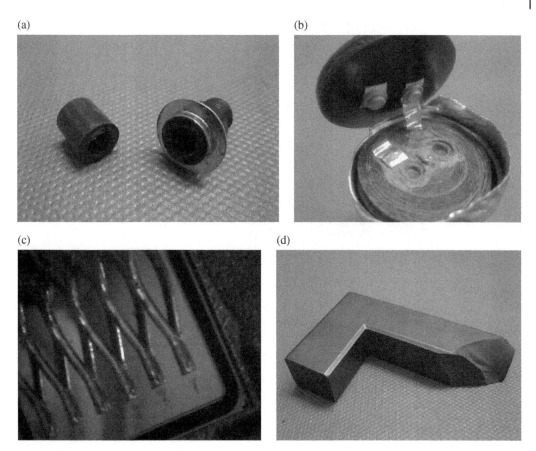

Figure 3.11 Effect of mechanical forces on the components: (a): broken connector, (b) disconnected lead of a capacitor, (c) internal wires of a switch module, (d) inductor ferrite core.

all of the elements in the system are compromised and/or the system becomes functionally disconnected from the source of its load. For example, under certain conditions a large power grid can collapse after the failure of a single transformer.

3.2.2.2 Environmental Factors

Various factors exist in the environment that affect the useful life of the converters. Dust/dirt can accumulate on a board surface, reducing the dielectric strength. Many electronic devices have humidity specifications, for example, 5–95%. At the top end of the range, moisture may increase the conductivity of permeable insulators, leading to malfunction. Too low humidity may make materials brittle or cause an electrostatic discharge. Figure 3.12 shows the effect of humidity on a transformer core.

3.2.3 Malfunctioning

One category of faults is one that does not cause a catastrophic failure but leads to malfunction of the system. One of the best examples is electromagnetic interference (EMI). Electromagnetic

Figure 3.12 Effect of humidity on the transformer core.

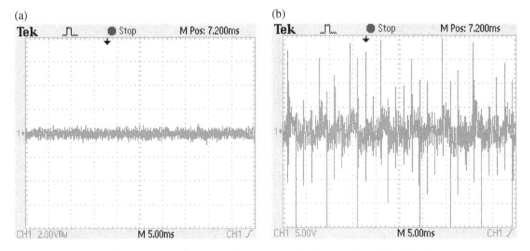

Figure 3.13 Effect of the noise on the signal of converter: (a) without noise, (b) with noise.

compatibility (EMC) is the branch of electrical sciences that studies the unintentional generation, propagation, and reception of electromagnetic energy with reference to the unwanted effects that such energy may induce. The goal of EMC is the correct operation, in the same electromagnetic environment, of different pieces of equipment that use electromagnetic phenomena, and the avoidance of any interference effects. Figure 3.13 shows an example of the noise effect on the reference signal of a power electronic system. In order to achieve this, EMC pursues two different kinds of issues. Emission issues are related to the unwanted generation of electromagnetic energy by some source and to the countermeasures that should be taken in order to reduce such generation and to avoid the escape of any remaining energy into the external environment. Susceptibility or immunity issues, in contrast, refer to the correct operation of electrical equipment, referred to as the victim, in the presence of unplanned electromagnetic disturbances. Interference mitigation and hence EMC is achieved by addressing both emission and susceptibility issues, i.e. diminishing the sources of interference and hardening the potential victims. The coupling path between the source and victim may also be separately addressed to increase its attenuation.

3.3 Availability

Availability means the probability that a system is operational at a given time, i.e. the amount of time a device is actually operating as the percentage of total time it should be operating. Availability features allow the system to stay operational even when faults do occur. A highly available system would disable the malfunctioning portion and continue operating at a reduced capacity. In contrast, a less capable system might crash and become totally non-operational. In the case of resilient power electronic systems, the availability means that the converter continues load support, even during the fault, with an acceptable value of the converter output.

3.3.1 Load Supporting During the Fault

Protection methods save the converter from non-catastrophic faults. For example, a fast protection system can save the converter against a load short-circuit fault. However, what happens if the load fault is temporary? Figure 3.14 shows a temporary overcurrent fault versus the time. A resilient power converter continues the load support. However, the protection system saves the converter but it also takes the converter out of service. This issue is about converters that are not damaged but cannot operate normally. The concept of availability was originally developed for repairable systems that are required to operate continuously. A system may be unavailable although none of its parts are damaged. In fact, there is an important difference between reliability and availability. A converter may be highly reliable but unavailable and vice versa.

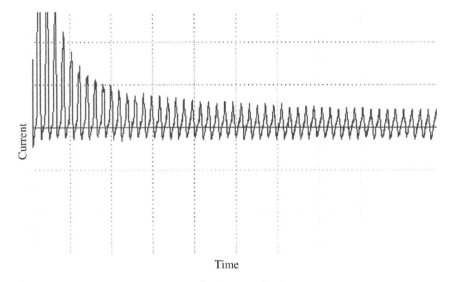

Figure 3.14 A temporary overcurrent fault versus the time.

3.3.2 Acceptable Limits of The Converter Output

The voltage and respective current of the power electronic converter outputs is defined for each load. In a resilient converter it is important to keep the value of the output within the defined range. If the output has a value beyond acceptable limits for a certain time interval, it is important to keep this time interval as low as possible. Figure 3.15a shows a voltage drop in the output of a non-resilient converter. It seems that the voltage drop and the time are considerable. Figure 3.15b shows the same voltage for a resilient converter. The voltage drop and its time are reduced significantly.

3.4 Road to Resilience

The resilience characteristics are achieved by various methods. These methods can be sorted with regards to the time of applying and the level of applying.

3.4.1 Level of Resilience

As described in Chapter 1, the system can be deposited to its subsystems or elements. In a power electronic system, the method of resilience can be applied at various levels: element level, converter level, and system level.

3.4.1.1 Element Level

Reliability of a converter element increases if it operates at a set point with low stress. The method for reducing an electric field are one of these methods. Low temperature operating conditions for an electric power converter is one method. Series connection for voltage sharing and parallel connection for current sharing is recommended. Novel control methods of power converters for reducing the complexity and reliable operation are presented. Control of an inrush current as a typical transient problem in electric power converters is recommended. Methods for preventing the

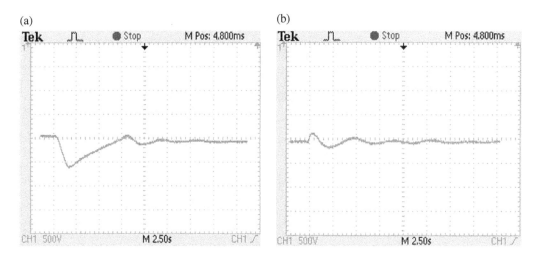

Figure 3.15 Comparison between the operation of a resilient converter and the non-resilient one: (a) non-resilient converter, (b) resilient converter.

(a)   (b)

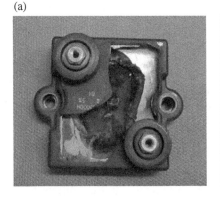 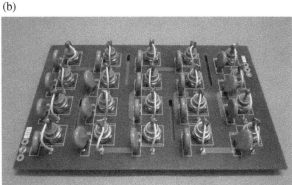

Figure 3.16 Resilience at the element level: (a): a non-resilient resistor, (b) increasing the resilience by series connection of the diodes.

overstress condition on the components in faulty cases helps to extend the resilience time interval. Techniques for reducing mechanical and environmental stress exist. Figure 3.16a shows a resistor that is not resilient against the short-circuit current. Figure 3.16b shows the series connection of the diodes for reducing voltage stress.

3.4.1.2 Converter Level
At this level, the converters are considered as the system of elements. Methods of redundancy, parallel operation, and sharing of the load are some of the methods in this category. Some of the above-mentioned methods can be applied in this category because the converters are the system elements at this level.

3.4.1.3 System Level
Establishing the paths for supporting the load during faults in the system is a method of resilience at the system level.

3.4.2 Time of Applying the Methods

The resilience methods can be applied with regard to the time of the fault. Thus, there are pre-fault or post-fault methods.

3.4.2.1 Pre-Fault Methods
These methods provide information about the fault before its occurrence or the conditions that the converters operate under a lower stress.

- Prediction of failure
  Implementation of the methods of reliability improvement needs an important tool: condition monitoring of the power converter. Condition monitoring allows information to be obtained about the growth of failure in power converters. A decision about this failure is related to the data that are obtained from the condition monitoring system. For example, monitoring of a long-term high output current beyond a nominal specification of a power electronic converter usually leads to operating the protection system and taking it out of service. As another alternative, monitoring a high temperature hotspot in this converter may lead to application of a derating

scenario and holding the converter in service. There are two general goals for monitoring the state of an electric power converter:

1) Monitoring for controlling a variable (usually output voltage of the power converter)
2) Condition monitoring for preventing a catastrophic failure

Monitoring is a key function during implementation of any control process. All closed-loop control schemes work based on monitoring the output variable of the process. However, converter condition monitoring methods are also used for informing about the state of the converter from a failure point of view. Many reliability improvement techniques need to have a view about the state of an electric power converter. Condition monitoring is the technique of monitoring a parameter in a power converter in order to identify a considerable change, which is an index of a developing fault. Conditional monitoring of power converters has many benefits for the converter. Condition monitoring is important in certain conditions that would shorten a normal lifespan and can be informed before they lead to a major failure. Condition monitoring allows scheduling of the maintenance to prevent failure and avoid its consequences. The monitoring is applied at the component level [1–4] and the system level [5–10].

- Stress reduction

  The failure factors act from the beginning of a converter application. In the long term, they cause the converter to age and its failure. The time interval for changing a stress factor to a failure factor is directly related to the value of stress. Higher stress leads to a shorter time to failure and vice versa. Thus, the first method of reliability improvement is to reduce the stress factors on the converter.

3.4.2.2 Post-Fault Methods

These methods are applied after the fault occurrence.

- Protection

  In general, we try to postpone the failure time by reducing the stress on the converter. However, the converter must operate and it is logically affected by stress. It is true that we reduce the stress but the stress still exists and causes failure in the long term. Suppose we are faced with a fault in the converter. What should we do? This section deals with one of the methods for preventing catastrophic damage in faulty converters: protection systems. Protection systems act when a fault occurs in the converter. Their performance is very important; isolation of the converter is not always the best choice because this strategy has a bad effect on availability of the converter. In the current chapter, we are not sensitive to this concern. However, we can consider this from another viewpoint; any failure factor needs to a time interval for damage to the converter. If the protection system is very fast, it protects the converter but conflict with noise is possible. Therefore, protection systems should be fast enough to protect the converter but not so fast as to cause incorrect operation. Protection techniques prevent catastrophic failures in electric power converters. However, protection systems cannot prevent any fault. There are two main approaches to a faulty converter. Isolating the faulty converter is one of the approaches and is used when the converter mission can be interrupted for repair and maintenance. If the mission of the converter is important at the time of failure, users should use other methods.

- Derating

  Derating is a method for allowing the faulty converter to continue its mission but with a lower performance. Derating increases the margin of safety between part design limits and applied

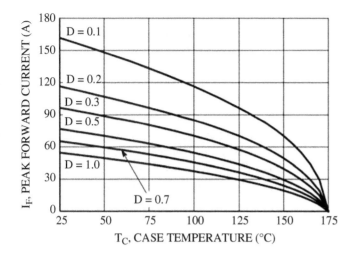

Figure 3.17 Derating curve of a diode. *Source:* Used with permission from SCILLC dba ansemi.

stresses, thereby providing extra protection for the part. By applying derating in an electrical or electronic component, its degradation rate is reduced. The reliability and life expectancy are improved. Figure 3.17 shows the derating curve for a diode.

3.5 Inherently Resilient Converters

A resilient system is a fault-tolerant system with mission critical considerations. There are some approaches to handle a fault in a power electronic converter. The methods of reliability improvement as well as reliability calculation techniques help to provide a safe converter without catastrophic failures. A derating method is usually an exclusive method of reliability improvement for a faulty converter. However, derating is derating! It usually means the derated converter continues to operate with new rated characteristics that are less than the converter original nominal specifications. In many cases, this is not acceptable and the original nominal rating of the converter is required. For example, consider a DC power distribution unit (PDU) with several output voltage levels. In a PDU, as presented for a satellite in Chapter 2, there are several voltage regulators that provide some output voltage levels from a common DC input voltage source such as a battery. In this system, failure in one of the output voltage levels causes failure in the subsystem related to the failed output voltage level. This is true when just one of the output channels fails but not the whole PDU, as the system does not operate properly even though other output channels operate normally.

Fault-tolerant methods are the solution for this drawback [11]. Fault tolerance is the property that enables a system to continue operating properly in the event of the failure of some of its components. A fault-tolerant design enables a system to continue its intended operation, possibly at a reduced level, rather than failing completely when some part of the system fails. There are two main approaches in a fault-tolerant system: using redundant systems and reconfiguration. In the field of power electronics, fault-tolerant techniques can be classified into three categories based on the type of hardware redundancy unit: element-level, converter-level, and system-level categories. Also, various fault-tolerant methods are assessed according to cost, complexity, performance, etc. It is important to note that the fault-tolerant methods are usually applied for internal faults.

3.6 Maintenance Scheduling

Proper planning of maintenance helps to find the system problems before the fault occurs. In general, the fault occurs when the level of the system stress reaches the level of the system strength. Figure 3.18a shows this concept. Proper maintenance time separates the strength distribution of the system from the stress distribution at the time before increasing the probability of the fault, as shown in Figure 3.18b.

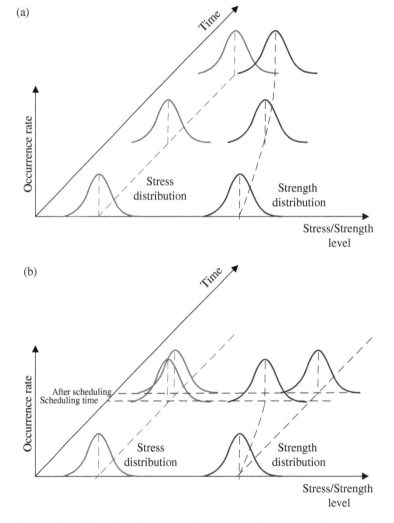

Figure 3.18 Effect of maintenance scheduling on the system fault: (a) without maintenance, (b) with maintenance.

3.7 Two Case Studies

The aim of this section is to clarify the difference between resilience and fault tolerance. We present two case studies: one with a description of a fault-tolerant system and the other that tries to achieve a resilient system.

3.7.1 Resilient Converter with a Low System Performance Index Drop

In this section, a fault-tolerant power electronic converter is explained. It will be shown that this power converter is fault tolerant against failure of its internal components. However, it may or may not be resilient. It depends on the defined mission profile [12]. If the recovery time of the output voltages in the converter is less than the critical defined time interval of the mission, the converter is resilient. This is a PDU and its output voltage value is considered as the system performance index. The PDU is one of the most critical units in an electric equipment since all the equipment subsystems require a continuous, regulated, and conditioned power supply to maintain their function. A DC PDU consists of some DC voltage regulators that receive input power from a DC voltage source and convert it into a number of regulated DC voltage levels. A DC PDU with two output voltage levels and a battery as the input voltage source is shown in Figure 3.19. The reliability of a PDU is very important in some applications, such as telecommunications systems in remote areas in which replacing the faulty components is difficult or impossible. There are some commonly used techniques to improve the reliability of systems. Some researchers have attempted to increase the reliability in a designing process by reducing the stress of components, using simpler circuit structures than the conventional structure and fast detecting the fault and preventing the catastrophic breakdown of the component. These techniques improve the reliability to some extent, but if the detected fault cannot be removed or if there is a catastrophic breakdown of components, then the PDU fails. Therefore, it is necessary to design a system where failure/damage of a component/subsystem does not result in the failure of the whole PDU. To achieve this goal, fault-tolerant operation methods are proposed. The fault-tolerant operation methods are recognized in two ways: use of control strategies and use of redundancy for components with the probability of more failures. Control strategies are used to reconfigure a failed structure into a new structure in order for the system to continue to work with its nominal or

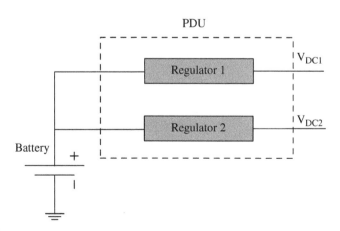

Figure 3.19 A basic structure for PDU with two output voltage levels.

de-rated power. Redundancy is defined as the use of additional components or subsystems beyond the number actually required for the system to operate reliably. Redundancies can be categorized as active (parallel) or passive (standby). In systems with active redundancy, all redundant components are in operation and with the failure of one component, the surviving components carry the load. In a passive or standby system with a switching device, the redundant component starts to operate only when one or more of the components fail. The conventional fault-tolerant techniques are classified into the component-level, module-level, and system-level solutions, based on the type of hardware redundancy. In all of these methods, exclusive redundancy is a common characteristic.

On the other hand, the failure rates of voltage regulators of a DC PDU are different because of single events or a different temperature rise. This means that each component or module with a high failure rate has only its exclusive redundant part. However, this characteristic leads to a serious drawback in a DC PDU: if one of the DC voltage regulators and its redundant units fail, then the DC PDU fails, while other redundant units from the other regulators are not used. This drawback and our proposed idea for solving it are presented in Figure 3.20 for a PDU with two DC

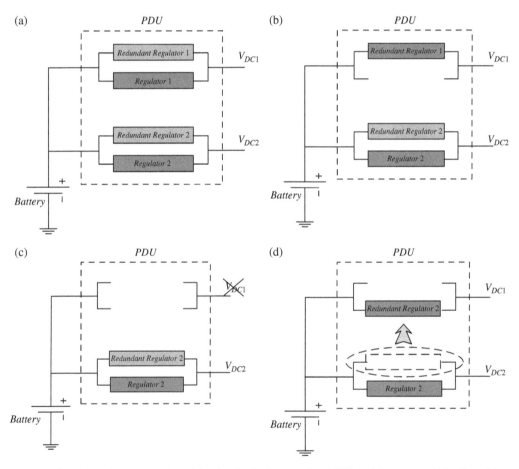

Figure 3.20 Illustrative presentation of the drawback of conventional PDU and the proposed idea for solving it: (a) normal operation, (b) main regulator fails, (c) main and redundant regulator fail, (d) the proposed idea.

output voltage levels. Each regulator module has one redundant module. In Figure 3.20a the PDU in its normal state is shown. If one of the voltage regulators fails, it is replaced by its exclusive redundant regulator, as shown in Figure 3.20b. If both the main and redundant voltage regulators of a DC voltage level fail, the respective voltage level is interrupted, which leads to the PDU malfunction, as shown in Figure 3.20c. In this section, in order to have a DC PDU with a high reliability, the use of non-exclusive adjustable standby regulators is suggested. In this approach, each redundant regulator can be used instead of any failed regulator. The idea of the proposed method is shown in Figure 3.20d for resuscitation of the failed DC PDU shown in Figure 3.20c. In Figure 2.2d, both the main and redundant regulators of a DC voltage level (VDC1) are damaged. Since the adjustable redundant regulator 2 is not exclusive, it can be adjusted to VDC1 and then placed in the faulty direction to provide VDC1. In this structure, the PDU fails only when all of the redundant units are used. The proposed method is presented for a DC PDU with standby redundancy. However, the proposed idea can also be used for active (parallel) redundancy. In a DC PDU with redundancy, input and output switches are used to isolate the faulty voltage regulator from the input voltage source and load, respectively. The combination of a regulator with input and output isolating switches makes a regulator module.

Figure 3.21 shows a regulator module, which consists of a regulator, an input isolating switch, Si, and an output isolating switch, So. In the conventional DC PDU, exclusive regulator modules are used as standby redundancies. In Figure 3.22, a conventional DC PDU with M output voltage levels is shown. In this study, we consider one redundant regulator for each main regulator. The number of main voltage regulators and the number of redundant voltage regulators are equal to the number of voltage levels of the PDU, M. In this PDU, the kth DC output voltage level (VDC-k in Figure 3.22) is normally produced by the *Main Regulator-k*. In a faulty condition, there are three steps that should be performed correctly in order to return to the normal condition. These steps are: fault Detection, failed regulator Identification, and PDU Reconfiguration (DIR). Reconfiguration means that the failed *Main Regulator-k* is isolated by the switches SiM-k and SoM-k and is replaced by exclusive *Redundant Regulator-k* via the switches SiR-k and SoR-k. In the conventional structure, if a main regulator and its standby regulator fail, then the PDU fails. In this structure, other unused exclusive standby regulators are not used instead of the faulty regulators. In this method, the use of the non-exclusive adjustable regulators as standby units is proposed. In this structure, the PDU fails only when all of the PDU standby units are used. In the proposed structure, the main regulator module is similar to the one in the conventional DC PDU. The key idea for a redundant module is that it must be able to connect to all of the output lines. Figure 3.23 shows a redundant module in the proposed structure.

The number of isolating output switches and the number of adjusting switches are equal to the number of PDU output DC voltage levels, M. Each of the isolating output switches operates

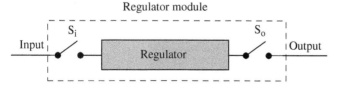

Figure 3.21 A regulator module.

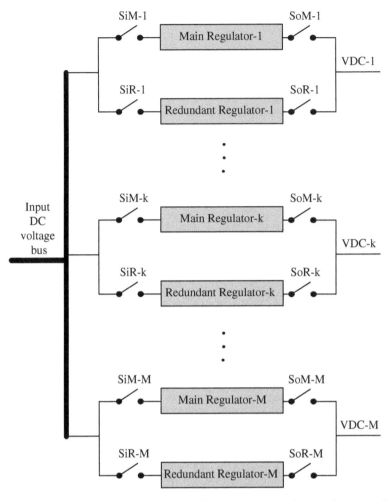

Figure 3.22 A conventional PDU with M DC output voltage level and one standby redundancy for each voltage regulator.

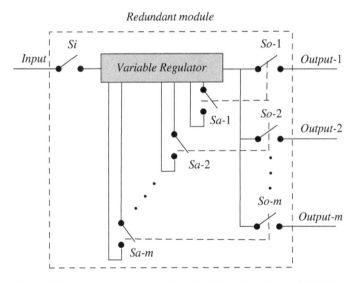

Figure 3.23 A redundant adjustable module in the proposed structure.

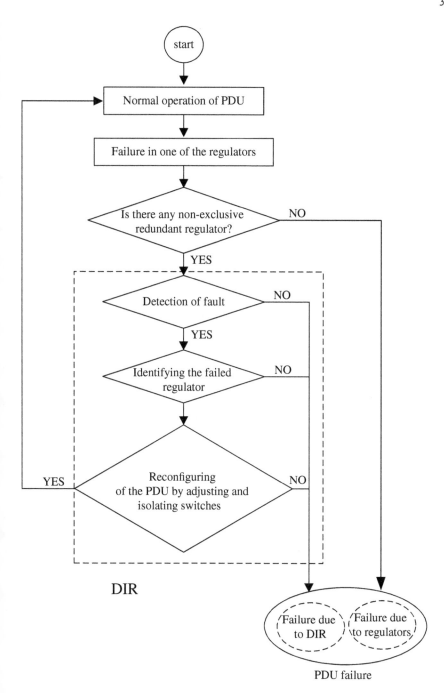

Figure 3.24 Flow diagram of the proposed PDU.

synchronously with one of the adjusting switches. The output voltage of the adjustable regulator is set to a desired DC voltage by one of the adjusting switches and then is connected to one of the output lines by its respective isolating switch. Figure 3.24 shows the flow diagram of the proposed PDU. A PDU with this proposed structure and M output DC voltage levels is shown in Figure 3.25. In this structure, before replacing the failed regulator, the redundant regulator is

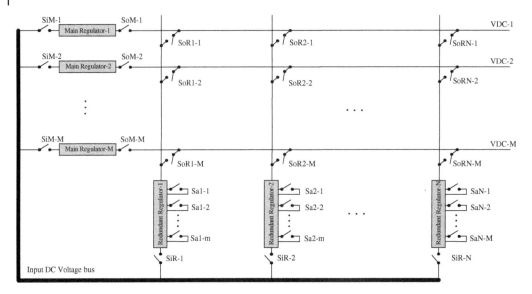

Figure 3.25 The proposed PDU with M DC output voltage level and N adjustable non-exclusive redundancy.

adjusted for the failed voltage level. If there is at least one redundant regulator, neither failure in the main regulators nor failure in the redundant regulators cause the failure of PDU, as shown in Figure 3.26. This gives more reliability using the proposed structure. Since the redundant regulators are not exclusive in the proposed structure, the number of redundant modules can be less than the number of main modules. In this study, switching DC/DC converters are used as adjustable regulators. The output voltage of the switching DC/DC converters is controlled by a duty cycle. Therefore, the output voltage of the PDU regulators can be tuned by adjusting the duty cycle. The most important parameters that affect the design of switching regulators are input voltage, output voltage, and load current. The input voltages of the converters are the same, so the output voltage and load current of the converters should be similar to each other. Great differences between the output voltage and the load current of these converters result in a non-optimized design with a high voltage ripple and poor efficiency. However, a low-quality PDU with normal operation is usually preferred to a PDU where some of its output voltages have stopped.

In order to compare the reliability of these two structures, we compare them numerically. In this analysis, LM2576 is used as the regulator. The most important element in the DIR process is the microprocessor, ATMEGA32. Table 3.1 shows the characteristics of the analyzed PDU. Detailed calculations of the component failure rate were performed based on MIL-HDBK-217. The reliability curves of a conventional PDU with $M = 2$, 6 and a proposed PDU with $M = N = 2$, 6 are shown in Figure 3.27. The assumed reliability for a safe operation time was considered to be greater than 0.99. Based on this definition, the reliability curves with a value greater than 0.99 are shown in Figure 3.28. In this figure, the safe operation time is shown for the conventional PDU with $M = 2$, 6 and the proposed PDU with $M = N = 2$, 6. The safe operation time of the proposed structure is remarkably longer than the conventional PDU. This is the most important advantage of using the proposed structure and means that we can guarantee that the PDU works correctly for a longer time.

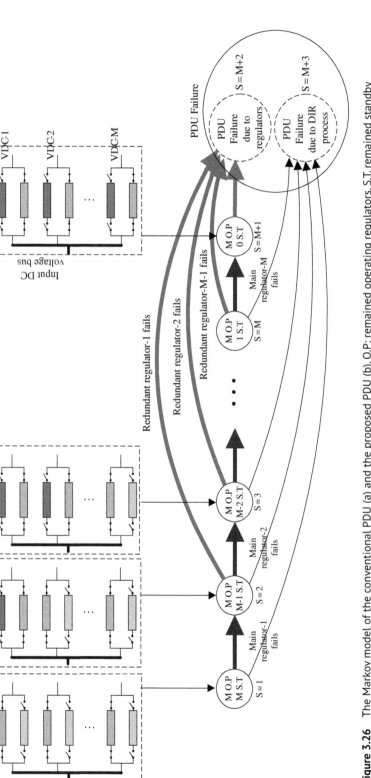

Figure 3.26 The Markov model of the conventional PDU (a) and the proposed PDU (b), O.P.: remained operating regulators, S.T.:remained standby regulators.

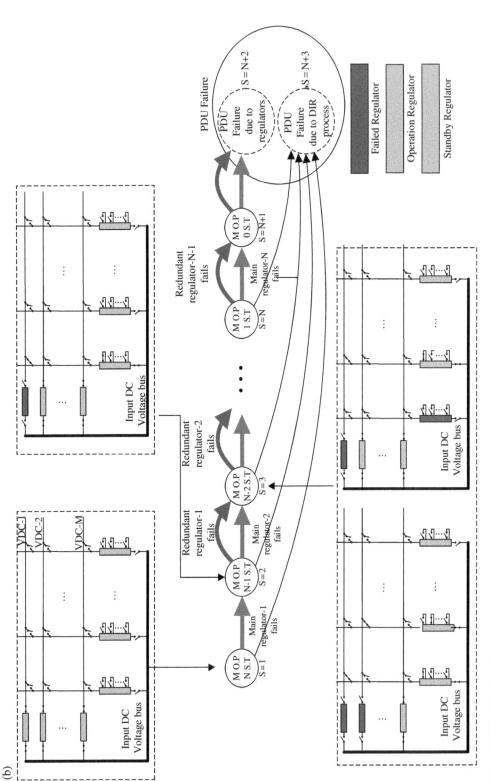

(b)

Figure 3.26 (Continued)

Table 3.1   Characteristics of the analyzed PDU.

Components	Failure rate (λ)
LM2576	2.83
Diode	0.013
Capacitor	0.18
ATMEGA32	0.2
Inductor	Negligible

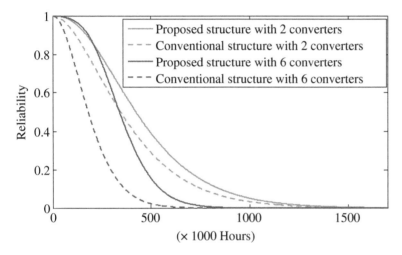

Figure 3.27   Reliability curves of the proposed structure and the conventional structure of PDU with two and six output voltage levels.

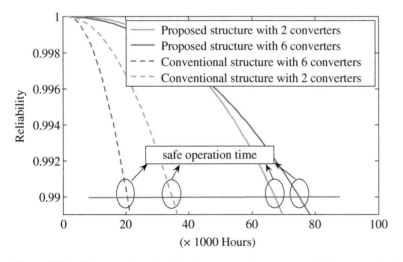

Figure 3.28   Safe operating time for the proposed structure and the conventional structure of PDU with two and six output voltage levels.

Figures 3.29 and 3.30 show the transient performance of the proposed PDU during a fault that leads to a decrease in voltage and in the output of the regulator.

In this test, the main regulator, 12 VDC, is subjected to a fault that leads to a decrease in voltage. This fault is detected by comparing the output voltage and its lower limit, 11.8 V. This is a very important value as it means that the PDU has a very short time for recovery because 12−11.8=0.2 V, which is a very low voltage drop. If we permit the system to drop to a lower voltage, for example 8 V, the PDU has more time recover. Then, the standby regulator was adjusted to 12 VDC and the faulty regulator was replaced by this adjusted regulator. Figure 3.31 shows the fast transient

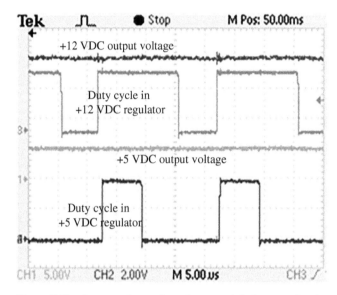

Figure 3.29 Output voltage of regulators and their respective duty cycles in continuous conduction mode.

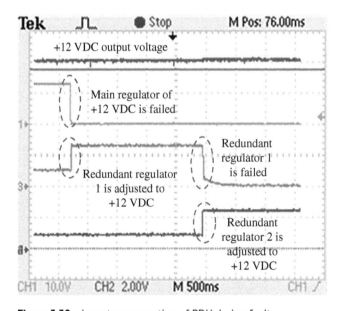

Figure 3.30 Long term operation of PDU during fault.

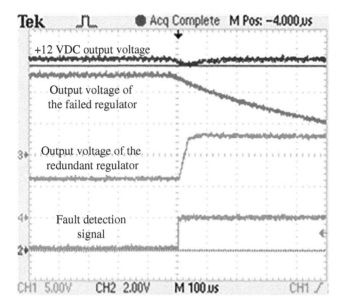

Figure 3.31 Transient performance of the proposed PDU during a fault that leads to voltage decreasing in output of the main regulator, 12V DC.

response of the PDU and the transient performance of the proposed PDU during the faults that leads to the voltage increasing in the output of the regulator. In this case, the output voltage of PDU, 5VDC, increases to its upper limit. Then the PDU isolates the faulty regulator and replaces it with a standby regulator.

3.7.2 Resilient Power Electronic Converter

In the following case study, the recovery time of a resilient converter is studied [13]. It is shown that this recovery time is long in the case of fault tolerance only, but is a short time in the case of resilience to meet the mission critical.

High-voltage power supplies are used in many applications, such as medical research, water treatment, materials processing, and telecommunication. One of the most important applications of this power supply is to serve as a power supply for high-power vacuum tubes. In this application, the high-voltage power supply (HVPS) is faced with the fault of an internal arc in the tube. To remove the internal arc fault in the vacuum tube, the tube HVPS should be cut off for a specified time interval. This time interval is named as the arc clearing time interval in this study. In addition, to prevent any damage to the tube during the internal arc fault, the released energy in the tube should be less than a specified limit in the order of 10J. On the other hand, a huge amount of the energy is usually stored in the HVPS output stage capacitors, which is much larger than the energy limit of the tubes. Therefore, a fast and reliable protection mechanism is essential to limit the fault energy that is deposited into the tube during the internal arcs. Application of a crowbar circuit is a common method used to limit the delivered energy to the tube during the fault. The diagram of this power supply is shown in Figure 3.32.When a fault occurs in the tube, a shunt crowbar device is closed quickly and diverts the energy from the high-voltage fault. Subsequently, the mechanical circuit breaker (CB) isolates the power supply from the grid.

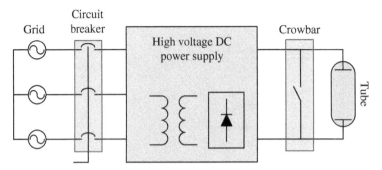

Figure 3.32 Diagram of the high voltage DC power supply.

Various kinds of devices have been used in a crowbar circuit, such as thyratron, ignitron, spark gap, and series-connected thyristors. They act effectively as a fast-closing switch. However, these devices do not have a turn-off capability through their command signal after triggering. This means that after detecting a fault and turning on the crowbar, it is impossible or difficult to turn it off. Therefore, the crowbar remains on until all the energy-storage components of the HVPS are completely discharged and the current of the crowbar reaches zero. When the fault is cleared, all energy storage components must be fully charged in order to return the HVPS to normal operation. Since the HVPS restoration time after the fault can reach several tenths of a second depending on the specifications of the crowbar device and HVPS, this time interval is larger than the required time to perform vacuum tube arc clearing. This disadvantage reduces power supply availability and resilience against an arc fault in the tube and false detection due to a transient phenomenon such as noise. In this section, a crowbar circuit is presented for fault resilience and fast restoration in the HVPS. The proposed crowbar structure built with series-connected insulated gate bipolar transistors (IGBTs) makes it feasible to employ a reopening mechanism for a crowbar in the HVPS. The crowbar reopening action is similar to a recloser that is used in a power system that is operated as a protection mechanism to improve the power supply reliability indices against a transient fault. The presented reopening mechanism effectively passes the transient faults, such as an arc fault and false detection due to noise, and restores the supply in a very short time. A conventional configuration of the power supply is shown in Figure 3.33. The crowbar device is constructed from series-connected thyristors. When the arc fault occurs in the tube, the protection unit detects the fault. The protection system produces a fault signal for triggering the crowbar and circuit breaker. By closing the crowbar, the fault energy that is supplied by energy storage elements and the grid is diverted through the crowbar. When the crowbar receives the trigger signal, all the series-connected thyristors are turned on after the turn-on delay of the thyristors. The circuit breaker, which is the mechanical switch, opens with a delay of about 25 ms and isolates the power supply from the grid. R1 and R2 are fault current limiter (FCL) resistors. The R1 resistance has the duty to limit the fault current before closing the crowbar and the crowbar current after it is closed. The R2 prevent tube arc completely shorts out the crowbar, so without this resistor, it would be very difficult to fire the crowbar and the crowbar also cannot divert the fault current because its impedance is comparable to the fault impedance. When this kind of thyristor crowbar is closed, the crowbar will remain closed until the thyristor current reaches zero. The crowbar will turn off after all energy storage

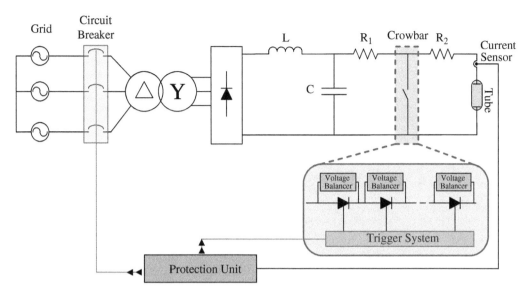

Figure 3.33 Conventional high voltage power supply with series-connected thyristor crowbar.

elements are discharged. Hence, if we want to restore the power supply after a fault, we should wait for a few hundred milliseconds before opening the crowbar. As a result, the power supply restoration time is extremely high.

In this section, the crowbar device with the intentionally opening capability is proposed. This type of crowbar is built with series-connected IGBTs. By use of this structure, we can implement the reopening mechanism to detect and pass the transient fault. A reopening is a mechanism that can open the crowbar after it has been closed due to a fault. This attempt repeats to a specified time, for if a fault is transient it is cleared; otherwise, if a fault is permanent, the crowbar switch and circuit breaker are held in the closed and open positions, respectively. In this section operation of the power supply in the presence of a reopening mechanism for the crowbar device is analyzed and the number of re-openings and restoration times are determined.

The design considerations are as follows:

1) The total open time should be such that, if the fault persists, the fault energy does not exceed its permissible limit.
2) Opened time (t_o) should be such that the protection system has the opportunity to detect the fault.
3) Closed time (t_c) must be greater than the maximum time that is needed to clear the transient faults.

Figure 3.34 shows the transient performance of the conventional structure during a fault in the load. After fault, the trigger unit sends a fault signal to the crowbar and CB, the crowbar is closed, and the output voltage descends to zero, as shown in Figure 3.34. Filter capacitors begin to discharge and the CB is opened after 25 ms. The crowbar current expands exponentially and after 150 ms it reaches 150 (mA), which is the thyristor holding current. The crowbar opens and the trigger unit resets the breaker fault signal and it closes again after 25 ms. As can be seen in

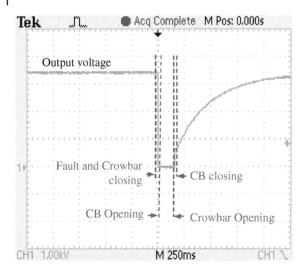

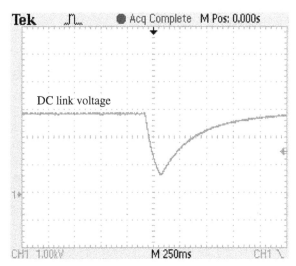

Figure 3.34 Waveforms of voltages variation under a transient fault with the thyristor crowbar. Upper figure, output voltage; lower figure, DC link voltage.

Figure 3.34, when the crowbar opens the DC bus voltage is equal to 800 (V). The output voltage of the HVPS is recovered 1 second after the fault occurs. Figure 3.35 shows the transient performance of the proposed structure during a fault in the load. When a fault is occurred, the trigger unit produces a fault signal that is a pulse with a width of 1 ms and the pulse transformer transfers the gate signal of the switch and provides the turn-on electrical charge of them. The crowbar closes and the output voltage descends to zero, as shown in Figure 3.35. After the 1 ms fault is cleared and the crowbar opens, the fault signals are reset to zero. Subsequently, the power supply is restored to its normal operation after a fault. As can be seen in lower figure of Figure 3.35, the DC bus voltage does not change much, since after the crowbar opens the output voltage immediately returns to a normal level. Figure 3.36 shows the converter output voltage in these two cases. It can be seen that the converter is out of service in the case of fault tolerance for about 200 ms.

Figure 3.35 Waveforms of a voltage variation under a transient fault with the IGBT crowbar. Upper figure, output voltage and DC link voltages, 250 ms/division; lower figure, the detailed trace of these variations, 1 ms/division.

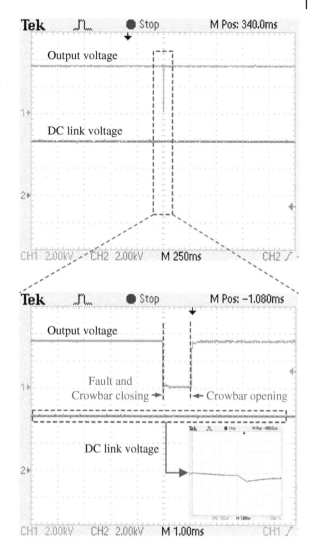

3.8 Summary and Conclusions

In this chapter, resilient power electronic converters were described. In these systems, continuous operation with usually nominal specifications have no interruption, even with the occurrence of a fault. The main topics of this chapter are summarized as follows:

1) The power electronic converters are faced with various faults. They can be sorted as internal faults, external faults, and malfunctioning.
2) The resilient characteristics can be performed at the component level, the converter level, and the system level.
3) Methods of resilience are applied both at the pre-fault period and the post-fault period.
4) A resilient converter keeps its performance index, such as its output voltage, in both a low drop value and a short recovery time.

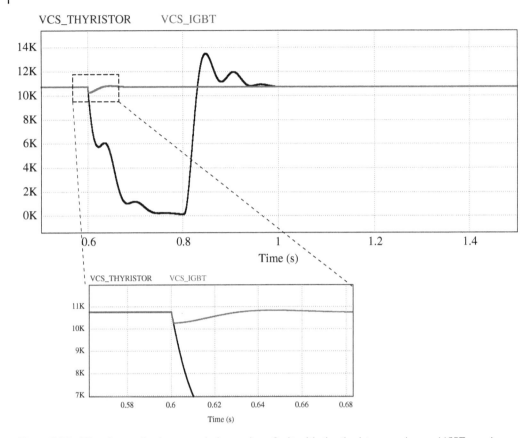

Figure 3.36 Waveforms of voltages variation under a fault with the thyristor crowbar and IGBT crowbar.

References

1 Kankanamalage, R.R., Foster, M.P., and Davidson, J.N. (2019). Online electrolytic capacitor prognosis system for PWM drives. In: *2019 IEEE 21st European Conference on Power Electronics and Applications (EPE '19 ECCE Europe)*, 1–8. Genova, Italy.

2 Wang, T., He, Y., Luo, Q. et al. (2017). Self-powered RFID sensor tag for fault diagnosis and prognosis of transformer winding. *IEEE Sensors Journal* 17 (19): 6418–6430.

3 Schwan, M., Schilling, K., Zickler, U., and Schnettler, A. (2006). Component reliability prognosis in asset management methods. In: *2006 IEEE International Conference on Probabilistic Methods Applied to Power Systems*, 1–6. Stockholm, Sweden.

4 Nasser, L. and Curtin, M. (2006). Electronics reliability prognosis through material modeling and simulation. In: *2006 IEEE Aerospace Conference*, 7. Big Sky, MT, USA.

5 Musallam, M., Johnson, C.M., Yin, C. et al. (2010). Real-time life consumption power modules prognosis using on-line rainflow algorithm in metro applications. In: *2010 IEEE Energy Conversion Congress and Exposition*, 970–977. Atlanta, GA, USA.

6 Nasser, L., Tryon, R., and Dey, A. (2005). Material simulation-based electronic device prognosis. In: *2005 IEEE Aerospace Conference*, 3579–3584. Big Sky, MT, USA.

7 Teng, W., Han, C., Hu, Y. et al. (2020). A robust model-based approach for bearing remaining useful life prognosis in wind turbines. In: *IEEE Access*, vol. 8, 47133–47143.

8 Sun, Q., Wang, Y., Jiang, Y., and Shao, L. (2017). Condition monitoring and prognosis of power converters based on CSA-LSSVM. In: *2017 IEEE International Conference on Sensing, Diagnostics, Prognostics, and Control (SDPC)*, 524–529. Shanghai, China.

9 Biswas, S.S., Srivastava, A.K., and Whitehead, D. (2015). A real-time data-driven algorithm for health diagnosis and prognosis of a circuit breaker trip assembly. *IEEE Transactions on Industrial Electronics* 62 (6): 3822–3831.

10 Spagnuolo, G., Xiao, W., and Cecati, C. (2015). Monitoring, diagnosis, prognosis, and techniques for increasing the lifetime/reliability of photovoltaic systems. *IEEE Transactions on Industrial Electronics* 62 (11): 7226–7227.

11 Zhang, W., Xu, D., Enjeti, P.N. et al. (2014). Survey on fault-tolerant techniques for power electronic converters. *IEEE Transactions on Power Electronics* 29 (12): 6319–6331.

12 Zarghany, M., Parvari, R., and Kaboli, S. (Nov. 2016). Fault-tolerant DC power distribution unit based on nonexclusive modules. *IEEE Transactions on Industrial Electronics* 63, 11: 6801–6811.

13 Pouresmaeil, K. and Kaboli, S. (2019). A reopened crowbar protection for increasing the resiliency of the vacuum tube high-voltage DC power supply against the vacuum arc. *IEEE Transactions on Plasma Science* 47 (5): 2717–2725.

4

State of the Art Resilient Power Converters

4.1   Mission Critical

Resilience finds its meaning in the mission critical applications. A mission critical application should be resilient against the faults to continue its mission at least for a certain time interval. Figure 4.1 shows the special characteristic of a mission critical system. Every reliable system is characterized with the concepts such as a long life time, its repairable capabilities, and relative fault tolerance. However, resilience is the exclusive characteristic of a mission critical system. The concept of mission critical is well known in engineering. Every system with an important and uninterrupted operation contains the concept of mission critical. Figure 4.2 shows a lamp with a critical application as an indicator in an aircraft. The lamp has two filaments to ensure its function during the flight. If one of the filaments is cut, the pilot still see the indicator but with reduced light intensity. Thus, the lamp is not only resilient against the fault but also sends an alarm to the pilot for replacing in a safe time.

In recent years, power electronics have been applied in various industrial processes. Some of these processes are mission critical, such as the petroleum process and healthcare instruments. Therefore, the mission critical concept enters the area of power electronics. Advances in power electronics enable efficient and flexible processing of electric power in the application of renewable energy resources, electric vehicles, adjustable-speed drives, etc. More and more efforts are being devoted to better power electronic systems in terms of reliability to ensure high availability, long lifetime, sufficient robustness, low maintenance costs, and low cost of energy. The reliability predictions are predominantly according to models and terms, such as MIL-HDBK-217H handbook models, Mean-Time-To-Failure (MTTF), and Mean-Time-Between-Failures (MTBF) analysis. A collection of methodologies based on the Physics-of-Failure (PoF) approach and mission profile analysis are presented in reference [1] to perform reliability-oriented design of power electronic systems. Power electronics are efficient for conversion and conditioning of electrical energy through a wide range of applications. Proper life consumption estimation methods applied for power electronics that can operate in real time under in-service mission profile conditions will not only provide an effective assessment of the product's life expectancy but can also deliver reliability design information. This is an important aid in manufacturing and thus helps in reducing costs and maximizing through-life availability. In reference [2], a mission profile-based approach for real-time life consumption estimation that can be used for reliability design of power electronics is presented. Reference [2] presents the use of electrothermal models coupled with PoF analysis by means of a real-time counting algorithm to provide accurate life consumption estimations for power modules operating under in-service conditions. These models, when driven by the actual

Resilient Power Electronic Systems, First Edition. Shahriyar Kaboli, Saeed Peyghami, and Frede Blaabjerg.
© 2022 John Wiley & Sons Ltd. Published 2022 by John Wiley & Sons Ltd.
Companion website: www.wiley.com/go/kaboli/resilientpower

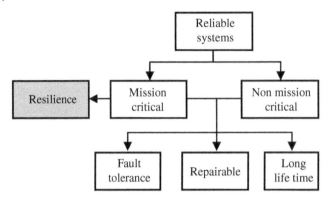

Figure 4.1 Resilience characteristic of a mission critical system.

Figure 4.2 A lamp with two filaments.

mission profiles, can be utilized to provide advanced warning of failures and thus deliver information that can be useful to meet particular application requirements for reliability at the design stage. In the following, to have a comprehensive insight about the concept of mission critical, this concept is reviewed in various applications.

4.1.1 Mission Critical in Network

The Internet of Things (IoT) is advancing, and the augmented role of architecture in automating processes is at its vanguard. Mission-critical applications are becoming a vital category of future IoT applications, and due to the advancements in the 5G, the design of a mission-critical application overcomes a big hurdle. Mission-critical applications must be reliable and the output should be known in advance. Therefore, to model such an application, architecture is considered to be the cornerstone. One of the major requirements is flexibility of the operation and adaptability to new devices. In [3], an optimal orchestration mechanism is proposed to automate the processes in a conventional multidevice and multitask mission-critical architecture for flexible and scalable operations. The central goal of this reference is threefold: firstly, to model tasks in such a way as to maximize the flexibility in the operation plane; secondly, to design a strongly correlated pair that has a maximum relation and thus the chance to hit the task of the devices that will be potentially maximized and also to ensure that the idle time among operations is minimal; and, lastly, to register devices in a network that is optimal for the group of this device in terms of correlation. Reference [3] proposes a multilayer particle swarm optimization for each of the optimization objectives. Results show that the operation plan is flexible and with scaling up the problem size, the orchestration is still graceful and within the requirements of mission-critical applications. The performance of multilevel particle swarm optimization is compared with conventional single-level particle swarm optimization and it has been learned that the latter is not only slower but also less accurate.

Network softwarization is a major paradigm shift, which enables programmable and flexible system operation in challenging use cases. In the fifth-generation (5G) mobile networks, the more advanced scenarios envision the transfer of high-rate mission-critical traffic. Achieving end-to-end reliability of these stringent sessions requires support from multiple radio access technologies and calls for dynamic orchestration of resources across both radio access and core network segments. Emerging 5G systems can already offer network slicing, multiconnectivity, and end-to-end quality provisioning mechanisms for critical data transfers within a single software-controlled network. Whereas these individual enablers are already in active development, a holistic perspective on how to construct a unified, service-ready system as well as understand the implications of critical traffic on serving other user sessions is not yet available. Against this background, reference [4] first introduces a softwarized 5G architecture for end-to-end reliability of mission-critical traffic. Then, a mathematical framework is contributed to model the process of critical session transfers in a softwarized 5G access network, and the corresponding impact on other user sessions is quantified. Finally, a prototype hardware implementation is completed to investigate the practical effects of supporting mission-critical data in a softwarized 5G core network, as well as substantiate the key system design choices.

Wireless Sensor Networks (WSNs) are generally designed to support applications in long-term deployments, and thus WSN protocols are primarily designed to be energy efficient. However, the research community has recently explored new WSN applications such as industrial process automation. These mission-critical applications demand not only energy efficient operation but also strict data transport performance. In particular, data must be transported to a sink in a timely and reliable fashion. Both WSN's data transport performance and energy consumption pattern are mainly defined by the employed medium access control (MAC) protocol. Therefore, reference [5] explores to what extent existing MAC protocols for WSNs can serve mission-critical applications. The reviewed protocols are classified according to data transport performance and suitability for mission-critical applications. The survey reveals that the existing solutions have a number of limitations and only a few recently developed MAC protocols are suitable for this application domain.

As sensor networks become widespread, security issues become a central concern, especially in mission-critical tasks. Reference [6] identifies the threats and vulnerabilities to WSNs and summarizes first the defense methods based on the networking protocol layer analysis. Then we give a holistic overview of security issues. These issues are divided into seven categories: cryptography, key management, attack detections and preventions, secure routing, secure location security, secure data fusion, and other security issues. Along the way reference [6] analyzes the advantages and disadvantages of current secure schemes in each category. In addition, we also summarize the techniques and methods used in these categories, and point out the open research issues and directions in each area.

Reference [7] presents a method for cost-benefit analysis of backup power systems for information and communications technology (ICT) applications. In recent years, ICT systems have played a very important role in modern society and life. For stable operation of the ICT systems without interruption, the electric power quality and reliability (PQR) must meet certain requirements for each type of load equipment. To improve and maintain PQR and to meet the requirements, various types of backup power supplies must be installed at the customer-side. It is advisable to estimate the cost benefits before introducing an expensive backup power supply. However, cost-benefit analysis is very difficult because of unknown and uncertain parameters, such as interruption frequency and economical damage. In reference [7], a general expression for cost-benefit analysis is reported. The expression uses discounted cash flow methods, such as the net present value, and interruption frequencies obtained through reports and field surveys in Japan.

4.1.2 Mission Critical in Computer Systems

If the performance of a computing system is "degradable," performance and reliability issues must be dealt with simultaneously in the process of evaluating system effectiveness. For this purpose, a unified measure, called "performability," is introduced in reference [8] and the foundations of performability modeling and evaluation are established. A critical step in the modeling process is the introduction of a "capability function," which relates low-level system behavior to user-oriented performance levels. A hierarchical modeling scheme is used to formulate the capability function and capability is used, in turn, to evaluate performability.

Reliability analysis of fault-tolerant (FT) computer systems for critical applications is complicated by several factors. Systems designed to achieve high levels of reliability frequently employ high levels of redundancy, dynamic redundancy management, and complex fault and error recovery techniques. Reference [9] describes dynamic fault-tree modeling techniques for handling these difficulties. Three advanced fault-tolerant computer systems are described: a fault-tolerant parallel processor, a mission avionics system, and a fault-tolerant hypercube. Fault-tree models are presented for their analysis. HARP (Hybrid Automated Reliability Predictor) is a software package developed at Duke University and NASA Langley Research Center that can solve those fault-tree models.

The techniques used to build highly available computer systems are sketched in reference [10]. Historical background is provided and terminology is defined. Empirical experience with computer failure is briefly discussed. Device improvements that have greatly increased the reliability of digital electronics are identified. Fault-tolerant design concepts and approaches to fault-tolerant hardware are outlined. The role of repair and maintenance and of design-fault tolerance is discussed. Software repair is considered. The use of pairs of computer systems at separate locations to guard against unscheduled outages due to outside sources (communication or power failures, earthquakes, etc.) is addressed.

4.1.3 Mission Critical in Industry

In the metal industry, rolling is the most widely used steel forming process to provide high production and control of the final product. Rolling mills must be able to change the speed of the strip at the same time as the speed is controlled within precise limits. Furthermore, this application has a severe load profile, with high torque variations during the lamination process. These characteristics include rolling mills among the classical mission-critical industry applications (MCIAs). In addition to the high cost/failure rate, rolling mills have a critical dynamic loading, making the design of a reliable system doubly challenging. Design for reliability (DFR) is the process conducted during the design of a component or system that ensures that they perform at the required reliability level. In the context of the power converters for rolling mills and other MCIAs, the DFR should be known and adopted in the design of the converter proper (component level) as well as in the specification of power converters (system level). Reference [11] contributes to the knowledge in the field by proposing a methodology covering the necessary steps for decision making during the design (component level) and selection (systems level) of power converters for MCIAs. A rolling mill system from a large steel plant in southeastern Brazil is adopted as the case study. The standard high-power converter solution is compared with two high-reliability converter topologies: the fault-tolerant active neutral point clamped (ANPC) and the triple-star bridge cells modular multilevel converter (MMC).

Mine hoist systems play a vital role in gold ore transportation. They are the connection between the underground mine and the beneficiation plant as shown in Figure 4.3. As a result, any hoist system interruption compromises the production.

In reference [12], a mission critical analysis and design methodology for power converters is proposed in order to achieve proper reliability of the hoist drive system. Since the failure of power devices is responsible for a significant portion of the total downtime of power converters, a detailed study of the lifetime of the insulated gate bipolar transistor (IGBT) power modules is carried out. The three-level neutral-point-clamped and ANPC IGBT-based topologies are considered for this application. The recently presented Fault-Tolerant (FT)-ANPC converter is proposed as a higher reliability solution, leading to longer power modules and drive system lifetimes.

Figure 4.3 Conveyor of the Sirjan iron mine, Iran.

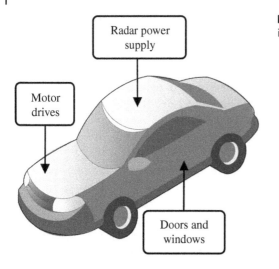

Figure 4.4 Applications of power electronics in an electrical vehicle.

Due to the increasing importance of power electronic components in automobiles, it becomes necessary to consider their reliability. Figure 4.4 shows some applications of power electronics in an electrical vehicle.

This applies especially to hybrid electrical vehicles (HEVs) where a malfunction of the power electronics may prevent the vehicle from operating. Of paramount importance for the reliability of power electronics is the component operating temperature and temperature cycling. Reference [13] deals with the development of an advanced simulation tool that is capable of determining the component temperature of a three-phase converter over long mission profiles. In addition, the expected converter reliability is calculated. To accomplish this, losses in the semiconductors and dc-link capacitors are determined first. Next, this loss data is fed into a thermal model to compute the component temperatures for the whole mission profile. As the basis for the reliability computation, failure-rate catalogs, such as Military Handbook 217F or RDF 2000, are used. Also an approach using simple formulas for a lifetime prediction is presented. According to failure-rate catalogs, temperature cycles are of particular importance for the reliability of power semiconductors. A novel algorithm, detecting all relevant temperature cycles within the computed temperature curve, has been developed. Finally, the applicability and significance of the presented reliability prediction methods are assessed.

4.1.4 Mission Critical in Devices

Mission-Critical Application is a general term for any system whose loss could cause downtime in crucial operations for a company. Given the importance assumed by power converters in many critical systems, reference [14] is conducted in the framework of the reliability engineering applied to a well-established medium voltage power converter in the industry environment. Condition monitoring (CM) techniques are used to detect changes in parameters of the IGBT power modules aiming to detect die and package degradation during operation. Figure 4.5 shows the IGBT package with die and its connections.

The main goal is to anticipate degradation failures in power modules in such a way that the power converter can be reconfigured before a fatal fault occurs. This fully proactive approach prevents the uncertainty related to different failure modes of power semiconductor devices. The premature detection of a failure event allows better equipment availability management.

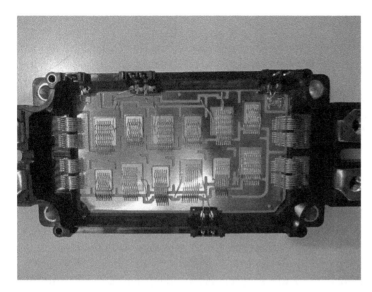

Figure 4.5 A typical IGBT package with die and its connections.

Mission-critical applications require that any failure that may lead to erroneous behavior and computation is detected and signaled as soon as possible in order not to jeopardize the entire system. Totally self-checking (TSC) systems are designed in reference [15] to be able to autonomously detect faults when they occur during normal circuit operation. Based on the adopted TSC design strategy and the goal pursued during circuit realization (e.g. area minimization), the circuit, although TSC, may not promptly detect the fault, depending on the actual number of input configurations that serve as test vectors for each fault in the network. If such a number is limited, although TSC, it may be improbable that the fault is detected once it occurs, causing detection and aliasing problems. The reference presents a design methodology, based on a circuit re-design approach and an evaluation function, for improving a TSC circuit promptness in detecting faults occurrence, a property we will refer to as TSC quality.

In the nanoelectronics era, multiple faults or failures in circuits and systems deployed in mission and safety-critical applications, such as space, aerospace, nuclear, etc., are known to occur. To withstand these, higher-order redundancy is suggested to be used selectively in the sensitive portions of a circuit or system. In reference [16], the distributed minority and majority voting-based redundancy (DMMR) scheme was proposed as an alternative to the N-modular redundancy (NMR) scheme for the efficient implementation of higher-order redundancy. However, the DMMR scheme is not self-healing. In this reference, we present a new self-healing redundancy (SHR) scheme that can inherently correct its internal faults or failures without any external intervention, which makes it ideal for mission/safety-critical applications. To achieve the same degree of fault tolerance, the SHR scheme requires fewer function blocks than the NMR and DMMR schemes. We present the architectures of the proposed SHR scheme, discuss the system reliability, and provide the design metrics estimated for example SHR systems alongside the corresponding NMR and DMMR systems using a 32/28-nm CMOS technology. From the perspectives of fault tolerance, self-healing capability, and optimizations in the design metrics, the SHR scheme is preferable to the NMR and DMMR schemes.

4.1.5 Mission Critical in Micro Grids

The increasing penetration of distributed renewable energy resources and increasing participation from a demand response from consumers make the microgrid (MG) a favorable option for future power system development. Since some loads are mission critical and require a 24/7 uninterruptable supply, a smart microgrid is expected as an effective architecture to serve this purpose. Reference [17] proposes the application of an electronic power transformer (EPT) to improve the controllability and reliability of a mission-critical microgrid. The application scenarios of EPT in the microgrids are analyzed in detail. Some case studies including applying the EPTs into a grid-connected microgrid, among autonomous microgrids and between two microgrids, are analyzed. The simulation results demonstrate the potential of the EPT in enhancing the controllability and reliability of the microgrid.

Mission profiles such as environmental and operational conditions together with the system structure, including energy resources, grid and converter topologies, induce stress on different converters and thereby play a significant role in power electronic systems reliability. Temperature swing and maximum temperature are two of the critical stressors on the most failure-prone components of converters, i.e. capacitors and power semiconductors. Temperature-related stressors generate electrothermal stress on these components, ultimately triggering high potential failure mechanisms. Failure of any component may cause converter outage and system shutdown. Reference [18] explores the reliability performance of different converters operating in a power system and indicates the failure-prone converters from a wear-out perspective. It provides a system-level reliability insight for design, control, and operation of a multiconverter system by extending the mission-profile-based reliability estimation approach. The analysis is provided for a dc microgrid due to the increasing interest that dc systems have been gaining in recent years; however, it can be applied for reliability studies in any multiconverter system. The outcomes can be worthwhile for maintenance and risk management as well as security assessment in modern power systems.

4.2 Resilient Systems

The resilience concept is applied in applications with a critical mission. In this section, some existing examples of the resilient systems are presented.

4.2.1 Resilient Control Systems

Since digital control systems were introduced to the market more than 30 years ago, the operational efficiency and stability gained through their use have fueled our migration and ultimate dependence on them for the monitoring and control of critical infrastructure. While these systems have been designed for functionality and reliability, a hostile cyber environment and uncertainties in complex networks and human interactions have placed additional parameters on the design expectations for control systems [19].

Common cause events are those specific groups of dependent events that might adversely affect the operation of a redundant system. Common cause failures (CCFs) are considered a subset of dependent failures with a major difference that they cannot be explicitly modeled. A CCF is a single point of failure (SPOF) causing a unit and its "perceived" redundant unit to fail simultaneously. During the first few years of commercial nuclear power plant operations, many probability risk assessment (PRA) studies revealed that CCF have significantly contributed to core damages. Consequently, the US Nuclear Regulation Commission (NRC) initiated various CCF analyses in the 1980s. Since then, CCF analyses have been synonymous with nuclear facilities, and a majority of CCF analysis efforts have rightfully been focused on the safety of nuclear plants. Given that many industrial and commercial mission-critical facilities and high-reliability organizations now require a high degree of availability, CCF therefore cannot be ignored. The objectives of this reference are to present (i) the importance of considering CCF in reliability, availability, and maintainability (RAM) analysis for industrial and commercial mission-critical facilities and high-reliability organizations, (ii) simple quantitative analysis methods of CCF for industrial and commercial mission-critical facilities, and (iii) the need for including CCF as part of future failure data collection efforts [20].

4.2.2 Resilient Microgrids

Reference [21] proposes a novel comprehensive operation and self-healing strategy for a distribution system with both dispatchable and non-dispatchable distributed generators (DGs). In the normal operation mode, the control objective of the system is to minimize the operation costs and maximize the revenues. A rolling-horizon optimization method is used to schedule the outputs of dispatchable DGs based on forecasts. In the self-healing mode, the on-outage portion of the distribution system will be optimally sectionalized into networked self-supplied microgrids (MGs) so as to provide a reliable power supply to the maximum loads continuously. The outputs of the dispatchable DGs will also be rescheduled accordingly. In order to take into account the uncertainties of DG outputs and load consumption, we formulate the problems as a stochastic program. A scenario reduction method is applied to achieve a tradeoff between the accuracy of the solution and the computational burden. A modified IEEE 123-node distribution system is used as a test system. The results of case studies demonstrate the effectiveness of the proposed methodology.

Microgrids with distributed generation (DG) provide a resilient solution in the case of major faults in a distribution system due to natural disasters. Reference [22] proposes a novel distribution system operational approach by forming multiple microgrids energized by DG from the radial distribution system in real-time operations to restore critical loads from the power outage. Specifically, a mixed-integer linear program is formulated to maximize the critical loads to be picked up while satisfying the self-adequacy and operation constraints for the microgrid formation problem by controlling the ON/OFF status of the remotely controlled switch devices and DG. A distributed multiagent coordination scheme is designed via local communications for the global information discovery as inputs of the optimization, which is suitable for autonomous communication requirements after a disastrous event. The formed microgrids can be further utilized for power quality control and can be connected to a larger microgrid before the restoration of the main grids is complete.

Continuously expanding deployments of distributed power-generation systems (DPGSs) are transforming the conventional centralized power grid into a mixed distributed electrical network. The modern power grid requires flexible energy utilization but presents challenges in the case of a high penetration degree of renewable energy, among which wind and solar photovoltaics (PVs) are typical sources. The integration level of the DPGS into the grid plays a critical role in developing sustainable and resilient power systems, especially with highly intermittent renewable energy resources. To address the challenging issues and, more importantly, to leverage the energy generation, stringent demands from both utility operators and consumers have been imposed on the DPGS. Furthermore, as the core of energy conversion, numerous power electronic converters employing advanced control techniques have been developed for the DPGS to consolidate the integration. In light of the above, reference [23] reviews the power-conversion and control technologies used for DPGSs.

Reference [24] proposes a novel approach to power wireless communication network base stations (BSs) in order to achieve higher resilience. In the proposed dc microgrid, resilience is improved with the coordinated operation of renewable energy sources, energy storage devices, and BS traffic using an integrated controller that adjusts BS traffic and distributes stored energy among cell sites based on weather forecasts and other operational conditions. The essential role played by power electronics and active power distribution nodes in this resilient microgrid is also explained. Resilience improvements are evaluated quantitatively with metrics analogous to that of availability, which considers energy storage levels, power generation, and load. The analysis shows that the proposed system allows reducing service restoration times by a factor of 3 or improving resilience by double-digit percentage points, while the battery life is extended by about 10%. Additionally, battery bank size needs and PV array footprints can be reduced without compromising resilience. The proposed microgrid is implemented for wireless communication networks because both energy and communications are identified in the US Presidential Policy Directive 21 as the two infrastructures that are especially critical for community resilience. Nevertheless, this same technology can also be used in other applications, such as residential neighborhoods or industrial campuses.

Following a major outage in the main grid due to natural disasters, microgrids have the ability to disconnect from the main grid and provide electricity to their consumers. However, integration of power electronic-based generation units and small-scale energy resources into MGs reduces the system inertia. Therefore, frequency deviations arising from loss of grid power or fluctuations of renewable energy resources and loads should be managed. In reference [25], a two-stage robust day-ahead optimization model for resilient operation of MGs is proposed in which the hierarchical frequency control structure of the microgrid is precisely formulated. Based on this model, the operation cost of a microgrid is minimized while sufficient primary and secondary reserves are scheduled to restrict frequency deviations and avoid load shedding under the worst-case realization of islanding events. A column-and-constraint generation (C&CG) algorithm is utilized to efficiently solve the problem. Numerical cases on a test system show the effectiveness of the proposed model and the solution algorithm. The obtained results verify that by applying the proposed model, the operating costs of a microgrid is minimized while the frequency deviation and load shedding can be successfully managed during islanding events.

Smart power grids are being enhanced by adding communication infrastructure to improve their reliability, sustainability, and efficiency. Despite all of these significant advantages, their open communication architecture and connectivity renders the power systems' vulnerability to a range of cyber attacks. Reference [26] proposes a novel resilient control system for a Load Frequency Control (LFC) system under false data injection (FDI) attacks. It is common to use encryption in data transfer links as the first layer of a defending mechanism; here, we propose a second defense layer that can jointly detect and mitigate FDI attacks on power systems. Here, we propose a new anomaly detection (AD) technique that consists of a Luenberger observer and an artificial neural network (ANN). Since FDI attacks can happen rapidly, the observer structure is enhanced by the Extended Kalman filter (EKF) to improve the ANN ability for online detection and estimation. The resilient controller is designed based on the attack estimation, which can eliminate the need for control reconfiguration. The resilience of the proposed design against FDI attacks has been tested on the LFC system.

Microgrids are seen as an effective way to achieve reliable, resilient, and efficient operation of the power distribution system. Core functions of the microgrid control system are defined by the

IEEE standard 2030.7; however, the algorithms that realize these functions are not standardized and are a topic of research. Furthermore, the corresponding controller hardware, operating system, and communication system to implement these functions vary significantly from one implementation to the next. In reference [27], we introduce an open-source platform, Resilient Information Architecture Platform for the Smart Grid (RIAPS), ideally suited for implementing and deploying distributed microgrid control algorithms. RIAPS provides a design-time tool suite for development and deployment of distributed microgrid control algorithms. With support from a number of run-time platform services, developed algorithms can be easily implemented and deployed into real microgrids. To demonstrate the unique features of RIAPS, we propose and implement a distributed microgrid secondary control algorithm capable of synchronized and proportional compensation of voltage unbalance using DGs.

Reference [28] discusses microgrid power supply resilience in extreme events and the impact of power electronic interfaces, energy storage, lifelines, and the characteristics of distribution architectures. Resilience is characterized based on metrics analogous to those of availability considering the presence of power electronic interfaces and energy storage. The effect of energy storage on microgrid resilience is analyzed and resilience improvement in the presence of diverse sources is discussed. Resilience metrics, which are time dependent in nature, are derived under natural disaster conditions. Resilience is calculated for microgrids containing PVs with batteries, diesel generators, and fuel storage with a discontinuous fuel supply. Hurricane conditions are considered to provide a practical context for the discussion. Microgrid resilience formulas are derived for radial, ring, and ladder architectures. It is shown that the resilience formula for radial architecture can be used as a building block to derive the formulas for the ring and ladder networks. Architecture resiliencies are compared in islanded and grid-tie modes. The results indicate that the effects of including energy storage and source diversity, such as combinations of renewable energy sources and diesel generators, can improve power supply resilience when the microgrid is in island mode, which is the most likely operating mode during extreme events.

4.3 Resilient Power Electronic Converters

With widespread application of power electronic converters in high power systems, there has been a growing interest in system reliability analysis and fault-tolerant capabilities. Figure 4.6 shows a magnetic beam former for radiopharmaceutical applications. Any interruption in the magnet power supply leads to failure of the process and radioactive issues of the processing room. Considering a mission is critical, a fault-tolerant power electronic converter changes to a resilient converter. In this section, existing resilient power electronic converters are presented.

4.3.1 Resilient Multievel Converters

In multilevel converters, there are many opportunities for fault tolerance because there are many unused states in the converter. Figure 4.7 shows a multilevel pulser for a medical linear accelerator (Linac). If one of the pulsers fails, the other pulsers compensate its voltage for a short time.

Reference [29] presents a resilient framework for fault-tolerant operation in MMCs to facilitate normal operation under internal and external fault conditions. This framework is realized by designing and implementing a supervisory algorithm and a post-fault restoration scheme. The supervisory algorithm includes monitoring and decision-making units to detect and identify faults

Figure 4.6 Magnetic beam former for radiopharmaceutical applications.

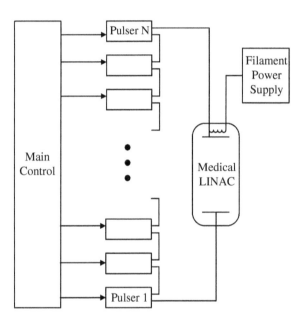

Figure 4.7 A multilevel high-voltage pulser for medical applications.

by analyzing the circulating current and submodule capacitor voltages in a very short time. The post-fault restoration scheme is proposed to immediately replace the faulty submodule with the redundant healthy one. The restoration is achieved by virtue of a multilevel modular capacitor-clamped dc/dc converter (MMCCC), which is redundantly aggregated to each arm of the MMC. This design effectively guarantees a smooth mode transition and handles the failure of multiple submodules in a short time interval. In addition, a modified modulation scheme is presented to ensure submodule capacitor voltage balancing of the MMC without implementing any additional hardware. Fast fault identification, a fully modular structure, and robust post-fault restoration are the main features of the proposed framework. Digital time-domain simulation studies are conducted on a 21-level MMC to confirm the effectiveness and resilience of the proposed fault-tolerant framework during internal and external faults.

A high failure rate of the semiconductor devices and capacitors, matched with their high requirement in the architecture of a multilevel inverter topology, results in low reliability of the inverter.

A low device count, preserving the output power upon fault occurrence, the ability to tolerate both open and short switch failures on all fault locations, handling both single and multiple switch failure, and achieving natural voltage balancing of the capacitors are the main challenges for the existing solutions to the fault-tolerant topologies. Thus, a highly resilient fault-tolerant topology, based on conventional inverters, is proposed in reference [30]. The proposed solution accomplishes all the challenges above without deteriorating the efficiency of the inverter during faulty conditions. The robustness and the effectiveness of the proposed topology are verified by the obtained experimental results. Multiple cases of switch failure are considered to cover all fault locations and types.

4.3.2 Resilient Motor Drive

Existing position sensorless methods rely on: (i) one-to-one correspondence between the magnetic characteristics of a switched reluctance motor (SRM) and rotor position and (ii) access to terminal quantities (i.e. voltage and current) for all phases. The occurrence of a fault in one or more phases will prohibit the necessary access to the phase/s and as such introduces a challenge to successful implementation of the existing position sensing techniques. Reference [31] investigates the existing sensorless methods with the occurrence of phase faults and introduces a family of generalized strategies for sensorless operation of SRM under single and multiphase faults.

The integrated gate-commutated thyristor and presspack power diodes have been successfully applied in medium-voltage neutral point-clamped converters in the power range from hundreds of kilowatts to tenths of a megawatt. Responsible for driving key processes in the industry, high reliability and availability are crucial for these converters, since their repair or replacement after failure events may take too long. Given the vital importance of such equipment for the drive systems, they are equipped with protection schemes that are usually reliable, but not infallible. If the protection scheme of the converter does not work properly in a short-circuit situation, serious damage may be expected on its power semiconductor devices. In Reference [32], the power semiconductors' thermal behavior is investigated using finite-element models in the COMSOL Multiphysics software. Three-dimensional thermal models of the power devices were raised by industrial radiography techniques, aiming to expand the information provided by the manufacturers. Reference [32] shows how these results can be used in a real equipment to attenuate the catastrophic effects of the protection scheme malfunction, so limiting the damage pattern within the converter to their least complex power devices.

Reference [33] presents a robust observer–controller scheme for sensor fault-resilient control in dc servomotor drive-based applications (such as antennae control for satellite tracking, radio telescopes, and conveyor belt systems). In contrast to the earlier works on abrupt faults, this reference considers incipient sensor faults and defects using the higher-order sliding mode (HOSM) observer, followed by a tracking controller, which maintains the acceptable drive performance. A robust output tracking controller based on a fractional integral terminal sliding mode surface with HOSM terms is developed to ensure faster and finite-time convergence of the error trajectory. Moreover, various slopes of incipient faults are considered to analyze the detection delay, and switching strategy reconfigures the system with the estimated speed whenever the residual crosses the threshold. The closed-loop performance in the presence of most common faults (abrupt, incipient, and intermittent) is experimentally validated on a dc motor-based industrial mechatronic drives unit with a belt-drive inertial load (which exhibits nonlinear friction, torque variations, and other disturbances).

Reliability is a fundamental requirement in aircraft safety-critical equipments. Its pursuit involves the adoption of protective design concepts such as fault-tolerant or redundant approaches, aiming to minimize mission failure probabilities. Multiphase motor drives are showing a growing interest to this approach, because they permit a boost in torque and power density, allowing the design of very compact high efficiency drives with intrinsic fault-tolerant capabilities. Reference [34] presents a five-phase permanent magnet brushless motor drive developed for an aircraft flap actuator application. The motor is designed to satisfy the load specifications with one or two phases open or with a phase short-circuited, while a failure in the rotor position sensors is remedied through a sensorless strategy. Design studies aiming to predict the faulty mode performance in case of different remedial strategies are presented.

4.3.3 Resilient HVDC Systems

Voltage source converter (VSC)-based high-voltage direct current (HVDC) transmission systems have attractive advantages compared to classical thyristor-based HVDC transmission systems. However, VSC-based HVDC transmission systems are vulnerable to dc side fault, and expensive dc circuit breakers are required to protect them against dc fault. Reference [35] proposes a control method of a dc fault-resilient VSC that can be protected against a dc fault without using expensive dc circuit breakers. In the VSC configuration, several H-bridge modules are connected in cascade, as shown in Figure 4.8, so the voltage balancing control of several floating dc capacitors is required. In this reference, an appropriate control structure with the capacitor voltage balancing controller is proposed.

In an effort to minimize the power disruption between a dc grid and ac grids that host power converters during ac and dc network faults, reference [36] proposes a novel converter station structure to improve ac and dc fault ride-through performance of the multi-terminal HVDC grid. The proposed structure consists of two independent ac and dc interfacing circuits, which are

Figure 4.8 Floating capacitors of a VSC.

a half-bridge MMC and a cascaded H-bridge (CHB) based energy storage system. Taking the advantages of high controllability and flexibility of the independent CHB converter and ease of integrating energy modules, a decoupled power relationship between the ac and dc sides is achieved, which is important for enhancing ac and dc fault performance. Operation of the proposed converter station under normal conditions and during ac and dc faults is explained, with the control system presented. Simulation validation of the proposed structure on a three-terminal HVDC grid confirms the enhanced performance, including the continuous operation during ac and dc faults with negligible power transfer disruption.

4.3.4 Resilient Converters in Space Applications

Reference [37] presents a remote, resilient dc microgrid for mission-critical space applications. This structured microgrid features resilience, energy balancing, autonomy, standard interface, fault tolerance, fault isolation, modularity, and scalability. A minute-day simulation is then presented to demonstrate the system-level energy balance and smooth dynamic behaviors when entering and exiting the lunar eclipse. Fuse clearing transient capability is also discussed. A dead-bus recovery circuitry is presented, which ensures self-resilience. Figure 4.9 shows the power distribution unit of a satellite with redundant modules for resilience.

4.3.5 Resilient Multiconverter Systems

A traditional droop-controlled system has assumed that generators can always generate the powers demanded from them. This is true with conventional sources, where fuel supplies are usually planned in advance. For renewable sources, it may also be possible if energy storage is available. Energy storage, usually as batteries, may however be expensive, depending on its planned capacity. Renewable sources are therefore sometimes installed as non-dispatchable sources without storage.

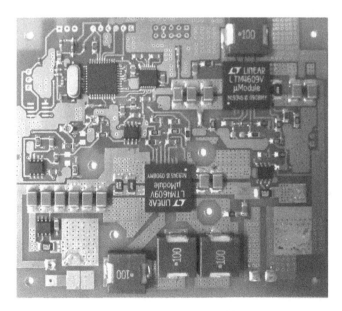

Figure 4.9 Power distribution unit of a satellite with redundant modules for resilience.

This may not be viable for remote grids, where renewable sources may be the only or major type of source. In those cases, a traditional droop scheme may not work well when its demanded power cannot be met by some renewable sources due to intermittency. When that happens, the system may become unstable with some sources progressively brought out of generation. To avoid such occurrence, an enhanced dual droop scheme is proposed in reference [38] for general two-stage converters with front rectifiers or dc–dc converters for conditioning powers from renewable sources and rear inverters for channeling powers to remote grids. Unlike the traditional droop scheme, the proposed dual droop scheme uses both dc-link voltage and generated powers for determining the required control actions, which have subsequently been proven stable by small-signal analysis.

4.3.6 Resilient Protection Systems

High-voltage power supplies (HVPSs) are widely used to supply vacuum tubes. The amount of delivered energy from the HVPS to the tube is an important issue during the vacuum arc in the tube. The conventional protection mechanism consists of a shunt crowbar that diverts the fault current from the tube to itself as a parallel path. The crowbar circuit is usually built of devices without the turn-off capability. It is a drawback since the output of the power supply is shortened for a long time, as shown in Figure 4.10. Thus, the restoration time of these power supplies is excessive. This demerit can have detrimental effects on mission-critical applications.

In reference [39], the IGBT-based crowbar structure is studied to overcome this issue. In the proposed protection mechanism, the crowbar can be reopened intentionally after closing. A theoretical analysis is presented to compare power supply performance in the presence of the conventional crowbar with the IGBT-based crowbar. The experimental results are presented to demonstrate the practical considerations of the proposed crowbar structure and its appropriate performance. This reference shows that HVPS has more fault resilience and less restoration time with the proposed IGBT-based crowbar.

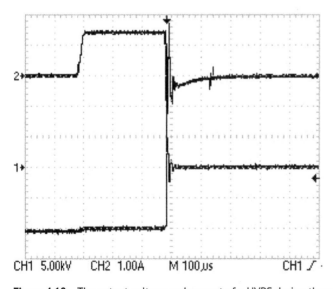

Figure 4.10 The output voltage and current of a HVPS during the crowbar operation.

4.3.7 Resilient Power Devices

Recent growth of the IGBT module market has been driven largely by increasing demand for an efficient way to control and distribute power in the field of renewable energy, hybrid/electric vehicles, and industrial equipment. For safety-critical and mission-critical applications, the reliability of IGBT modules is still a concern. Understanding the PoF of IGBT modules has been critical to the development of effective CM techniques as well as reliable prognostic methods. Figure 4.11 shows the infrared photo of a power converter. Detection of the over-temperature in the converter switch prevents the unwanted shutdown.

Reference [40] attempts to summarize past developments and recent advances in the area of CM and prognostics for IGBT modules. The improvement in material, fabrication, and structure is described. The CM techniques and prognostic methods proposed in the literature are presented. Reference [40] concludes with recommendations for future research topics in the CM and prognostics areas.

Precursor parameters have been identified to enable development of a prognostic approach for IGBTs [41]. The IGBTs were subjected to thermal overstress tests using a transistor test board until device latch-up. The collector-emitter current, transistor case temperature, transient and steady state gate voltages, and transient and steady state collector-emitter voltages were monitored in-situ during the test. Pre- and post-aging characterization tests were performed on the IGBT. The aged parts were observed to have shifts in capacitance-voltage (C-V) measurements as a result of trapped charge in the gate oxide. The collector-emitter ON voltage $V_{CE(ON)}$ showed a reduction with aging. The reduction in the $V_{CE(ON)}$ was found to be correlated to die attach degradation, as observed by scanning acoustic microscopy (SAM) analysis. The collector-emitter voltage and transistor turn-off time were observed to be precursor parameters to latch-up. The monitoring of these precursor parameters will enable the development of a prognostic methodology for IGBT failure. The prognostic methodology will involve trending precursor data and using physics of failure models to predict the remaining useful life of these devices.

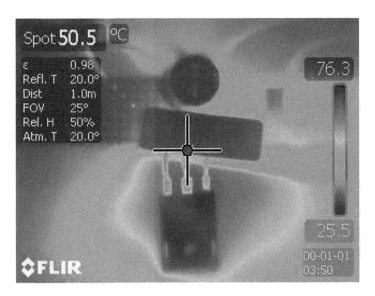

Figure 4.11 Infrared photo of a converter shows over-temperature in the switch.

4.3.8   Resilient Renewable Energy Systems

As a key component in the wind turbine systems and photovoltaic plants, the power electronic converter and its power semiconductors suffer from complicated power loadings related to the environment, as shown in Figure 4.12.

Therefore, correct lifetime estimation of a wind power converter is crucial for the reliability improvement and also for cost reduction of wind power technology. Unfortunately, the existing lifetime estimation methods for the power electronic converter are not yet suitable in the wind power application, because the comprehensive mission profiles are not well specified and included. Consequently, a relative more advanced approach is presented in reference [42], which is based on the loading and strength analysis of devices and takes into account different time constants of the thermal behaviors in a power converter. With the established methods for loading and lifetime estimation for power devices, more detailed information of the lifetime-related performance in a wind power converter can be obtained.

Reference [43] presents the reliability analysis of a push–pull converter intended for connection to a 125-W PV panel. Four prototypes of the converter were built and tested, using transistors with different ratings. Failure rates were calculated using the MIL HDBK 217 and the IEC TR 62380 procedures. In the latter case, the prediction was performed taking into account an annual mission profile obtained from the intended installation site, in an area with desert climate temperatures. Failure rate results obtained with MIL HDBK 217 show small differences among the converters, the best performance obtained being from the prototype with the lowest on-resistance. Results obtained with IEC TR 62380 indicate that thermal cycles have a significant effect in reliability performance, and should be considered carefully, because PV systems often see large temperature variations. With both procedures, the failure rate contributions from magnetic devices were higher than expected.

Figure 4.12   Partial shadow on a solar plant.

4.4 Summary and Conclusions

In this chapter, a survey of the concept of resilience in power electronic converters was presented. Old applications of this concept in network and computer science were reviewed. Existing examples of resilient power electronic converters were presented. The chapter can be summarized as follows:

1) Resilience is a well-known concept in some areas of engineering. The old application of resilience can be found in computer science and earlier in networks. The reason for this old application is the importance of the service in those areas. Uninterrupted operation of a network or computer system is obvious. Therefore, they must be resilient against faults.
2) Mission-critical applications need to be resilient against faults. In the area of power engineering, microgrids are examples of the application of the resilience concept. A microgrid may handle a critical mission. Therefore, resilient microgrids were studied in recent years.
3) As the penetration rate of the power electronic converters increase in industries, critical missions of these converters increase. Therefore, resilience in power electronic converters design is considered. *A resilient power electronic converter is a fault-tolerant converter with mission critical considerations.*

References

1 Huai Wang, Ke Ma, and Frede Blaabjerg. (2012). Design for reliability of power electronic systems. *IECON 2012 – 38th Annual Conference on IEEE Industrial Electronics Society*, 25–28 October 2012.
2 Musallam, M., Yin, C., Bailey, C., and Johnson, M. (2015). Mission profile-based reliability design and real-time life consumption estimation in power electronics. *IEEE Transactions on Power Electronics* 30 (5): 2601–2613.
3 Ahmad, S., Khudoyberdiev, A., and Kim, D. (2018). Towards the task-level optimal orchestration mechanism in multi-device multi-task architecture for Mission-critical IoT applications. *IEEE Journal on Selected Areas in Communications* 36 (3): 140922–140935.
4 Petrov, V., Lema, M.A., Gapeyenko, M. et al. (2012). Achieving end-to-end reliability of mission-critical traffic in softwarized 5G networks. *IEEE Communication Surveys and Tutorials* 14 (2): 485–501.
5 Suriyachai, P., Roedig, U., and Scott, A. (2009). A survey of MAC protocols for mission-critical applications in wireless sensor networks. *IEEE Communication Surveys and Tutorials* 11 (2): 240–264.
6 Chen, X., Makki, K., Yen, K., and Pissinou, N. (2009). Sensor network security: a survey. *IEEE Communication Surveys and Tutorials* 11 (2): 52–73.
7 Hirose, K., Matsumura, T., and Yamasaki, M. (2010). Cost-benefit analysis of emergency backup power systems for mission critical applications. *Intelec* 2010: 6–10.
8 Meyer, J.F. (1980). On evaluating the performability of degradable computing systems. *IEEE Transactions on Computers* C-29 (8): 720–731.
9 Dugan, J.B., Bavuso, S.J., and Boyd, M.A. (1992). Dynamic fault-tree models for fault-tolerant computer systems. *IEEE Transactions on Reliability* 41 (3): 363–377.
10 Gray, J. and Siewiorek, D.P. (1991). High-availability computer systems. *Computer* 24 (9): 39–48.
11 de Nazareth Ferreira, V., Cupertino, A.F., Pereira, H.A. et al. (2018). Design and selection of high reliability converters for mission critical industrial applications: A Rolling Mill Case Study. *IEEE Transactions on Industry Applications* 54 (5): 4938–4947.

12 de Nazareth Ferreira, V., Mendonça, G.A., Rocha, A.V. et al. (2017). Mission critical analysis and design of IGBT-based power converters applied to mine hoist systems. *IEEE Transactions on Industry Applications* 53 (5): 5096–5104.

13 Hirschmann, D., Tissen, D., Schroder, S., and De Doncker, R.W. (2007). Reliability prediction for inverters in hybrid electrical vehicles. *IEEE Transactions on Power Electronics* 22 (6): 2511–2517.

14 Ferreira, V.N., Cadoso Filho, B.J., and Rocha, A.V. (2017). Proactive fault-tolerant IGBT-based power converters for mission critical applications in MW range. *IEEE Applied Power Electronics Conference and Exposition (APEC)*, 26–30 March 2017.

15 Bolchini, C., Pomante, L., Salice, F., and Sciuto, D. (2003). The design of reliable devices for mission-critical applications. *IEEE Transactions on Instrumentation and Measurement* 52 (6): 1703–1712.

16 Balasubramanian, P. and Maskell, D.L. A self-healing redundancy scheme for mission/safety-critical applications. *IEEE Access* 6: 69640–69649.

17 Wang, D., Mao, C., Jiming, L., and Lee, W.-J. (2014). Electronic power transformer to secure the power supply of a mission critical microgrid. *IEEE Industry Application Society Annual Meeting*, 5–9 October 2014.

18 Peyghami, S., Wang, H., Davari, P., and Blaabjerg, F. (2019). Mission-profile-based system-level reliability analysis in DC microgrids. *IEEE Transactions on Industry Applications* 55 (5): 5055–5067.

19 Rieger, C.G., Gertman, D.I., and McQueen, M.A. (2009). Resilient control systems: Next generation design research. *2nd Conference on Human System Interactions*, 21–23 May 2009.

20 Pourali, M. (2014). Incorporating common cause failures in mission-critical facilities reliability analysis. *IEEE Transactions on Industry Applications* 50 (4): 2883–2890.

21 Wang, Z. and Wang, J. (2015). Self-healing resilient distribution systems based on Sectionalization into microgrids. *IEEE Transactions on Power Systems* 30 (6): 3139–3149.

22 Chen, C., Wang, J., Qiu, F., and Zhao, D. (2016). Resilient distribution system by microgrids formation after natural disasters. *IEEE Transactions on Smart Grid* 7 (2): 958–966.

23 Blaabjerg, F., Yang, Y., Yang, D., and Wang, X. (2017). Distributed power-generation systems and protection. *Proceedings of the IEEE* 105 (7): 1311–1331.

24 Kwon, Y., Kwasinski, A., and Kwasinski, A. (2016). Coordinated energy management in resilient microgrids for wireless communication networks. *IEEE Journal of Emerging and Selected Topics in Power Electronics* 4 (4): 1158–1173.

25 Mohiti, M., Monsef, H., Anvari-Moghaddam, A., and Lesani, H. (2019). Two-stage robust optimization for resilient operation of microgrids considering hierarchical frequency control structure. *IEEE Transactions on Industrial Electronics* 67 (7): 9439–9449.

26 Abbaspour, A., Sargolzaei, A., Forouzannezhad, P. et al. (2019). Resilient control design for load frequency control system under false data injection attacks. *IEEE Transactions on Industrial Electronics* 67 (9): 7951–7962.

27 Hao, T., Yuhua, D., Hui, Y. et al. (2019). Resilient information architecture platform for the smart grid (RIAPS): a novel open-source platform for microgrid control. *IEEE Transactions on Industrial Electronics* 67 (11): 9393–9404.

28 Krishnamurthy, V. and Kwasinski, A. (2016). Effects of power electronics, energy storage, power distribution architecture, and lifeline dependencies on microgrid resiliency during extreme events. *IEEE Journal of Emerging and Selected Topics in Power Electronics* 4 (4): 1310–1323.

29 Ghazanfari, A. and Mohamed, Y.A.-R.I. (2016). A resilient framework for fault-tolerant operation of modular multilevel converters. *IEEE Transactions on Industrial Electronics* 63 (5): 2669–2678.

30 Jalhotra, M., Sahu, L.K., Gupta, S., and Gautam, S.P. (2019). Resilient fault tolerant topology of single phase multilevel inverter. *IEEE Journal of Emerging and Selected Topics in Power Electronics* 9 (2): 1915–1922.

31 Wang, W. and Fahimi, B. (2014). Fault resilient strategies for position sensorless methods of switched reluctance motors under single and multiphase fault. *IEEE Journal of Emerging and Selected Topics in Power Electronics* 2 (2): 190–200.

32 Rocha, A.V., Bhattacharya, S., Moghaddam, G.K. et al. (2016). Thermal stress and high temperature effects on power devices in a fault-resilient NPC IGCT-based converter. *IEEE Transactions on Power Electronics* 31 (4): 2800–2807.

33 Kommuri, S.K., Rath, J.J., and Veluvolu, K.C. (2018). Sliding-mode-based observer–controller structure for fault-resilient control in DC servomotors. *IEEE Transactions on Industrial Electronics* 65 (1): 918–929.

34 Villani, M., Tursini, M., Fabri, G., and Castellini, L. (2012). High reliability permanent magnet brushless motor drive for aircraft application. *IEEE Transactions on Industrial Electronics* 59 (5): 2073–2081.

35 Yousefpoor, N., Narwal, A., and Bhattacharya, S. (2015). Control of DC-fault-resilient voltage source converter-based HVDC transmission system under DC fault operating condition. *IEEE Transactions on Industrial Electronics* 62 (6): 3683–3690.

36 Wang, S., Ahmed, K., Adam, G. et al. (2019). A novel converter station structure for improving multi-terminal HVDC system resiliency against AC and DC faults. *IEEE Transactions on Industrial Electronics* 67 (6): 4270–4280.

37 Tan, D., Baxter, D., Foroozan, S., and Crane, S. (2016). A first resilient DC-dominated microgrid for mission-critical space applications. *IEEE Journal of Emerging and Selected Topics in Power Electronics* 4 (4): 1147–1157.

38 Liu, H., Yang, Y., Wang, X. et al. (2017). An enhanced dual droop control scheme for resilient active power sharing among paralleled two-stage converters. *IEEE Transactions on Power Electronics* 32 (8): 6091–6104.

39 Pouresmaeil, K. and Kaboli, S. (2019). A reopened crowbar protection for increasing the resiliency of the vacuum tube high-voltage DC power supply against the vacuum arc. *IEEE Transactions on Plasma Science* 47 (5): 2717–2725.

40 Hyunseok, O., Han, B., McCluskey, P. et al. (2015). Physics-of-failure, condition monitoring, and prognostics of insulated gate bipolar transistor modules: a review. *IEEE Transactions on Power Electronics* 30 (5): 2413–2426.

41 Patil, N., Celaya, J., Das, D. et al. (2009). Precursor parameter identification for insulated gate bipolar transistor (IGBT) prognostics. *IEEE Transactions on Reliability* 58 (2): 271–276.

42 Ma, K., Liserre, M., Blaabjerg, F., and Kerekes, T. (2015). Thermal loading and lifetime estimation for power device considering mission profiles in wind power converter. *IEEE Transactions on Power Electronics* 30 (2): 590–602.

43 De León-Aldaco, S.E., Calleja, H., Chan, F., and Jiménez-Grajales, H.R. (2013). Effect of the mission profile on the reliability of a power converter aimed at photovoltaic applications – a case study. *IEEE Transactions on Power Electronics* 28 (6): 2998–3007.

Part II

Useful Life of the Power Electronic Systems

5

Useful Life Modeling

5.1 Failure Rate

There are reliability prediction techniques that depend on knowledge about design. As more details of the design are known, more accurate methods become available. These methods use part failure rate models, which predict the failure rates of parts based on various part parameters, such as technology, complexity, package type, quality level, and stress levels. Predictive methods attempt to predict the reliability of a part based on some models typically developed through empirical studies and/or testing. An attempt is made to identify critical variables such as materials, application environmental and mechanical stresses, application performance requirements, duty cycle and manufacturing techniques. Typically, a base failure rate for the component is assigned, which is then multiplied by factors for each critical variable identified. Some predictive models assume a constant failure rate over the lifetime of a product. This ignores higher failure rates typically seen at the beginning and the end of the component's useful life, infant mortality, and wear-out, respectively. Predictive methods can provide a relatively accurate reliability estimate in cases where good studies have been done to analyze field failures. In a power electronic converter, there are some components such as switch, diode, inductor, capacitor, and controller IC, which are in a series or parallel relationship from a reliability point of view. For evaluation of converter reliability, the failure rate of components must be determined. Then the converter should be analyzed systematically and the reliability model of the converter should be obtained.

Reliability is the probability of performing adequately to achieve the desired aim of the system. This can be mentioned as a time-dependent function. The reliability concept has more importance in specific applications such as space, healthcare, and military equipment where in that mission a failed part can hardly be replaced or be done by another system. In order to improve the system reliability, different research has been done and several methods have been introduced. Many of these methods need information about the failure rate of the system. Each system contains a number of components. One method to enhance the converter reliability is to improve its component reliability. This goal may be achieved by component-specific derating or by improvement of component specifications. At this level, the failure rate of the components should be modeled properly.

In the general form, the relation between reliability and the failure rate can be deduced as:

$$R(t) = e^{-\int_0^t \lambda(\tau)\,d\tau} \tag{5.1}$$

Resilient Power Electronic Systems, First Edition. Shahriyar Kaboli, Saeed Peyghami, and Frede Blaabjerg.
© 2022 John Wiley & Sons Ltd. Published 2022 by John Wiley & Sons Ltd.
Companion website: www.wiley.com/go/kaboli/resilientpower

where λ is the system failure rate. The mean time to failure (MTTF) is the mean time that is expected for the component or system to work adequately before failure. It can be noted that MTTF is equal to the integral of reliability of the component, as shown in the following:

$$\text{MTTF} = \int_0^\infty R(t)\,dt \tag{5.2}$$

The mean time to repair (MTTR) is the mean time which takes to repair the failed component that in electronic reliability evaluations is generally negligible with respect to MTTF. The mean time between failure (MTBF) is the summation of MTTF and MTTR, which can be assumed to equal the MTTF.

Based on the above-mentioned equations, if the failure rate is clear then the other reliability and useful life indices can be calculated. The Weibull distribution is one of the most widely used lifetime distributions in reliability engineering, as shown in Figure 5.1 [1, 2]. The failure process in a converter can be divided into three intervals. This figure shows that the failure rate has a considerable value at the beginning of the system operation (infant mortality period), an almost constant value at most times (random failures), and a high value at the end of life (EOL).

With regards to this curve, the failure rate is a nonlinear function of the time. Therefore, the reliability calculation based on Equation (5.2) is difficult. Here, it is noted that the early failures are clear in a short period after the system operation. In addition, they can be detected in a shorter period by the accelerated aging tests (AATs) [3]. Therefore, the failure rate can be represented by:

$$\lambda(t) = \lambda_0 + \lambda_v(t) \tag{5.3}$$

where λ_0 is the constant failure rate and λ_v is the time-variant failure rate that models the EOL. Equation (5.2) for the constant failure rate reduces to Equation (5.3) and the MTTF can be derived as

$$R(t) = e^{-\lambda t} \tag{5.4}$$

$$\text{MTBF} = \frac{1}{\lambda} \tag{5.5}$$

5.1.1 Early Failures

The infant mortality period is characterized by an initially high failure rate. This is normally the result of poor design, the use of substandard components, or lack of adequate controls in the manufacturing process. When these mistakes are not caught by quality control inspections, an

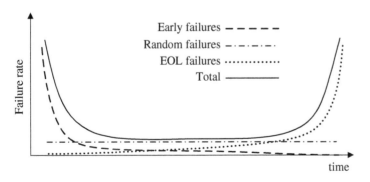

Figure 5.1 Weibull distribution of the failure rate.

early failure is likely to occur. Early failures can be eliminated from the customer by "burn-in," during which time the equipment is operated at stress levels equal to the intended actual operating conditions. The equipment is then released for actual use only when it has passed through the "burn-in" period.

5.1.2 Random Failures

The useful life period is characterized by an essentially constant failure rate. This is the period dominated by chance failures [4]. Chance failures are those failures that result from strictly random or chance causes. They cannot be eliminated by either lengthy burn-in periods or good preventive maintenance practices. Equipment is designed to operate under certain conditions and up to certain stress levels. When these stress levels are exceeded due to random unforeseen or unknown events, a chance failure will occur. While reliability theory and practice are concerned with all three types of failures, its primary concern is with chance failures, since they occur during the useful life period of the equipment. The time when a chance failure will occur cannot be predicted; however, the likelihood or probability that one will occur during a given period of time within the useful life can be determined by analyzing the equipment design. If the probability of a failure chance is too great, either design changes must be introduced or the operating environment must be made less severe.

5.1.3 End of Life

The wear-out period is characterized as a result of equipment deterioration due to age or use. For example, mechanical components such as transmission bearings will eventually wear out and fail, regardless of how well they are made. Early failures can be detected and the useful life of the equipment can be extended by good design and maintenance practices. The only way to prevent failure due to wear-out is to replace or repair the deteriorating component before it fails.

5.1.4 Effect of Mission Profile

The power electronic converter reliability depends on its loading profile, which induces thermal stresses on the failure-prone components, hence limiting the converter lifetime. The loading profile of different converters in a power system is relevant to the mission profiles (including environmental and operational conditions), which can be controlled by a power management strategy. For example, as a key component in the wind turbine system, the power electronic converter and its power semiconductors suffer from complicated power loadings related to the environment and are prone to having high failure rates. Therefore, correct lifetime estimation of wind power converters is crucial for reliability improvement and also for cost reduction of wind power technology. The existing lifetime estimation methods for the power electronic converter are not yet suitable in these applications, because the comprehensive mission profiles are not well specified and included. Consequently, a relative more advanced approach is proposed in reference [5], which is based on the loading and strength analysis of devices and takes into account different time constants of the thermal behaviors in the power converter. With the established methods for loading and lifetime estimation for power devices, more detailed information of the lifetime-related performance in the wind power converter can be obtained. Proper life consumption estimation methods applied for power electronics that can operate in real time under in-service mission profile conditions will not only provide an effective assessment of the product's life expectancy but can also deliver reliability design information. This is an important aid in manufacturing and thus helps in reducing costs and maximizing through-life availability. In reference [6], a mission profile-based approach for real-time life consumption estimation that can be used for reliability design of power electronics is presented. The reference presents the use of electrothermal models coupled with physics-of-failure (PoF) analysis by means of a real-time counting algorithm to provide accurate life consumption estimations for power modules operating under in-service conditions. These models, when driven by the actual mission profiles, can be utilized to provide advanced warning of failures, thus delivering information that can be useful to meet particular application requirements for reliability at the design stage. Existing converter-level reliability analysis methods have two major limitations: being based on constant failure rate models and lack of consideration of long-term operation conditions (i.e. mission profile). Although various studies have been presented on power electronic component-level lifetime predictions based on wear-out failure mechanisms and mission profiles, it is still a challenge to apply the same method to the reliability analysis of converters with multiple components. A component lifetime prediction based on associated models provides only a B_X lifetime information (i.e. the time when X% items fail) but the time-dependent reliability curve is still not available. In reference [7], a converter-level reliability analysis approach is proposed based on time-dependent failure rate models and long-term mission profiles. Reference [8] presents the reliability analysis of a push–pull converter intended for connection to a 125-W photovoltaic (PV) panel. Four prototypes of the converter were built and tested, using transistors with different ratings. Failure rates were calculated using the MIL HDBK 217 and the IEC TR 62380 procedures. In the latter case, the prediction was performed taking into account an annual mission profile obtained from the intended installation site, in an area with desert climate temperatures. Failure rate results obtained with MIL HDBK 217 show small differences among the converters, the best performance being obtained from the prototype with the lowest on-resistance. Results obtained with IEC TR 62380 indicate that thermal cycles have a significant effect in reliability performance, and should be considered carefully, because PV systems often see large temperature variations. With both procedures, the failure rate contributions from magnetic devices were higher than expected.

5.2 Accelerated Aging Tests

The performance of many products degrades as the product ages. Such degradation is usually slow but can be accelerated by a high stress. For example, the breakdown strength of electrical insulation depends on age and temperature. In some tests, the performance of a test unit is measured only once at a chosen age. For example, an insulation specimen yields only one breakdown measurement [9]. Common ways to determine a life stress relationship are:

- Arrhenius model
- Eyring model
- Inverse power law model
- Temperature–humidity model
- Temperature non-thermal model

The Arrhenius equation is a simple but remarkably accurate formula for the temperature dependence of reaction rates. Arrhenius provided a physical justification and interpretation for the formula. Currently, it is best seen as an empirical relationship. It can be used to model the temperature variation of diffusion coefficients, population of crystal vacancies, creep rates, and many other thermally-induced processes. A historically useful generalization supported by the Arrhenius model is that, for many common chemical reactions at room temperature, the reaction rate doubles for every 10 °C increase in temperature.

Life data analysis involves analyzing times-to-failure data obtained under normal operating conditions in order to quantify the life characteristics of a product, system, or component. For many reasons, obtaining such life data (or times-to-failure data) may be very difficult or impossible. The reasons for this difficulty can include the long lifetimes of today's products, the small time period between design and release, and the challenge of testing products that are used continuously under normal conditions. Given these difficulties and the need to observe failures of products to understand their failure modes and life characteristics better, reliability practitioners have attempted to devise methods to force these products to fail more quickly than they would under normal use conditions. In other words, they have attempted to accelerate their failures.

The purpose of accelerated life testing (ALT) is to induce field failure in the laboratory at a much faster rate by providing a harsher, but nonetheless representative, environment. In such a test, the product is expected to fail in the lab just as it would have failed in the field – but in much less time. The main objective of an accelerated test is either of the following:

- To discover failure modes
- To predict the normal field life from the high stress lab life

In reliability tests, at each test one or some parameters of environmental conditions are stressed more than the typical state to reduce the test time less than the real state. Then, there are some determined relations between these accelerated test results and typical condition results that are used to find failure rates and reliability evaluations. One of the main concerns in these tests is this question: does the over stress test have an improper effect on the equipment [10]? This question becomes more important when we want to evaluate the remaining useful life of a converter. In this case, the reliability tests must have no destructive effect on the equipment. In destructive testing, tests are carried out to the specimen failure, in order to understand a specimen's structural performance or material behavior under different loads. These tests are generally much easier to carry out, yield more information, and are easier to interpret than non-destructive testing (NDT). Destructive testing is most suitable, and economic, for objects that will be mass-produced, as the cost of destroying a small number of specimens is negligible. NDT includes a wide group of analysis techniques used in science and industry to evaluate the properties of a material, component, or system without causing damage. Because NDT does not permanently alter the objects being inspected, it is a highly valuable technique that can save both money and time in product evaluation, troubleshooting, and research. Common NDT methods include ultrasonic, magnetic-particle, liquid penetrant, radiographic, remote visual inspection, eddy-current testing, and low coherence interferometry. NDT is commonly used in forensic engineering, mechanical engineering, electrical engineering, civil engineering, systems engineering, aeronautical engineering, and medicine.

To show the difference between destructive physical analysis (DPA) and NDT an insulator is a good example. Figure 5.2 shows the nonlinear I–V characteristic of a typical isolator. If the test voltage causes a nonlinear breakdown of the isolator, a DPA occurs. If the applied voltage is less than the breakdown voltage, the isolator is not damaged during the test, as shown in Figure 5.3.

5.2.1 Stressors

Accelerated aging tests (AATs) are usually performed in chambers. An environmental chamber is an enclosure used to test the effects of specified environmental conditions on biological items, industrial products, materials, and electronic devices and components. Temperature

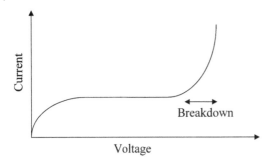

Figure 5.2 V-I characteristics of an electric insulator.

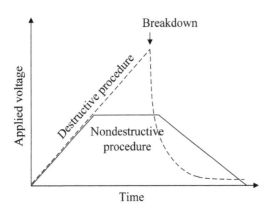

Figure 5.3 Difference between NDT and DT in an electric withstand test.

Figure 5.4 A thermal cycling chamber. *Source*: Garouk Co. with permission.

cycling is the process of cycling through two temperature extremes, typically at relatively high rates of change. It is an environmental stress test used in evaluating product reliability as well as in manufacturing to catch early-term, latent defects by inducing failure through thermal fatigue. Figure 5.4 shows a typical test chamber for a thermal cycling test of electronic boards. Figures 5.5 and 5.6 show typical test scenarios for thermal cycling and thermal shock tests.

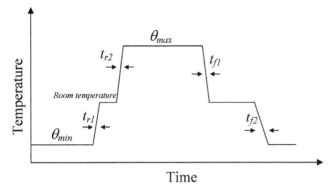

Figure 5.5 A scenario for thermal cycling.

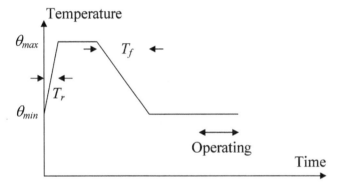

Figure 5.6 A scenario for thermal shock.

An electrical field is an important failure factor in power converters. Under normal conditions, any electrical device will produce a minimal amount of leakage current due to the voltages and internal capacitance present within the product. However, due to design flaws or other factors, the insulation in a product can break down, resulting in excessive leakage current flow.

5.2.2 Types of Tests

Acceptance testing and qualification testing are engineering processes that products undergo before they are deemed safe and ready to use. Many types of equipment are subject to a realm of rigorous tests to meet industry and government safety standards. There is typically a range of tests that occur between the qualification of an electrical product and its final acceptance. The first step in the testing process is usually a materials test to ensure each piece of the finished electrical product meets or exceeds expected standards. The next step is the qualification testing phase, which determines whether the overall design is credible. Again, this is an important phase to complete before you even begin manufacturing, because a flawed design will result in a flawed finished product.

Before an electrical equipment can pass the acceptance testing stage, it goes through exhaustive field testing. The final stage involves acceptance testing, which is performed after installation and field testing are complete. The product runs through its normal application. The acceptance testing process is performed primarily to test the installation and ferret out any bugs. Acceptance testing should not exceed the designed capabilities of the product.

Industrial standards in the electrical industry guide the qualification testing process and usually include aging tests as well as comparisons to similar products. Manufacturers of different cables or software platforms must comply with a set of industry qualification testing standards. Once the product passes the materials and qualification tests, it can then go to production. Production testing occurs periodically during and after the manufacturing process. Industry standards are used to measure the viability of the product as it goes through production. Random Failure Models

The failure rate shown in Figure 5.1 is almost constant for a long period. In this period, the failure rate is mainly due to random failures. Although a constant failure rate is used in most calculations of reliability (mainly owing to the simplified derivations), there are cases for which this simplifying assumption is inappropriate, especially during the "infant mortality" and "wear-out" phases of a component's life. However, assuming a constant failure rate simplifies the reliability relations and is valid for the period before EOL.

5.2.3 MIL-HDBK-217

Maintaining reliability and providing reliability engineering is an essential need with modern electronic systems. Reliability engineering for electronic equipment requires a means for a quantitative baseline, or a reliability prediction analysis. There are some technical standards for numerical evaluation of power electronic components failure rate. One of the most widely used reliability prediction methodology handbooks is MIL-HDBK-217, the Military Handbook for "Reliability Prediction of Electronic Equipment." The MIL-HDBK-217 handbook contains failure rate models for the various part types used in electronic systems, such as ICs, transistors, diodes, resistors, capacitors, relays, switches, connectors, etc. These failure rate models are based on the best field data that could be obtained for a wide variety of parts and systems; this data is then analyzed, with many simplifying assumptions thrown in, to create usable models. The purpose for developing this handbook was to establish and maintain consistent and uniform methods for estimating the inherent reliability of military electronic equipment and systems.

MIL-217 uses the part stress analysis method for characterizing the component's reliability. The part stress analysis method is used the majority of times and is applicable when the design is near completion and a detailed parts list, or bill of materials (BOM), plus component stresses are available. By component stresses, the standard is referring to the actual operating conditions, such as environment, temperature, voltage, current, and power levels applied. The MIL-217 standard groups components or parts by major categories and then has subgroups within the categories. An example is a "fixed electrolytic (dry) aluminum capacitor," which is a subcategory of the "capacitor" group. Each component or part category and its subgroups have a unique formula or model applied to it for calculating the failure rate for that component or part. The failure rate formulas include a base failure rate for the category and subgroup selected. The base failure rates apply to components and parts operating under normal environmental conditions, with power applied, performing the intended functions, using base component quality levels, and operating at the design stress levels. The standard then applies many correction factors to the base failure rates in order to factor in the actual operating conditions, environment, and stress levels. Base failure rates are adjusted by applying the correction factors to the underlying equation or model provided for each component category.

The over temperature is one of the most important parameters affecting the failure rate of a component. Ambient and operating temperatures have a major impact on the failure rate prediction results of electronic equipment, especially equipment involving semiconductors and integrated circuits. The MIL-217 standard requires an input of ambient temperatures and more definitive data required for the calculation of junction temperatures in semiconductors and microcircuits.

A thermal analysis should be a part of the design and reliability analysis process for electronic equipment.

The voltage stress factor accounts for the acceleration of failure mechanisms associated with abnormally high supply voltages.

The design quality of the component utilized has a direct effect on the part failure rate and appears in the models as a correction factor. Many of the components covered by the MIL-217 specification are available in several quality levels and each has an associated correction factor.

Environmental stress is of major concern in establishing the failure rate for components and parts included in a system. Environmental stresses can be quite different from one application environment to another and can subject the equipment to a controlled environment with constant temperature and humidity, or an environment with rapid temperature changes, high humidity, high vibration, and high acceleration.

In MIL-HDBK-217, the following equation is determined for the MOSFET failure rate:

$$\lambda_p = \lambda_b \times \pi_T \times \pi_A \times \pi_Q \times \pi_E \quad \text{Failure} / 10^6 \text{ hours} \tag{5.6}$$

where π_T is the temperature factor and is deduced from

$$\pi_T = \exp\left\{-1925 \times \left(\frac{1}{T_j + 273} - \frac{1}{298}\right)\right\} \tag{5.7}$$

where T_j is the junction temperature that is related to MOSFET power losses. Therefore, for determination of π_T, one important step is a MOSFET power loss evaluation. The reduction of power losses results in a decrement of π_T and therefore the failure rate comes down. The quality factor π_Q is determined by a packaging and covering method, where for commercial applications it is "lower" in quality and is equal to 5.5. Other quality factors that have less value are for specific applications, such as for the military and aerospace. For example, using the JANTX package quality, which has π_Q equal to 1.0, results in a lower switch failure rate.

The diode failure rate is expressed as

$$\lambda_p = \lambda_b \times \pi_T \times \pi_S \times \pi_C \times \pi_Q \times \pi_E \quad \text{Failure} / 10^6 \text{ hours} \tag{5.8}$$

where π_T relates to diode power losses. Lower power losses cause a smaller π_T and the failure rate

$$\pi_T = \exp\left\{-3091 \times \left(\frac{1}{T_j + 273} - \frac{1}{298}\right)\right\} \tag{5.9}$$

while π_S is the electrical stress factor that is related to the rated and applied voltages:

$$\pi_S = \left(\frac{\text{Applied voltage}}{\text{Rated voltage}}\right)^{2.43} \tag{5.10}$$

Whenever the portion of applied voltage to rated voltage becomes smaller, this factor also gets smaller. Hence, selection of a diode with a higher rated voltage results in a lower failure rate, but there is one important fact that should be considered for diode selection. In fact, choosing diodes that have a larger rated voltage may have a larger forward voltage, which can increase diode conduction loss and affect π_T. Therefore, optimizing between different parameters is required to choose an appropriate diode.

For an inductor, the following equation is used to calculate the failure rate:

$$\lambda_p = \lambda_b \times \pi_T \times \pi_Q \times \pi_E \quad \text{Failure} / 10^6 \text{hours} \tag{5.11}$$

where λ_b is equal to 0.00003 for inductors that have a fixed inductance and π_T is the temperature factor:

$$\pi_T = \exp\left(\frac{-0.11}{8.617 \times 10^{-5}} \left(\frac{1}{T_{HS} + 273} - \frac{1}{298}\right)\right) \tag{5.12}$$

where

$$T_{HS} = T_A + \frac{1.1 \times 11.5 \times P_{loss}}{W_L^{0.6766}}$$

It is clear from the above equations that the inductor failure rate has a direct relationship with power losses. However, the inductor failure rate is much lower than that of other components and there have not been any reports that the inductor failure rate produces a problem for reliability of power electronic converters.

The other component is the capacitor, which does give a failure rate:

$$\lambda_p = \lambda_b \times \pi_T \times \pi_C \times \pi_V \times \pi_Q \times \pi_E \quad \text{Failure} / 10^6 \text{hours} \tag{5.13}$$

where π_T is determined by the ambient temperature:

$$\pi_T = \exp\left\{\frac{0.15}{8.617 \times 10^{-5}} \times \left(\frac{1}{298} - \frac{1}{T_A + 273}\right)\right\}$$

and

$$\pi_C = C^{0.09} \quad \left(C \text{ is in } \mu f \text{ units}\right) \tag{5.14}$$

As presented, a greater rated voltage leads to a smaller π_V, which results in a lower failure rate. Consequently, one method to provide a failure rate reduction uses a capacitor that has a higher rated voltage:

$$S = \frac{\text{Applied voltage to capacitor}}{\text{Rated voltage}}$$

with

$$\pi_V = \left(\frac{S}{0.6}\right)^5 + 1 \tag{5.15}$$

Figures 5.7–5.14 show the comparison between the various failure factors for different elements.

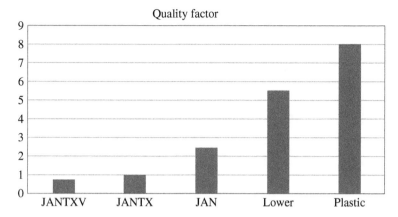

Figure 5.7 Different values of quality factor in MIL-217.

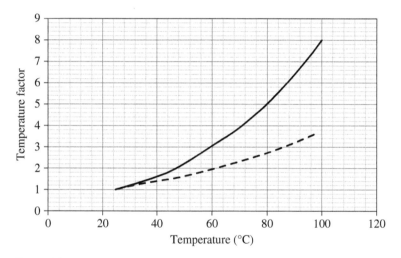

Figure 5.8 A comparison between the temperature factor in MIL-217 between diode (dashed) and MOSFET (solid).

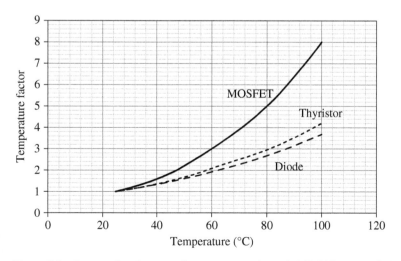

Figure 5.9 A comparison between the temperature factor in MIL-217 among diode and MOSFET and thyristor.

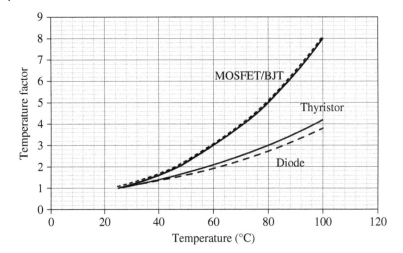

Figure 5.10 A comparison between the temperature factor in MIL-217 between different switches.

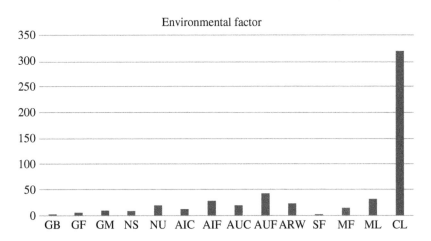

Figure 5.11 Different values of environmental factor in MIL-217.

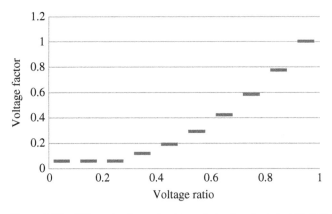

Figure 5.12 Different values of voltage factor for diode in MIL-217.

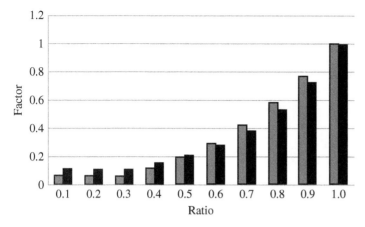

Figure 5.13 A comparison between values of voltage factor for diode (left) and BJT (right) in MIL-217.

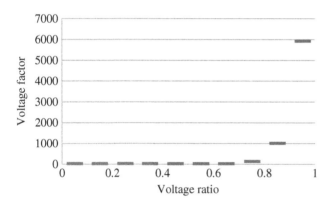

Figure 5.14 Different values of factor for a capacitor in MIL-217.

5.2.4 FIDES Standard

The FIDES Guide is a global methodology for reliability engineering in electronics. It contains two parts:

- A reliability prediction guide
- A reliability process control and audit guide

The FIDES Guide aims to enable a realistic assessment of the reliability of electronic equipment, including systems operating in severe environments (aeronautics, defense systems, industrial electronics, transport, etc.). The FIDES Guide also aims to provide a concrete tool to develop and control this reliability. Its key features are:

- Providing models both for electrical, electronic, and electromechanical components, and for some subassemblies.
- Revealing and taking into consideration all technological and physical factors that play an identified role in a product's reliability.
- Taking into precise consideration the mission profile.
- Taking into consideration the electrical, mechanical, and thermal overstresses.

- Taking into consideration the failures linked to the development, production, field operation, and maintenance processes.
- The possibility of distinguishing several suppliers of the same component.

By identifying the factors contributing to reliability, whether technological, physical, or process-based, the FIDES Guide makes it possible to revise product definition and intervene throughout the product lifecycle, to improve and control reliability.

5.3   End of Life Models

Reliability prediction in power electronic converters is of paramount importance for converter manufacturers and operators. Conventional approaches employ generic data provided in hand-books for random chance failure probability prediction within a useful lifetime. However, the wear-out failures affect the long-term performance of the converters. Reference [11] proposes a comprehensive approach for estimating the converter reliability within a useful lifetime and wear-out period. Moreover, this article proposes a wear-out failure prediction approach based on a structural reliability concept. The proposed approach can quickly predict the converter wear-out behavior, unlike conventional Monte Carlo-based techniques. Hence, it facilitates reliability modeling and evaluation in large-scale power electronic-based power systems with a huge number of components. The proposed comprehensive failure function over the useful lifetime and wear-out phase can be used for optimal design and manufacturing by identifying the failure-prone components and end-of-life prediction. Moreover, the proposed reliability model can be used for optimal decision-making in design, planning, operation, and maintenance of modern power electronic-based power systems. The proposed methodology is exemplified for a PV inverter by predicting its failure characteristics.

5.3.1   Empirical Models

These models are based on the experiments and the statistical analysis of the results. For many applications, the impact on the component lifetime of temperature swings generated by low output frequency is of special interest. A simple form of the EOL empirical model is

$$N_f = \alpha.\Delta T^{-n} \tag{5.16}$$

where ΔT is the temperature swing and α and n are the empirical coefficients.

Therefore, active power cycling tests with a variation of load pulse durations were conducted. The requirements for the active power cycling test equipment related to load pulse durations down to 70 ms is discussed in reference [12]. The analysis of the test results leads to the proposal of a new empirical lifetime model for advanced power modules, which explicitly contains the impact of the load pulse duration. These reliability results will affect the design of advanced inverters with low-frequency output currents.

Temperature cycle profiles at various stress levels are investigated for accelerated reliability testing of electronic device packaging. A failure analysis is conducted for test failures to determine their root cause failure mechanisms and failure modes. Weibull analysis is conducted for failures with the main failure mechanisms (such as solder fatigue, mechanical cracks). The Coffin-Manson model has been used to model crack growth due to repeated temperature cycling [13]. In reference [14], a physical model is proposed to estimate the TRIAC solder joints fatigue during power cycling. The lifetime prediction is based on the following assumptions: the case temperature swing (ΔT_{case}) is the main acceleration factor, the solder joints are the weakest materials in the TRIAC assembly, and the plastic strain within the solder layer due to shearing is the failure cause.

Stippich et al. [15] investigate how thermal buffers in power electronic modules can reduce thermal cycles and thus mitigate fatigue and improve the lifetime of power modules. Safe and reliable power module operation over an extended lifetime is crucial for many power electronic applications, e.g. automated driving, aerospace, and energy supply. To address this objective, this work adds passive thermal buffer elements made of copper to the power module to extend its lifetime. For the analysis of different thermal buffer geometries, detailed 3-D thermal models are derived. The junction temperature responses of the power module geometries are simulated with 3-D thermal models of typical mission profiles from electric vehicles. Consequently, the occurring thermal cycles are extracted by a rainflow-counting algorithm and used to compute the lifetime of the power modules on the basis of empirical models. The results show a significant influence of thermal buffers on the lifetime of the modules, as they mitigate thermal cycles that drive thermally-induced degradation. It could be demonstrated that this technology can increase the lifetime of the power modules by up to a factor of six in electric vehicles. Thus, thermal buffers provide a simple, effective, and cost-efficient way to increase the lifetime and thus reliability of power electronics modules.

5.3.2 Physics of Failure Models

A simple form of the EOL empirical model is the number of cyclings to failure, N_f:

$$N_f = W_{crit} \cdot \left(\Delta W_{hys} \right)^{-n} \tag{5.17}$$

where W_{crit} is the energy leading to the failure and ΔW_{hys} represents the accumulated deformation energy per cycle.

A collection of methodologies based on the PoF approach and mission profile analysis are presented in reference [16] to perform a reliability-oriented design of power electronic systems. The corresponding design procedures and reliability prediction models are provided. Further on, a case study on a 2.3 MW wind power converter is discussed with emphasis on the reliability critical components of insulated gate bipolar transistors (IGBTs). Different aspects of improving the reliability of the power converter are mapped. The challenges and opportunities to achieve more reliable power electronic systems are addressed. Understanding the PoF of IGBT modules has been critical to the development of effective condition monitoring (CM) techniques as well as reliable prognostic methods. Reference [17] summarizes past developments and recent advances in the area of CM and prognostics for IGBT modules. The improvements in material, fabrication, and structure are described. The CM techniques and prognostic methods proposed in the literature are presented. This reference concludes with recommendations for future research topics in the CM and prognostics areas. Reference [4] serves to give an overview of the major aspects of reliability in power electronics and to address future trends in this multidisciplinary research direction. The ongoing paradigm shift in reliability research is presented first. Then, the three major aspects of power electronics reliability are discussed, respectively, which cover a PoF analysis of critical power electronic components, state-of-the-art design for the reliability process and robustness validation, and intelligent control and CM to achieve improved reliability under operation. Finally, the challenges and opportunities for achieving more reliable power electronic systems in the future are discussed.

5.4 Thermomechanical Models

Over temperature is one of the main reasons for failure in electric power converters. In addition, some of other failure factors, such as dielectric breakdown, act as over temperature in a damaging process of a converter. In previous sections, it is emphasized that the temperature factor is a key index in a reliability calculation. Unlike fully electrical variables, thermal calculations require details of geometry of the system and its environment. In this chapter, thermal analysis as the most important factor in the failure of converters is presented. Two main approaches for this goal are presented: numerical and lumped models.

5.4.1 Analytical Models

Finite element method (FEM) techniques for analyzing the temperature profile are usually accurate tools and yield the temperature of many nodes in the studied system. However, these methods usually have a long run time and can suffer from divergence problems. Especially in the optimization process with iterative solving techniques, it is an important limitation. Another alternative technique for analyzing the temperature is usage of a thermal model. A thermal model is a summarized model of the one that is used in FEM. In this summarized model, the number of nodes is limited to a few nodes. Therefore, the set of the thermal equation becomes small and can be solved rapidly. Thus, the run time decreases and the temperature profile is achieved in a reasonable time interval. The cost paid for this short convergence time is decreasing in the accuracy of the solutions. Regarding these characteristics, thermal models are used in the design process to reach an acceptable design and then the accuracy of the design is raised with an FEM-based analysis. Thermal models facilitate the design and analysis of cooling systems. The underlying idea of such models is the formal similarity between heat transfer and electric circuits. Thermal quantities and their electrical counterparts are listed in Table 5.1.

Figure 5.15 shows a typical heat transfer problem, in which a lossy system dissipates some energy into its ambient. The thermal equivalent for this typical problem is shown in Figure 5.16.

In this circuit, T_2 and T_1 are the ambient and system temperatures, respectively, P_{loss} is the power loss in the system and R_{th} is the thermal resistance of the heat transfer path. Using this equivalent circuit, the following relationship can be readily obtained:

$$T_1 - T_2 = P_{loss}.R_{th} \tag{5.18}$$

It is worth noting that the thermal capacity is not considered in this study, as it does not affect the steady-state temperature. However, the transient swing of the junction temperature is important in the EOL models. There are two approaches for thermal transient modeling [18]. The Foster

Table 5.1 Thermal and electrical equivalent pairs.

Thermal quantity	Electrical quantity
Heat current (power), P (W)	Current, I (A)
Temperature, T (°K)	Voltage, V (V)
Thermal resistance, R (°K/W)	Resistance, R (Ω)
Thermal capacity, C (J/°K)	Capacitance, C (F)

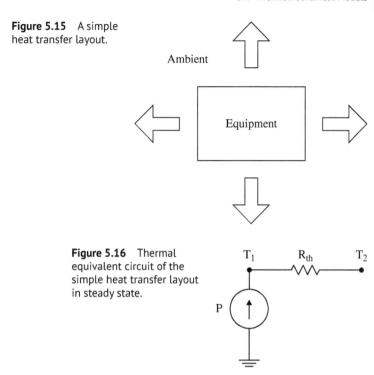

Figure 5.15 A simple heat transfer layout.

Figure 5.16 Thermal equivalent circuit of the simple heat transfer layout in steady state.

Thermal Model block represents heat transfer through a semiconductor module. Figure 5.17 shows an equivalent circuit for a third-order Foster Thermal Model block. T_j is the junction temperature and T_c is the base plate or ambient temperature. The Cauer Thermal Model block represents heat transfer through multiple layers of a semiconductor module. A Cauer Thermal Model contains multiple Cauer Thermal Model Element components. Figure 5.18 shows an equivalent circuit for a Cauer Thermal Model. Figure 5.19 shows the equivalent thermal impedance of an IGBT, which is based on the Foster thermal model.

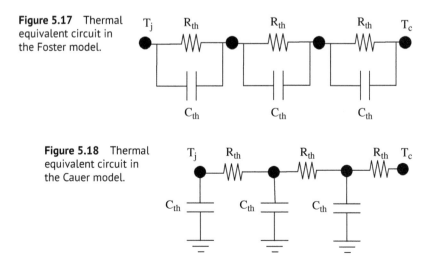

Figure 5.17 Thermal equivalent circuit in the Foster model.

Figure 5.18 Thermal equivalent circuit in the Cauer model.

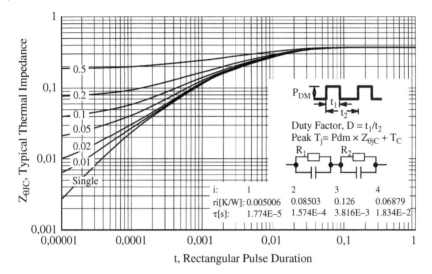

Figure 5.19 Thermal impedance of an IGBT. *Source:* Used with permission from SCILLC dba onsemi.

5.4.2 Numerical Models

In mathematics, the FEM is a numerical technique for finding approximate solutions to boundary value problems for differential equations. It uses the calculation methods to minimize an error function and produce a stable solution. FEM encompasses all the methods for connecting many simple element equations over many small subdomains, named finite elements, to approximate a more complex equation over a larger domain.

A typical work using this method is dividing the domain of the problem into a collection of subdomains, with each subdomain represented by a set of element equations to the original problem, systematically recombining all sets of element equations into a global system of equations for the final calculation. The global system of equations has known solution techniques and can be calculated from the initial values of the original problem to obtain a numerical answer. The subdivision of a whole domain into simpler parts has several advantages:

- Accurate representation of complex geometry
- Inclusion of dissimilar material properties
- Easy representation of the total solution
- Capture of local effects

Numerical methods provide a prediction of heat flows. This method is more appropriate for precise modeling to optimization. Numerical methods can give an insight into patterns that are difficult, expensive, or impossible to study using experimental methods. Experiments can give a quantitative description of flow phenomena using measurements for one quantity at a time, at a limited number of points and time instances. If a full-scale model is not available or is not practical, scale models or dummy models can be used. The experiments can have a limited range of problems and operating conditions. Simulations can give a prediction of flow phenomena using software for all desired quantities, with a high resolution in space and time and virtually any problem, and realistic operating conditions.

At the first step, the element equations are simple equations that locally approximate the original complex equations to be studied, where the original equations are often partial differential equations. The process is to construct an integral of the inner product of the residual and the weight functions and set the integral to zero. In simple terms, it is a procedure that minimizes the error of approximation by fitting trial functions into the partial differential equations. The residual is the error caused by the trial functions, and the weight functions are polynomial approximation functions that project the residual.

In the second step, a global system of equations is generated from the element equations through a transformation of coordinates from the subdomain's local nodes to the domain's global nodes. This spatial transformation includes appropriate orientation adjustments as applied in relation to the reference coordinate system. The process is often carried out by FEM software using coordinate data generated from the subdomains. Figure 5.20 shows the COMSOL software as a FEM analyzer. This software can be used to thermal study. In this figure, an IGBT module is studies. This investigation can be done as steady state or transient study. The results are shown in Figure 5.21.

FEM is a good choice for analyzing problems over complicated domains, when the domain changes, when the desired precision varies over the entire domain, or when the solution lacks smoothness. FEM software provides a wide range of simulation options for controlling the complexity of both modeling and analysis of a system. Similarly, the desired level of accuracy required and associated computational time requirements can be managed simultaneously to address most engineering applications. FEM allows entire designs to be constructed, refined, and optimized before the design is manufactured.

This powerful design tool has significantly improved both the standard of engineering designs and the methodology of the design process in many industrial applications. The introduction of FEM has substantially decreased the time to take products from concept to the production line. It is primarily through improved initial prototype designs using FEM that testing and development have been accelerated. In summary, the benefits of FEM include increased accuracy, enhanced design, and better insight into critical design parameters, virtual prototyping, fewer hardware prototypes, a faster and less expensive design cycle, increased productivity, and increased revenue.

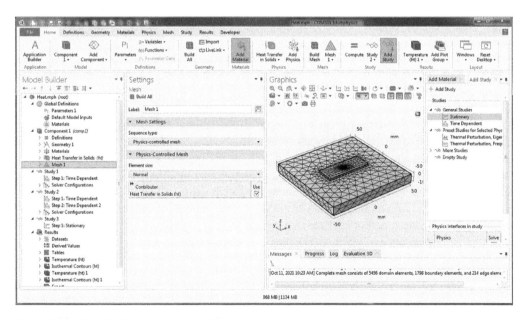

Figure 5.20 The COMSOL software as a FEM analyzer. *Source*: COMSOL INC.

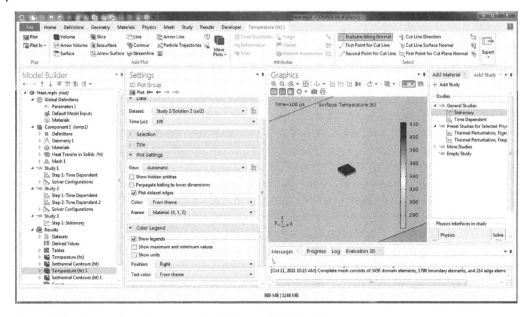

Figure 5.21 The results of the thermal study. *Source*: COMSOL INC.

5.5 System Life Modeling

Fault occurrence is a relatively random phenomenon. Randomness means lack of pattern or predictability in events. Therefore, essential methods of a lifetime prediction are based on a probability analysis. The fields of probability and statistics use formal definitions of randomness. In statistics, a random variable is an assignment of a numerical value to each possible outcome of an event space. This association facilitates the identification and the calculation of probabilities of the events. Random variables can appear in random sequences. A random process is a sequence of random variables describing a process whose outcomes do not follow a deterministic pattern, but follow an evolution described by probability distributions. These and other constructs are extremely useful in probability theory and the various applications of randomness. Probabilistic design is a discipline within engineering design. It deals primarily with the consideration of the effects of random variability upon the performance of an engineering system during the design phase. Typically, these effects are related to quality and reliability. Thus, probabilistic design is a tool that is mostly used in areas that are concerned with quality and reliability. Probability theory is applied in everyday life in risk assessment and in trade on financial markets.

A random variable x is, in the simplest term, a variable that randomly takes some values in a specific range. In the probability field, a random variable is expressed by its statistical moments. The probability with which different values are taken by a random variable is quantified by the probability distribution function $F(x)$, defined by $F(x) = \Pr(x \leq x)$, that is, the probability that the random variable x has a value less than or equal to x. The probability density function (PDF) $p(x)$ is the derivative given by $p(x) = dF(x)/dx$.

The expectation of a random variable x, denoted as $E\{x\}$, is given by

$$\bar{X} = E\{X\} = \int_{-\infty}^{\infty} x p(x) dx \tag{5.19}$$

This is also called the mean value of x or the first moment of x.

A discrete random process may be thought of as a collection of real or discrete sequences of time, any of which might be observed on any trial of an experiment. The mean or expected value of a random process $x[n]$ at the time index n is defined as

$$\bar{x}[n] = E\{X[n]\} = \sum_{i=1}^{n} x_i P_i \tag{5.20}$$

The autocorrelation of a discrete random process between two sample times indices n_1 and n_2 is defined as

$$R_{xx}[n_1, n_2] = E\{x[n_1] x*[n_2]\} \tag{5.21}$$

where x is the conjugate of x. Equation (5.21) is the engineering definition for autocorrelation, as first suggested by Wiener. The autocorrelation function of a random process is the appropriate statistical average that will be concerned with characterizing random signals in the time domain. A random process is wide-sense stationary (WSS) if its mean is constant for all time indices (i.e. independent of time) and its autocorrelation depends only on the time index difference $m = n_1 - n_2$. Therefore, autocorrelation of a WSS discrete random process $x[n]$ is given by

$$R_{xx}[n_1, n_2 - m] = E\{x[n+m] x*[n]\} \tag{5.22}$$

The power spectral density of a discrete random process is defined as the discrete time Fourier transform (DFT) of the autocorrelation sequence

$$P_{xx}(f) = \sum_{m=-\infty}^{+\infty} R_{xx}[n_1, n_2 - m] e^{-j2 fm} \tag{5.23}$$

5.5.1 Markov Tool

There are many approaches for a reliability analysis of systems. The Markov model is a powerful tool for analyzing these systems. However, the failure probability of a process is not a constant value. If the time-variable failure probability is considered, the traditional homogeneous Markov models are not valid. Time dependence of the detection–identification–reconfiguration (DIR) process failure probability leads to use of a non-homogeneous Markov model.

If X is a stochastic process, then X is a Markov chain if, for all $j \in E$,

$$\Pr\{X_{n+1} = j \mid X_0, X_1, \ldots, X_n\} = \Pr\{X_{n+1} = j \mid X_n\} \tag{5.24}$$

Here $E = \{1, 2, \ldots, s\}$ and $n \in \mathbb{N}$. The $p_n(i, j)$, $(n \in \mathbb{N}, i, j \in E)$ is the transition probability function of chain X in the nth subinterval. Multiple-step transition probabilities are defined by

$$
p_{n,m}(i,j) = \Pr\{X_m = j \mid X_n = i\} =
\begin{cases}
\left[\prod_{k=n}^{m-1} P_k \right](i,j), \text{for } m > n \geq 0, \\
1, \\
\quad \text{for } i = j, m = n \geq 0, \\
0, \\
\quad \text{for } i \neq j, m = n \geq 0
\end{cases}
\tag{5.25}
$$

$$p_{n,n+1}(i,j) = p_n(i,j) \tag{5.26}$$

If $p_n(i, j)$ does not depend on n, the Markov chain is time homogeneous.

The state probability vector in the Markov chain of X is defined by its α and $p_n(n \in \mathbb{N})$, where α is the initial distribution of X and p_n is the transition probability matrix at the nth subinterval

$$P_j(n) = \Pr\{X_n = j\} = \left[\prod_{k=0}^{n-1} p_k\right](j)$$

$$P(n) = \alpha . p_{0,n} \tag{5.27}$$

Also, MTTF is equal to

$$\text{MTTF} = \sum_{n=1}^{\infty} R(n) . \Delta t \tag{5.28}$$

5.5.2 Monte Carlo Tools

One alternative method can be to use the Monte Carlo simulation technique. The increase in the number of states does not increase the computation complexity in this method, since it treats the system as a series of real experiments. In other words, the increase in the number of elements and corresponding increase in the number of states only increase the number of random experiments. In the next subsections, the reliability evaluation of diode-fuse configurations is discussed and also one special framework is proposed to obtain the reliability indices of these configurations. Because of the sequential inherit of the failure of the diodes in the aforementioned configurations, the sequential Monte Carlo simulation (SMCS) method is employed to conduct the reliability evaluations. In SMCS, first, for each component, two random numbers are generated at interval (0, 1); a number for its short-circuit failure and a number for its open-circuit failure, and then these numbers are converted to random TTFs. Afterwards, the minimum value of the TTFs is selected and its corresponding components encounter the failure. Subsequently, the voltage and current stresses of the other components are changed, and, therefore, their failure rates should be updated. Next, new random TTFs are produced for the other components in accordance with the new failure rates of components. This process will continue until the experimented system fails. At the time, the whole time passed before the failure of the system will be saved as a TTF sample. This experiment is repeated over and over to obtain an ample number of TTF samples. Finally, a normalized cumulative frequency histogram of the resulting TTFs will be drawn. This histogram demonstrates the time-dependent probability of the system's well-being – the reliability of the system. The flowchart, which is shown in Figure 5.22, includes the above steps.

The conventional framework for evaluating the reliability of power electronic system by SMCS necessitates calculating a complex function that can quantify the effects of the failure of components on the voltage and current stresses of other components. Achieving this function for the aforementioned configurations is very formidable and time consuming. To deal with this issue, facilitating the issue of obtaining the reliability indices of very complex configurations, a special framework is used. In this framework, the simulation environment is divided into separate environments, namely, a state generation environment and an electrical circuit simulation environment. The random numbers and corresponding random TTFs are generated in the first environment, and the component that first encounters a failure is determined. Afterwards, a fault signal is produced for this component. At the second environment, each component has one normally open parallel switch, modeling its short-circuit failure, and one normally closed series switch, modeling its short-circuit failure. The fault signal produced is the state generation environment imposed on

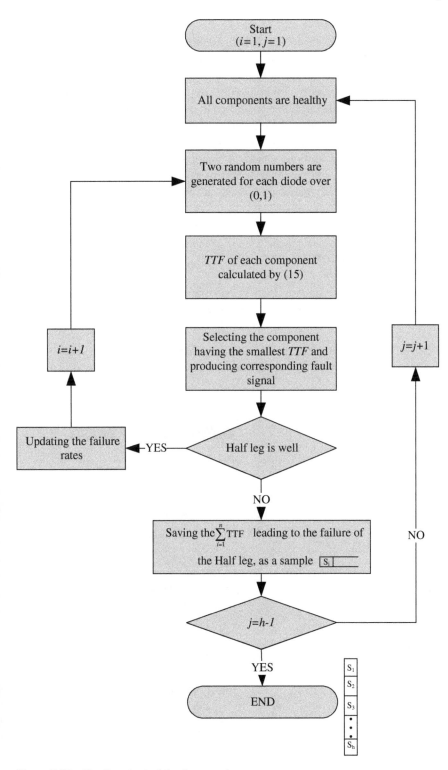

Figure 5.22 The flowchart of the framework.

the corresponding component. If the fault is of the type of a short-circuit failure, the parallel switch is closed. Otherwise, the parallel switch is open. Afterwards, the new voltage and current stresses are measured, and corresponding failure rates are calculated and sent to the state generation environments for the next experiment. By taking advantage of the framework, there is no need to model the system behavior and construct the functions including the voltage and current stresses of each component. Furthermore, this is a universal method that can be applied to complex power electronic and electronic systems to calculate the reliability indices of these systems without modeling the system from the aspect of the reliability. It needs only one environment for random state generation, such as MATLAB Editor, and one electrical circuit simulator environment, such as MATLAB Simulink. Figure 5.23 shows different parts of the proposed framework while Figure 5.24 shows a comparison between the study performed based on the Markov model and the Monte Carlo model.

5.6 Summary and Conclusions

In this chapter, a probability calculation was used to make a useful life prediction. Reliability models from component to system levels were described. The results of this chapter are summarized as follows:

1) Reliability predictions provide a quantitative basis for evaluating product reliability. You can use the information from a reliability prediction to guide design decisions throughout

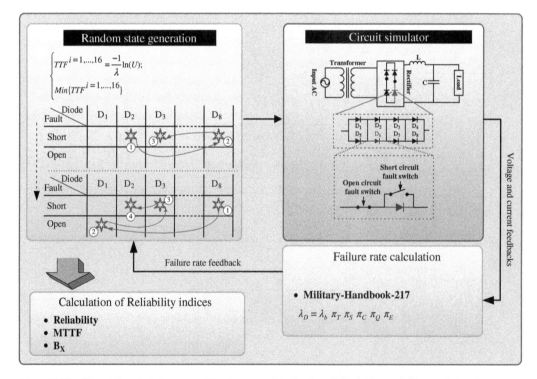

Figure 5.23 The elaborated diagram of the framework including both state generation and circuit simulator environments.

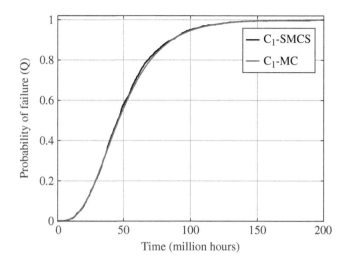

Figure 5.24 A comparison between the results from Markov or Monte Carlo modeling methods.

the development cycle. When an initial design concept is proposed, a reliability prediction can indicate the design feasibility from a reliability standpoint. Even though these early-stage predictions are based on limited design information and are approximate at best, they can give direction to your design decisions, which may be critical to product success. Reliability design begins with the development of a system model. Reliability is defined in terms of probability. Probabilistic design is a discipline within an engineering design. It deals primarily with the consideration of the effects of random variability upon the performance of an engineering system during the design phase. Typically, these effects are related to quality and reliability. Thus, probabilistic design is a tool that is mostly used in areas that are concerned with quality and reliability.

2) One of the most widely used reliability prediction methodology handbooks is MIL-HDBK-217, the Military Handbook for "Reliability Prediction of Electronic Equipment." MIL-HDBK-217 is published by the Department of Defense. It contains failure rate models for numerous electronic components. These models may incorporate predictions based on failure rates taken from historical data. Various parts of a converter have different rates of failure. A failure in a system occurs in the part with highest rate of failure.

3) Series and parallel models are two main types of reliability model of a converter. In a series model, the successful operation of the system depends on the proper functioning of all the system components. A component failure represents a total system failure. A parallel configuration, or the use of redundancy, is one of the design procedures used to achieve extremely high system reliability, greater than the individual component reliabilities.

4) The purpose of ALT is to induce field failure in the laboratory at a much faster rate by providing a harsher, but nonetheless representative, environment. In this method, at each test one or some parameters of environmental conditions are stressed more than a typical state in order to reduce the test time to less than the real state. Then, there are some determined relations between these accelerated test results and typical condition results that are used to find failure rates and reliability evaluations. The failure rate of the product can then be determined.

5) A theoretical thermal model is normally used as a first-order estimate. Thermal analysis can be carried out by numerical analysis or lumped models. The numerical method is not fast but is very accurate. FEM is a tool for a numerical thermal analysis. FEM is best understood from its

practical application and is applied in engineering as a computational tool for performing engineering analysis. It includes the use of mesh generation techniques for dividing a complex problem into small elements, as well as the use of a software program coded with the FEM algorithm. Lumped models are very fast and can be used for optimization, but their accuracy is less than that of numerical methods. Thermal models facilitate the design and analysis of cooling systems. The underlying idea of such models is the formal similarity between heat transfer and electric circuits.

References

1 Mudholkar, G.S. and Srivastava, D.K. (1993). Exponentiated Weibull family for analyzing bathtub failure-rate data. *IEEE Transactions on Reliability* 42 (2): 299–302.

2 Wu, E.Y. and Vollertsen, R. (2002). On the Weibull shape factor of intrinsic breakdown of dielectric films and its accurate experimental determination. Part I: theory, methodology, experimental techniques. *IEEE Transactions on Electron Devices* 49 (12): 2131–2140.

3 Nelson, W. (1980). Accelerated life testing – step-stress models and data analyses. *IEEE Transactions on Reliability* R-29 (2): 103–108.

4 Wang, H., Liserre, M., Blaabjerg, F. et al. (2014). Transitioning to physics-of-failure as a reliability driver in power electronics. *IEEE Journal of Emerging and Selected Topics in Power Electronics* 2 (1): 97–114.

5 Ma, K., Liserre, M., Blaabjerg, F., and Kerekes, T. (2015). Thermal loading and lifetime estimation for power device considering mission profiles in wind power converter. *IEEE Transactions on Power Electronics* 30 (2): 590–602.

6 Musallam, M., Yin, C., Bailey, C., and Johnson, M. (2015). Mission profile-based reliability design and real-time life consumption estimation in power electronics. *IEEE Transactions on Power Electronics* 30 (5): 2601–2613.

7 Zhou, D., Wang, H., and Blaabjerg, F. (2018). Mission profile based system-level reliability analysis of DC/DC converters for a backup power application. *IEEE Transactions on Power Electronics* 33 (9): 8030–8039.

8 De León-Aldaco, S.E., Calleja, H., Chan, F., and Jiménez-Grajales, H.R. (2013). Effect of the mission profile on the reliability of a power converter aimed at photovoltaic applications – a case study. *IEEE Transactions on Power Electronics* 28 (6): 2998–3007.

9 Nelson, W. (1981). Analysis of performance-degradation data from accelerated tests. *IEEE Transactions on Reliability* R-30 (2): 149–155.

10 Baird, P.J., Herman, H., Stevens, G.C., and Jarman, P.N. (2006). Non-destructive measurement of the degradation of transformer insulating paper. *IEEE Transactions on Dielectrics and Electrical Insulation* 13 (2): 309–318.

11 Peyghami, S., Wang, Z., and Blaabjerg, F. (2020). A guideline for reliability prediction in power electronic converters. *IEEE Transactions on Power Electronics* 35 (10): 10958–10968.

12 Scheuermann, U., Schmidt, R., and Newman, P. (2014). Power cycling testing with different load pulse durations. In: *7th IET International Conference on Power Electronics, Machines and Drives (PEMD 2014)*, Manchester, UK, 1–6.

13 Cui, H. (2005). Accelerated temperature cycle test and Coffin-Manson model for electronic packaging. In: *Proceedings of the Annual Reliability and Maintainability Symposium, 2005*, Alexandria, VA, USA, 556–560.

14 Jacques, S., Caldeira, A., Batut, N. et al. (2011). A Coffin–Manson model to predict the TRIAC solder joints fatigue during power cycling. In: *Proceedings of the 2011 14th European Conference on Power Electronics and Applications*, Birmingham, UK, 1–8.

15 Stippich, A., van der Broek, C.H., and De Doncker, R.W. (2020). Enhancing lifetime of power electronic modules via thermal buffers. In: *2020 IEEE Energy Conversion Congress and Exposition (ECCE)*, Detroit, MI, USA, 3178–3185.

16 Wang, H., Ma, K., and Blaabjerg, F. (2012). Design for reliability of power electronic systems. In: *IECON 2012 – 38th Annual Conference on IEEE Industrial Electronics Society*, Montreal, QC, Canada, 33–44.

17 Oh, H., Han, B., McCluskey, P. et al. (2015). Physics-of-failure, condition monitoring, and prognostics of insulated gate bipolar transistor modules: a review. *IEEE Transactions on Power Electronics* 30 (5): 2413–2426.

18 Schütze, T. (2008). *AN2008-03: Thermal Equivalent Circuit Models*. Application Note V1.0. Germany: Infineon Technologies AG.

6

Internal Faults

Converter Level

6.1   Converter Level Faults

Reliability prediction in power electronic converters is very important for converter manufacturers and operators. In the reliability study of the power electronic systems, the main focus of the studies is related to the wear-out and random failures. Conventional approaches employ generic data provided in handbooks for random chance failure probability prediction within a useful lifetime. However, the wear-out failures affect the long-term performance of the converters. In modern reliability studies, it is mentioned that the expected lifetime of a power electronic converter is limited by its weak components, such as semiconductor devices and capacitors. These components are prone to wear-out failures depending on the converter mission profile and component thermal characteristics. The operation modes and the loading profile affect the system reliability. Reference [1] explores the wear-out failure of the power converter semiconductor devices in inverting and rectifying modes of the converter using different power factors. The obtained results show the different wear-out characteristics of a converter under rectification, inversion, and also partial-inversion/rectification modes, as well as when providing reactive power support. Moreover, this reference presents B10 lifetime curves to estimate the converter reliability based on the operating conditions. Reference [2] proposes a comprehensive approach for estimating the converter reliability within a useful lifetime and wear-out period. Moreover, the article proposes a wear-out failure prediction approach based on a structural reliability concept. The proposed approach can quickly predict the converter wear-out behavior, unlike conventional Monte Carlo-based techniques. Hence, it facilitates reliability modeling and evaluation in large-scale power electronic-based power systems with a huge number of components. The proposed comprehensive failure function over the useful lifetime and wear-out phase can be used for optimal design and manufacturing by identifying the failure prone components and end-of-life prediction. Moreover, the proposed reliability model can be used for optimal decision-making in design, planning, operation, and maintenance of modern power electronic-based power systems. Reference [3] explores the impacts of non-exponentially distributed failures on reliability of microgrids. The failure rate of some components such as power electronic converters is not constant, although they play a major role in microgrids. Hence, the conventional reliability evaluation approaches based on mean time to failure may introduce inaccurate inputs for decision making in planning and operation of microgrids. In this reference, the obtained results indicate that the system reliability

Resilient Power Electronic Systems, First Edition. Shahriyar Kaboli, Saeed Peyghami, and Frede Blaabjerg.
© 2022 John Wiley & Sons Ltd. Published 2022 by John Wiley & Sons Ltd.
Companion website: www.wiley.com/go/kaboli/resilientpower

remarkably depends on the failure characteristics, and considering mean or steady-state probabilities instead of failure statistics may introduce erroneous reliability prediction results. Reference [4] proposes a reliability prediction approach for converters modeling the hardware random failures within the useful life and the wear-out period. The methodology is exemplified for a single-phase photovoltaic (PV) inverter and its reliability is predicted during a useful lifetime and the wear-out phase under two operating conditions. At the power system level, dc microgrids have been gaining continually increasing interest over the years. The most important advantages of dc microgrids include higher reliability and efficiency, simpler control and natural interface with renewable energy sources, and electronic loads and energy storage systems. With the rapid emergence of these components in modern power systems, the importance of dc in today's society is gradually being brought to a whole new level. A broad class of traditional dc distribution applications, such as traction, telecom, vehicular, and distributed power systems can be classified under the dc microgrid framework and ongoing development, and expansion of the field is largely influenced by concepts used there [5]. Since the dc microgrids are based on power electronics, internal faults of the power electronic converters are important.

At the element level, the wear-out failures have also been considered in reliability studies. Reliability is one of the key issues for the application of solid-state switches in power electronic converters. Many efforts have been devoted to the reduction of switches wear-out failure induced by accumulated degradation and catastrophic failure triggered by single-event overstress. The wear-out failure under field operation could be mitigated by scheduled maintenances based on lifetime predictions and condition monitoring. However, the catastrophic failure is difficult to predict and thus may lead to serious consequences concerning power electronic converters. To obtain a better understanding of catastrophic failure, the state-of-the-art research on the failure behaviors and failure mechanisms of insulated gate bipolar transistors (IGBTs) is presented in reference [6].

Reliability modeling of power electronic converters is of importance for optimal design, control, and operation of power electronic-based power systems. Suitable topology selection, converter component sizing, and a proper control strategy adoption in a single unit converter, together with operation of a multiconverter system, require a prediction to be made of the converter reliability within its useful life and wear-out phases. In all of the above-mentioned researches, proper design and construction are assumed. This means that the designer and the manufacturer must be certain about the design and construction process of the converters in the power electronic system. It is usually supposed that the failures originate from the internal faults inside the elements or from external faults such as a short circuit at the converter outputs. In this assumption, poor design, mounting issues, and the environmental effect on the converter construction have not been considered. However, a series of failures in the power electronic systems may be due to problems in the building process of the converter. In this chapter, the main important issues at the design and montage stages of the converters are presented.

The meaning of failure in this chapter is the type of failure that leads to damage in a converter. Based on this view, four main reasons of failure are:

- Electrical considerations including overvoltage
- Thermal considerations including thermal shock and overtemperature
- Mechanical considerations including mechanical forces on the various parts of the converters
- Environmental considerations including the effects of humidity and dust on the converters

6.2   Electrical Considerations

The electric field is one of the stressors in electronic devices. Various types of fault exist in this category.

6.2.1   Electric Breakdown

Electrical breakdown or dielectric breakdown refers to a rapid reduction in the resistance of an electrical insulator when the voltage applied across it exceeds the breakdown voltage. This results in a portion of the insulator becoming electrically conductive.

Under sufficient electrical stress, electrical breakdown can occur within solids, liquids, gases, or a non-ideal vacuum. Electrical breakdown is often associated with the failure of solid or liquid insulating materials used inside high voltage transformers or capacitors in the electricity distribution grid, usually resulting in a short circuit. Dielectric breakdown within a solid insulator can permanently change its appearance and properties. Thus, we will talk more about the breakdown of solid insulators. Figure 6.1 shows an electric breakdown in the winding of an inductor. The distance for withstanding against an electric breakdown is named the clearance distance.

In solid dielectrics, the breakdown strength is high. The highest breakdown strength is obtained under carefully controlled conditions. Dielectrics usually fail at stresses well below the intrinsic strength, usually due to one of the following causes:

- Electromechanical breakdown
- Breakdown due to internal discharges
- Thermal breakdown

Reference [7] reviews surface flashover (i.e. voltage breakdown along the surface of insulators), primarily in vacuum. It discusses theories and models relating to surface flashover and pertinent experimental results. Surface flashover of insulators in vacuum generally is initiated by the emission of electrons from the cathode triple junction (the region where the electrode, insulator, and vacuum meet). These electrons usually then multiply as they traverse the insulator surface, either as a surface secondary electron-emission avalanche or as an electron cascade in a thin surface

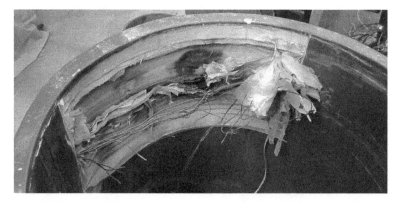

Figure 6.1   Electric breakdown in the winding of an inductor.

layer, causing desorption of gas that- had been adsorbed on the insulator surface. This desorbed gas is then ionized, which leads to surface flashover of the insulator.

6.2.2   Creepage

The creepage distance for insulators is the shortest distance along the insulator surface between the metal parts at each end of the insulator. A proper and adequate creepage distance protects against tracking, a process that produces a partially conducting path of localized deterioration on the surface of an insulating material as a result of the electric discharges on or close to an insulation surface. Figure 6.2 shows a high voltage solid-state switch with an increased creepage distance on its outer side.

Tracking that damages the insulating material normally occurs because of one or more of the following reasons:

- Humidity in the atmosphere
- Presence of contamination
- Corrosive chemicals
- Altitude at which the equipment is to be operated

Figure 6.3 shows the fault due to the creepage phenomenon in a high voltage bushing and Figure 6.4 shows the clearance and creepage distances on the terminals of a power module.

Figure 6.2   Increasing the creepage distance on the outer side of a high voltage switch.

Figure 6.3   Creepage phenomenon on a high voltage bushing.

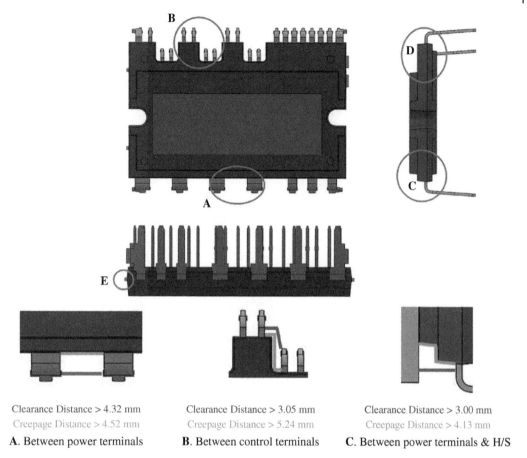

Clearance Distance > 4.32 mm
Creepage Distance > 4.52 mm

A. Between power terminals

Clearance Distance > 3.05 mm
Creepage Distance > 5.24 mm

B. Between control terminals

Clearance Distance > 3.00 mm
Creepage Distance > 4.13 mm

C. Between power terminals & H/S

Figure 6.4 Clearance and creepage distances on the terminals of a power module: *Source:* ON semiconductor. Used with permission from SCILLC dba onsemi.

As the voltage and current in power electronic converters are high, the printed circuit boards used for these converters should meet standards for minimum spacing between traces and also the width of traces. Figure 6.5 shows the minimum spacing between traces versus their voltage difference in various standards.

6.2.3 Voltage Spikes

In industrial applications, high voltage power supply spikes with durations ranging from a few microseconds to hundreds of milliseconds are commonly encountered. The electronic devices used in these systems must not only survive transient voltage spikes, but in many cases also operate reliably throughout an event. In systems where power is distributed over long wires severe transients are generated by load steps. These transients pose a difficult challenge for engineers trying to protect sensitive electronic devices. Figure 6.6 shows a sample of the voltage spikes on the drain terminal of a MOSFET.

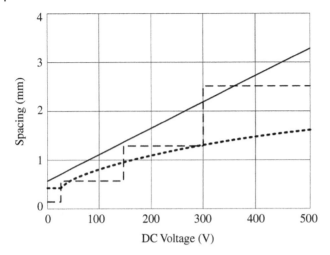

Figure 6.5 PCB minimum spacing in different standards.

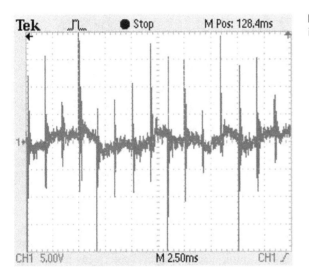

Figure 6.6 Voltage spikes in a power converter.

6.2.4 Partial Discharge

The partial discharge (PD) occurs due to the uneven voltage distributions of the insulator. This unbalanced voltage distribution has mainly two reasons: unavoidable existing voids through the insulator and the PD on the insulator surface. Voids usually consist of air and have a different permittivity from the insulator. After applying the voltage, the voltage across the void exceeds the break down voltage and a PD occurs. This process occurs whenever the total applied voltage is more than a predetermined level known as the inception voltage. In this condition, one or several PDs occur and in each PD a level of heat is released. The released energy through the insulator in PDs is the main factor deteriorating the insulators. The power of the PDs can be written as

$$P_{PD} \approx f_{sw} n_{PD} Q_{PD} V_{inc} \tag{6.1}$$

where, f_{sw} is the switching frequency, n_{PD} is the number of PD occurrence in each switching period, Q_{PD} is the electrical charge releasing during each PD, V_{inc} is the inception voltage of the insulator, and P_{PD} is the power loss caused by the PDs. The PDs in DC voltages are not prevailed. In low-frequency applications, due to the limited number of PDs in each second, the insulator lifetime is considerable enough to ignore the effect of insulators on the HVDC (high voltage DC) links lifetime. Inversely, in high frequency applications according to (6.1), due to high number of the cases in which the insulator voltage violates inception voltage, the insulation lifetime maybe less than several hours or even minutes. In order to highlight the issue, a commercial insulator with the part number of DuPont Nomex 410 is tested experimentally. The nominal voltage of the insulator is reported in the product datasheet as 6 kV. The test setup performs the standard sphere insulation test under a 50 kHz voltage stress with the amplitude equals 2 kV as shown in Figure 6.7. A sample of the measured PD current is shown in Figure 6.7. The PD current was measured with the high frequency current transformer (CT). It was observed that the insulator was damaged in 20 minutes under the test as shown in Figure 6.8. This figure indicates the importance of the insulator deterioration in high-frequency applications. In this regard, even with reduced applied voltage (equals to

Figure 6.7 The PD current in the tested insulator, CH1: applied voltage, CH2: measured current.

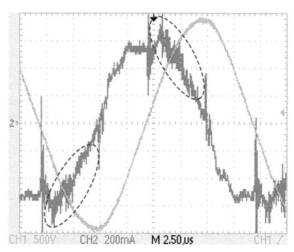

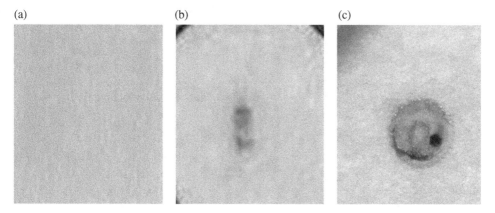

Figure 6.8 The insulator tests result: (a) before voltage applying, (b) after 10 minutes of the applied voltage, and (c) after 20 minutes of the applied voltage.

0.25 nominal voltage), the insulator is fully deteriorated after 20 minutes. This operation time is not acceptable for the reliable design.

6.3 Thermal Considerations

Overtemperature is the most important factor of failure in all of systems. Thermal damage is a very important factor in the fault of electric power converters. Figure 6.9 shows a sample damaged switch due to the overtemperature. There are two scenarios for thermal damage: overtemperature and thermal shock. Thermal shock occurs when a thermal gradient causes different parts of an object to expand by different amounts. At some point, this stress can exceed the strength of the material, causing a crack to form.

The basic thermal issues come from power losses in the power electronic converters. A power electronic converter consists of solid-state active devices and some passive devices. Therefore, the power loss of a power electronic converter can be listed as follows.

6.3.1 Power Loss Characteristics of Active Devices

Active devices, i.e. solid-state switches, are the key elements of power electronic systems. Every solid state device is a resistive element, so it dissipates power as current is conducted through the device. For the MOSFETs, these conduction losses are inversely proportional to the size of the MOSFET; the larger the switching device, the lower it is on resistance [2]. Another term of losses in solid-state devices is switching losses. Switching losses are created as a result of a simultaneous high voltage and current in a solid-state device during a transition between the open and closed states.

For power semiconductor devices, a safe operating area (SOA) is defined as the voltage and current conditions over which the device can be expected to operate without damage. The SOA specification combines the various limitations of the device: maximum voltage, current, and power, allowing a simplified design of protection circuitry. Often, in addition to the continuous rating, separate SOA curves are plotted for short duration pulse conditions (1 ms pulse, 10 ms pulse, etc.). SOA specifications are useful to the design engineer working on power circuits such as amplifiers and power supplies as they allow a quick assessment of the limits of the device performance, the design of appropriate protection circuitry, or selection of a more capable device. Figure 6.10 shows the SOA for two solid-state switches.

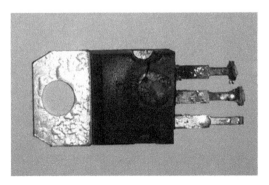

Figure 6.9 A damaged solid-state switch due to the overtemperature.

(a)

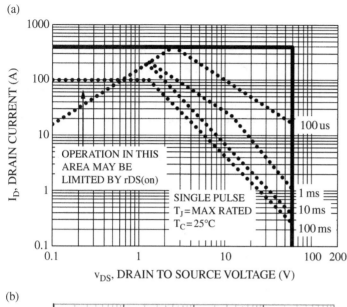

(b)

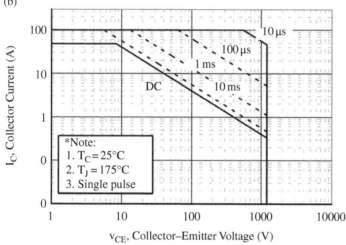

Figure 6.10 SOA of two switches: (a) MOPSFET, (b) IGBT: *Source:* ON semiconductor. Used with permission from SCILLC dba onsemi.

- Thermal loading of power devices

This is closely related to the reliability performance of the whole converter system. The electrical loading and device rating are both important factors that determine the loss and thermal behaviors of power semiconductor devices. In the existing loss and thermal models, only the electrical loadings are focused and treated as design variables, while the device rating is normally pre-defined by experience with limited design flexibility. Consequently, a more complete loss and thermal model is proposed in reference [8], which takes into account not only the electrical loading but also the device rating as input variables. The quantified correlation between the power loss, thermal impedance, and silicon area of IGBT is mathematically established. By this new modeling approach, all factors that have impacts to the loss and thermal profiles of the power devices can accurately be mapped, enabling more design freedom to optimize the efficiency and thermal loading of the power converter.

● Diode

The conduction losses in a diode appear when the diode is in the forward conduction mode due to the on-state voltage drop. Most of the time the conduction losses are the main contributor to the total diode power losses and the junction rise in temperature. Figure 6.11 shows the effect of temperature rise on the conduction voltage drop of diodes. It can be seen that the on-state voltage drop of some solid-state switches increases with temperature and leads to an increase in conduction losses of the switches. SiC diodes belong to wide bandgap devices as the new generation of the semiconductor switches. Wide bandgap semiconductors show superior material properties, enabling potential power device operation at higher temperatures, voltages, and switching speeds than current Si technology. As a result, a new generation of power devices is being developed for power

(a)

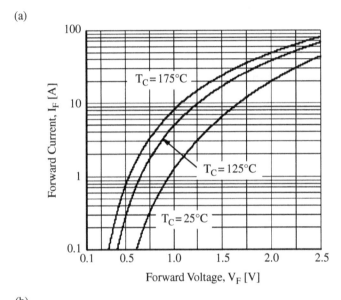

(b)

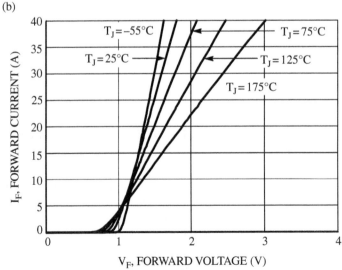

Figure 6.11 The diode I–V characteristic: (a) fast recovery diode, (b) SiC diode, (c) reverse leakage current of a SiC diode. *Source:* ON semiconductor. Used with permission from SCILLC dba onsemi.

(c)

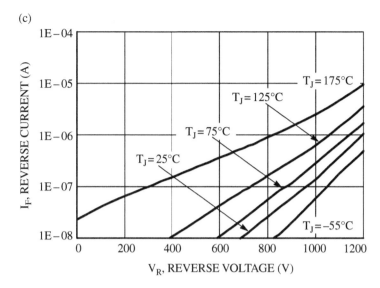

Figure 6.11 (Continued)

converter applications in which traditional Si power devices show limited operation. The use of these new power semiconductor devices will allow both an important improvement in the performance of existing power converters and the development of new power converters, accounting for an increase in the efficiency of the electric energy transformations and a more rational use of electric energy. At present, SiC and GaN are the more promising semiconductor materials for these new power devices as a consequence of their outstanding properties, commercial availability of starting material, and maturity of their technological processes. Reference [9] presents a review of recent progress in the development of SiC- and GaN-based power semiconductor devices together with an overall view of the state of the art of this new device generation.

- Thyristor

Reference [10] presents investigations on losses in the thyristors that are essential parts of the inductive current limiter. After the detailed switching characteristics of these thyristors are described, the power losses are calculated in the system. Analysis of the switching process in the inductive superconducting fault current limiter (SFCL) with the DC reactor is given and the on-state, switching, and snubber losses are also estimated under normal conditions. The proposed model is suitable for investigations with a system loss analysis and overall system cost. Figure 6.12 shows the voltage drop of a sample thyristor.

- MOSFET

An accurate analytical model is presented in reference [11] to calculate the power loss of a metal-oxide semiconductor field-effect transistor. The nonlinearity of the capacitors of the devices and the parasitic inductance in the circuit, such as the source inductor shared by the power stage and driver loop, the drain inductor, etc. are considered in the model. In addition, the ringing is always observed in the switching power supply, which is ignored in the traditional loss model. In this reference, the ringing loss is analyzed in a simple way with a clear physical meaning. Based on this model, the circuit power loss can be accurately predicted. Figure 6.13 shows the I–V characteristics of classic Si MOSFETs and new SiC MOSFETs while Figure 6.14 shows the switching losses of an SiC MOSFET.

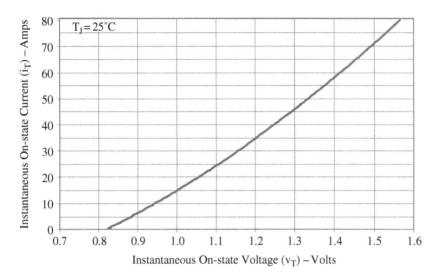

Figure 6.12 The thyristor I–V characteristic: TDK (with permission).

(a)

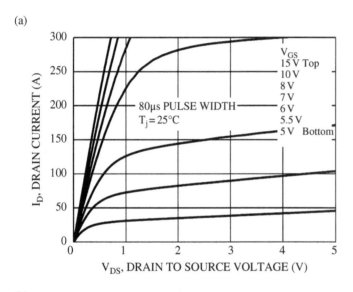

(b)

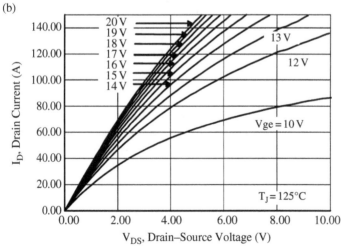

Figure 6.13 MOSFET I-V characteristic: (a) Si MOSFET, (b) SiC MOSFET. *Source:* ON semiconductor. Used with permission from SCILLC dba onsemi.

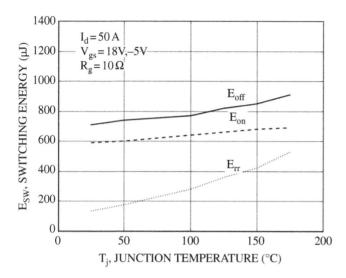

Figure 6.14 SiC MOSFET switching losses. *Source:* ON semiconductor. Used with permission from SCILLC dba onsemi.

- IGBT

In reference [12], the mechanisms of destructive failure of an IGBT in the short-circuit state are discussed. Results from a two-dimensional numerical simulation of p-channel and n-channel IGBTs are presented. It is found that there are two types of destructive failure mechanisms: a secondary breakdown and a latchup. Which type is dominant in the p-channel and n-channel IGBTs depends on an absolute value of a forward voltage. At a moderately low mod, the p-channel IGBT is destroyed by secondary breakdown, and the n-channel IGBT by latchup. This is due to the difference in the type of flowing carrier crossing a base-collector junction of a wide-base transistor and ionization rates of electrons and holes. Figure 6.15 shows the IV characteristic of an IGBT, while Figures 6.16 and 6.17 show the conduction and switching losses of a sample IGBT.

One of the future challenges in high-power converters is how to size key components with compromised costs and design margins, while fulfilling specific reliability targets. It demands better thermal modeling compared to the state-of-the-art in terms of both accuracy and simplicity. Reference [13] thus proposes a simple analytical thermal modeling method, which adopts equivalent periodic power loss profiles. More importantly, time-domain simulations are not required in the proposed method. Benchmarking of the proposed methods with the prior-art solutions is performed in terms of parameter sensitivity and model accuracy with a case study on a 30-MW power electronic system.

6.3.2 Power Loss in Passive Devices

For discrete electrical circuit components, power loss includes inductor winding, core loss, and capacitors loss. Pure capacitors and inductors do not dissipate energy; any process that dissipates energy must be treated as one or more resistors in the component model. Figure 6.18 shows a damaged resistor because of overcurrent.

- Losses in capacitors

A capacitor is typically made of a dielectric placed between conductors. The lumped element model of a capacitor includes a lossless ideal capacitor in series with a resistor termed the

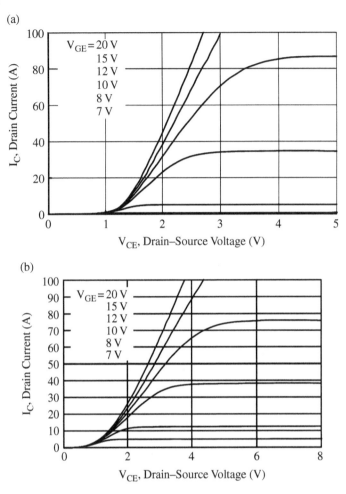

Figure 6.15 The I–V characteristic of an IGBT: (a) 25 °C, (b) 175 °C. *Source:* ON semiconductor. Used with permission from SCILLC dba onsemi.

equivalent series resistance (ESR). The ESR represents losses in the capacitor. Figure 6.19a, b shows the frequency response of two types of capacitor. The minimum value of the frequency response represents the ESR of the capacitor. Figure 6.19c, d shows the variation of ESRs of two capacitors versus the frequency. Figure 6.19c–f shows the failure rate and life time of a sample capacitor.

- Losses in magnetic devices

Switching circuits, operating at high frequencies, have led to considerable reductions in the size of magnetic components and power supplies. Non-sinusoidal voltage and current waveforms and high-frequency skin and proximity effects contribute to power transformer losses. Traditionally, power transformer design has been based on sinusoidal voltage and current waveforms operating at low frequencies. The physical and electrical properties of the transformer form the basis of a new design methodology while taking full account of the current and voltage waveforms and high-frequency effects.

(a)

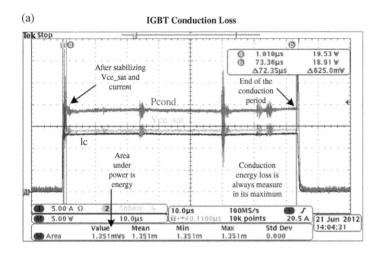

(b)

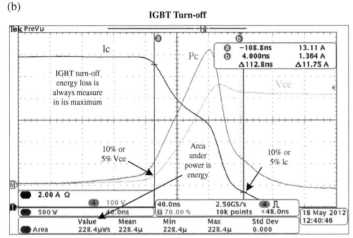

(c)

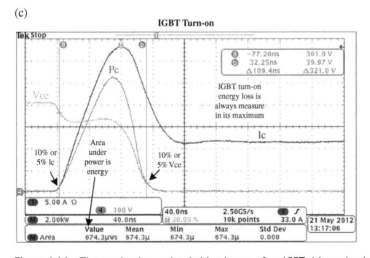

Figure 6.16 The conduction and switching losses of an IGBT: (a) conduction loss, (b) and (c) switching loss.
Source: ON semiconductor. Used with permission from SCILLC dba onsemi.

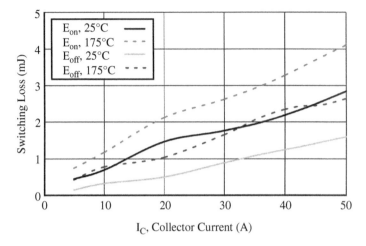

Figure 6.17 The effect of a temperature rise on the switching losses of an IGBT. *Source:* ON semiconductor. Used with permission from SCILLC dba onsemi.

Figure 6.18 A damaged resistor due to the overcurrent fault.

Skin effect is the tendency of an alternating electric current to become distributed within a conductor such that the current density is largest near the surface of the conductor and decreases with greater depths in the conductor. The skin effect causes the effective resistance of the conductor to increase at higher frequencies where the skin depth is smaller, thus reducing the effective cross-section of the conductor. The skin effect is due to opposing eddy currents induced by the changing magnetic field resulting from the alternating current. Application of litz-wire helps to control the power losses. The number and diameter of strands to minimize loss in a litz-wire transformer winding is determined in reference [14]. With fine stranding, the AC resistance factor can be decreased, but DC resistance increases as a result of the space occupied by insulation. A power law to model insulation thickness is combined with a standard analysis of proximity-effect losses to find the optimal stranding. Suboptimal choices under other constraints are also determined. A graphical and numerical method of calculating and minimizing losses in windings, which generalizes previous findings, has been introduced in reference [15]. Using electromagnetic theory and magneto-motive force (MMF) diagrams in both space and time, a method is proposed that provides insight into the mechanism of skin and proximity effect losses and also yields quantitative results. Using this method, several winding geometries for various topologies are covered. The

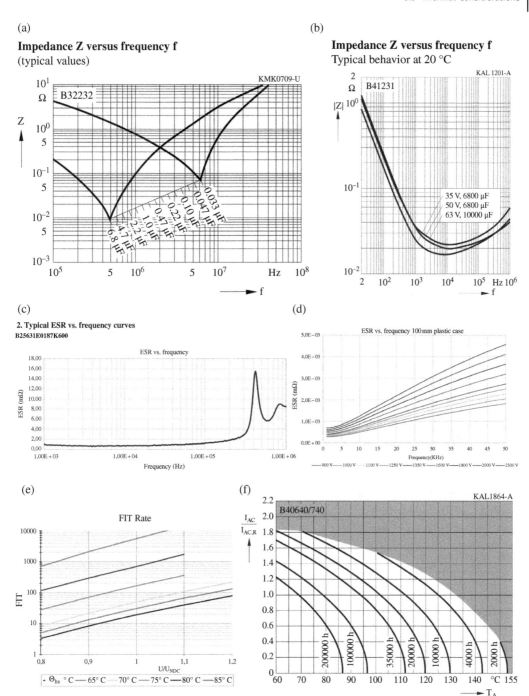

Figure 6.19 The capacitor loss characteristics: (a) frequency response of film capacitors, (b) frequency response of electrolyte capacitors, (c) and (d) ESR of two capacitors, (e) effect of temperature rise on the failure rate of capacitors, (f) effect of voltage ripple on the lifetime of the capacitors. *Source:* TDK (with permission).

analysis and optimization process is experimentally verified using an interleaved flyback transformer. The mathematical treatment justifying the use of the field method, and which is essential in arriving at any numerical result, is presented and more general equations for the calculation of copper losses are derived. The relation between the fields in the transformer and copper losses is emphasized. Also, the tools necessary to derive optimization diagrams are provided.

Core selection is based on the optimum throughput of energy with minimum losses. In reference [16], the optimum core is found directly from the following transformer specifications: frequency, power output, and temperature rise. The design methodology is illustrated with a detailed design of a push–pull power converter.

Each time the magnetic field is reversed, a small amount of energy is lost due to hysteresis within the core. Ferromagnetic materials are also good conductors and a core made from such a material also constitutes a single short-circuited turn throughout its entire length. Eddy currents therefore circulate within the core in a plane normal to the flux, and are responsible for resistive heating of the core material. The eddy current loss is a function of the square of the supply frequency. Eddy current losses can be reduced by making the core of a stack of plates electrically insulated from each other, rather than a solid block; all transformers operating at low frequencies use laminated or similar cores. In reference [17], an improved calculation of ferrite core loss for non-sinusoidal waveforms separates a flux trajectory into major and minor loops via a new recursive algorithm. It is highly accurate and outperforms two previous methods for our measured data. The only characteristics of the material required are the standard Steinmetz equation parameters. Reference [18] discusses the influence of non-sinusoidal flux waveforms on the remagnetization losses in ferro- and ferrimagnetic materials of inductors, transformers, and electrical machines used in power electronic applications. The non-sinusoidal changes of flux originate from driving these devices by non-sinusoidal voltages and currents at different switching frequencies. A detailed examination of a dynamic hysteresis model shows that the physical origin of losses in magnetic material is the average rate of remagnetization rather than the remagnetization frequency. This principle leads to a modification of the most common calculation rule for magnetic core losses, i.e. to the "modified Steinmetz equation" (MSE). In the MSE, the remagnetization frequency is replaced by an equivalent frequency, which is calculated from the average remagnetization rate. This approach allows, for the first time, the calculation of the losses in the time domain for arbitrary waveforms of flux while using the available set of parameters of the classical Steinmetz equation. DC premagnetization of the material, having a substantial influence on the losses, can also be included.

- Printed circuit boards

Determining appropriate trace sizes for current requirements is an important aspect of circuit board development. Since copper is not a perfect conductor, it presents a certain amount of impedance to current flowing through it, and some of the energy is lost in the form of heat. For many applications it is necessary to predict the temperature rise caused by this loss, which has been accomplished traditionally by using a chart created over 50 years ago by the National Bureau of Standards, or by using one of several calculators based upon it. The chart shows the relationship between current, conductor temperature rise, and conductor cross-sectional area. If any two of these are known, the third can be approximated. Figure 6.20 shows permissible current in traces of a printed circuit board (PCB). The trace current increases if a higher temperature rise is acceptable.

- Soldering

The applied heat in the soldering process of a power electronic system should be controlled to prevent an unacceptable temperature rise. There is usually a soldering profile, as shown in Figure 6.21, that each manufacturer provides for the customers.

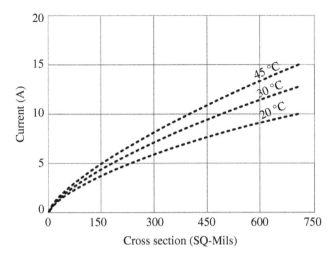

Figure 6.20 PCB trace width for different temperature rises.

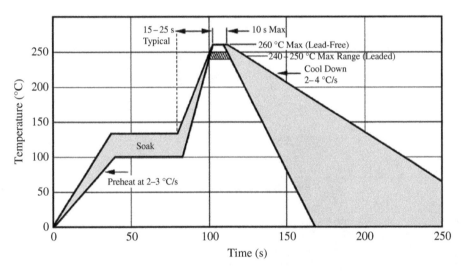

Figure 6.21 Soldering profile for a power electronic switch. *Source:* ON semiconductor. Used with permission from SCILLC dba onsemi.

6.4 Mechanical Considerations

The mechanical problems of power electronic systems at converter levels consist of two main problems: poor mounting from the crack generation viewpoint and poor installing from thermal considerations.

6.4.1 Mounting Considerations

One of the most likely failure modes is PCB flexing, which can cause the solder joints on the PCB to crack, causing intermittent or complete failure of the connection. An attempt is made to keep the PCB as compact as possible. The smaller the PCB, the smaller is the overall flex of the board. The user can use a conformal coating on the PCB, which will hold surface mount components in

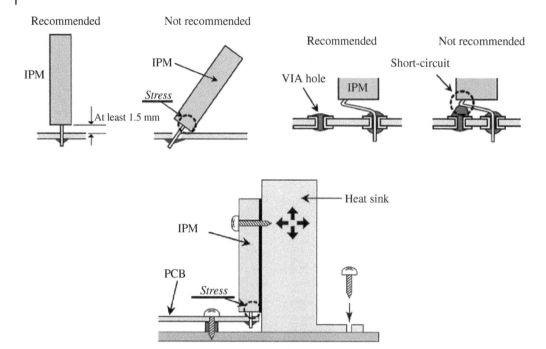

Figure 6.22 Examples of poor mounting of the devices in power electronic systems. *Source:* ON semiconductor. Used with permission from SCILLC dba onsemi.

Figure 6.23 Application of conformal coating for fixing the devices.

place as well as add some stiffness to the PCB. Figure 6.22 shows the examples of the poor mounting problems and Figure 6.23 shows an application of conformal coating for fixing the elements on a power converter.

6.4.2 Thermal Considerations

In electronic systems, a heat sink is a passive heat exchanger that cools a device by dissipating heat into the surrounding medium. In computers, heat sinks are used to cool central processing units or graphics processors. Heat sinks are used with high-power semiconductor devices such as power transistors, where the heat dissipation ability of the basic device is insufficient to moderate its temperature. A heat sink is designed to maximize its surface area in contact with the cooling medium surrounding it, such as the air. Air velocity, choice of material, protrusion design, and surface treatment are factors that affect the performance of a heat sink. Heat sink attachment methods and thermal interface materials also affect the die temperature of the integrated circuit. Thermal adhesive or thermal grease improve the heat sink's performance by filling air gaps between the heat sink and the device. Thermal contact resistance occurs due to the voids created by surface roughness effects, defects, and misalignment of the interface. The voids present in the interface are filled with air. Heat transfer is therefore due to conduction across the actual contact area and to conduction and radiation across the gaps. If the contact area is small, as it is for rough surfaces, the major contribution to the resistance is made by the gaps. To decrease the thermal contact resistance, the surface roughness can be decreased while the interface pressure is increased. However, these improving methods are not always practical or possible for electronic equipment. Thermal interface materials are a common way to overcome these limitations. Properly applied thermal interface materials displace the air that is present in the gaps between the two objects with a material that has a much-higher thermal conductivity. Figure 6.24 shows the standard method for providing a heatsink with thermally conductive coating. Figure 6.25 shows the effect of proper pressure during the mounting of a device on a heatsink, where it is seen that the thermal resistance has a minimum value for the properly applied force. Figure 6.26 shows the standard mounting equipment with controlled pressure on the device.

6.5 Environmental Considerations

Humidity is the amount of water vapor in the air. Water vapor is the gaseous state of water and is invisible. There are three main measurements of humidity: absolute, relative, and specific. Absolute humidity is the water content of air. Relative humidity, expressed in percent, measures the current absolute humidity relative to the maximum for that temperature. Specific humidity is the ratio of the water vapor content of the mixture to the total air content on a mass. Many electronic devices have humidity specifications, for example, 5–95%. At the top end of the range, moisture may increase the conductivity of permeable insulators leading to malfunction. Too low

Figure 6.24 The standard method for providing a heatsink with thermally conductive coating. *Source:* Used with permission from SCILLC dba onsemi.

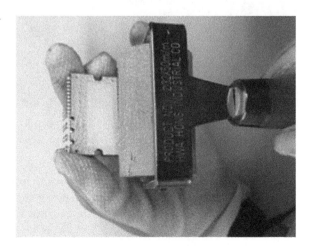

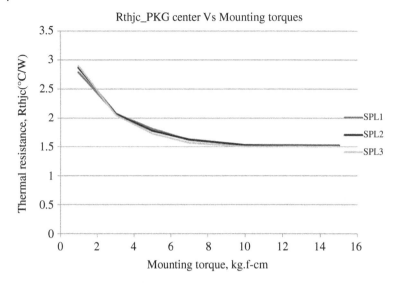

Figure 6.25 The effect of proper pressure during the mounting of a device on a heatsink. *Source:* ON semiconductor. Used with permission from SCILLC dba onsemi.

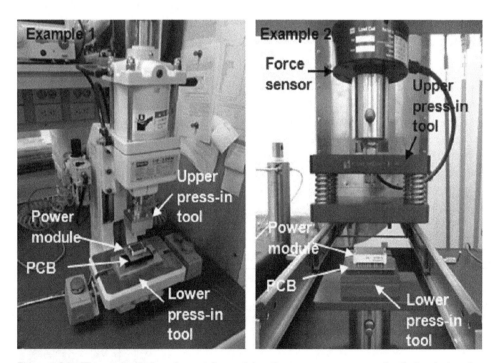

Figure 6.26 The mechanical equipment for applying the proper pressure on the device during the mounting process. *Source:* Used with permission from SCILLC dba onsemi.

humidity may make materials brittle or give an electrostatic discharge. A particular danger to electronic items, regardless of the stated operating humidity range, is condensation. When an electronic item is moved from a cold place to a warm humid place, condensation may coat circuit boards and other insulators, leading to a short circuit inside the equipment. Figure 6.27 shows the

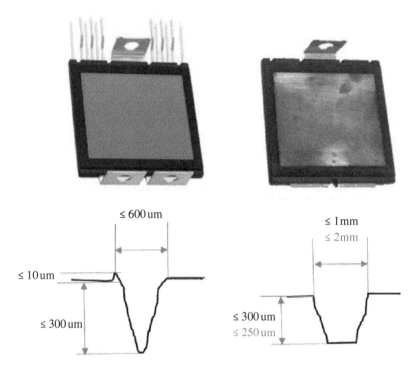

Figure 6.27 Effect of humidity on the surface quality of a power electronic switch. *Source:* Used with permission from SCILLC dba onsemi.

effect of humidity on the elements of power converters. In this figure, the surface of a switch is oxidized and leads to poor surface topography. This low-quality surface affects the thermal conductivity of the device.

In contrast, a very low humidity level favors the buildup of static electricity, which may result in spontaneous shutdown of computers when discharges occur. In addition, electrostatic discharges can cause dielectric breakdown in solid-state devices, resulting in irreversible damage.

6.6 Summary and Conclusions

Some of the main reasons for failure in power electronic converters were presented in this chapter. All these factors cause catastrophic damage in the systems. Overtemperature, overvoltage, mechanical forces, and environmental effects are the main factors of failure in power converters, which have been described in detail. The results of this chapter are summarized as follows:

1) Electrical breakdown or dielectric breakdown is a rapid reduction in the resistance of an electrical insulator when the voltage applied across it exceeds a specified voltage. Electric breakdown has a fast effect on the failure of converters. It acts to break down inside the body of an insulator or shows a creepage on the surface of the insulator. Environmental factors have a direct effect on the creepage type of insulator breakdown. A breakdown in a solid insulator usually leads to permanent damage. Therefore, special attention to this failure factor is very important because low power and low voltage power converters use elements with a solid insulator.

2) Overtemperature is the most important failure factor in electric power converters. Most failure factors finally lead to overtemperature. Since the power conversion process is not ideal, heat generation is a common problem in all power converters. Of course, this is true for all electrical systems and not only for electric power converters. However, it is a serious problem in these converters because the amount of energy transfer is high and, therefore, their energy loss is considerable. Conduction loss in conductors and semiconductors, switching loss in switching devices, and core loss are the main mechanisms of energy loss in electric power converters. Damage due to overtemperature is a long-term factor in comparison to damage due to electric breakdown.

3) Mechanical factors are affected over the long term in comparison to electrical factors. Vibration is an important mechanical factor in the failure of rotary power converters. In addition, outside generated vibration causes mechanical damage to leads in electronic elements. In the mechanical world, transient conditions like mechanical shock occur in a similar way to transient phenomena in the electrical world. Improper mounting of the elements can lead to mechanical damage. Modal analysis is a tool for analyzing the mechanical behavior of structures and elements.

4) Operation in a harsh environment accelerates the failure due to the above-mentioned factors. Operation in a high ambient temperature accelerates the failure due to overtemperature. Humidity and dust act by reducing insulator resistance and cause electrical breakdown at lower voltages than the designed nominal values.

References

1 Peyghami, S., Davari, P., Zhou, D. et al. (2019). Wear-out failure of a power electronic converter under inversion and rectification modes. In: *2019 IEEE Energy Conversion Congress and Exposition (ECCE)*, Baltimore, MD, USA, 1598–1604.

2 Peyghami, S., Wang, Z., and Blaabjerg, F. (2020). A guideline for reliability prediction in power electronic converters. *IEEE Transactions on Power Electronics* 35 (10): 10958–10968.

3 Peyghami, S., Fotuhi-Firuzabad, M., and Blaabjerg, F. (2020). Reliability evaluation in microgrids with non-exponential failure rates of power units. *IEEE Systems Journal* 14 (2): 2861–2872.

4 Peyghami, S., Wang, Z., and Blaabjerg, F. (2019). Reliability modeling of power electronic converters: a general approach. In: *2019 20th Workshop on Control and Modeling for Power Electronics (COMPEL)*, Toronto, ON, Canada, 1–7.

5 Dragičević, T., Lu, X., Vasquez, J.C., and Guerrero, J.M. (2016). DC microgrids – Part II: A review of power architectures, applications, and standardization issues. *IEEE Transactions on Power Electronics* 31 (5): 3528–3549.

6 Wu, R., Blaabjerg, F., Wang, H. et al. (2013). Catastrophic failure and fault-tolerant design of IGBT power electronic converters – an overview. In: *IECON 2013 – 39th Annual Conference of the IEEE Industrial Electronics Society*, Vienna, Austria, 507–513.

7 Miller, H.C. (1989). Surface flashover of insulators. *IEEE Transactions on Electrical Insulation* 24 (5): 765–786.

8 Ma, K., Bahman, A.S., Beczkowski, S., and Blaabjerg, F. (2015). Complete loss and thermal model of power semiconductors including device rating information. *IEEE Transactions on Power Electronics* 30 (5): 2556–2569.

9 Millán, J., Godignon, P., Perpiñà, X. et al. (2014). A survey of wide bandgap power semiconductor devices. *IEEE Transactions on Power Electronics* 29 (5): 2155–2163.

10 Joo, M. (2004). Losses of thyristor on modified bridge type high-temperature superconducting fault current limiter. *IEEE Transactions on Applied Superconductivity* 14 (2): 835–838.

11 Ren, Y., Xu, M., Zhou, J., and Lee, F.C. (2006). Analytical loss model of power MOSFET. *IEEE Transactions on Power Electronics* 21 (2): 310–319.

12 Iwamuro, N., Okamoto, A., Tagami, S., and Motoyama, H. (1991). Numerical analysis of short-circuit safe operating area for p-channel and n-channel IGBTs. *IEEE Transactions on Electron Devices* 38 (2): 303–309.

13 Zhang, Y., Wang, H., Wang, Z. et al. (2019). Simplified thermal modeling for IGBT modules with periodic power loss profiles in modular multilevel converters. *IEEE Transactions on Industrial Electronics* 66 (3): 2323–2332.

14 Sullivan, C.R. (1999). Optimal choice for number of strands in a litz-wire transformer winding. *IEEE Transactions on Power Electronics* 14 (2): 283–291.

15 Vandelac, J.P. and Ziogas, P.D. (1988). A novel approach for minimizing high-frequency transformer copper losses. *IEEE Transactions on Power Electronics* 3 (3): 266–277.

16 Hurley, W.G., Wolfle, W.H., and Breslin, J.G. (1998). Optimized transformer design: inclusive of high-frequency effects. *IEEE Transactions on Power Electronics* 13 (4): 651–659.

17 Venkatachalam, K., Sullivan, C.R., Abdallah, T., and Tacca, H. (2002). Accurate prediction of ferrite core loss with nonsinusoidal waveforms using only Steinmetz parameters. In: *Proceedings of the 2002 IEEE Workshop on Computers in Power Electronics, 2002*, Mayaguez, PR, USA, 36–41.

18 Reinert, J., Brockmeyer, A., and De Doncker, R.W.A.A. (2001). Calculation of losses in ferro- and ferrimagnetic materials based on the modified Steinmetz equation. *IEEE Transactions on Industry Applications* 37 (4): 1055–1061.

7

Internal Faults

Element Level

7.1 Element Level Faults

Power electronic converters are increasingly used in power systems. However, they are vulnerable components and prone to aging failures, thus affecting overall system reliability. Therefore, their availability modeling, especially in large-scale power electronic-based power systems, is very important. In the previous chapter, we discussed the faults at the converter level. They are normally the result of poor design, the use of substandard components, or lack of adequate controls in the manufacturing process. In this chapter, the faults of the power electronic system are studied during a useful lifetime and the wear-out phase. As shown in Figure 7.1, the useful lifetime is considered as the period with a constant failure rate. In this period, the failures happen by chance. They cannot be eliminated by either lengthy burn-in periods or good preventive maintenance practices. Equipment is designed to operate under certain conditions and up to certain stress levels. When these stress levels are exceeded due to random not anticipated or unknown events, a random failure will occur. The wear-out phase begins at the system end of life. Wear-out studies allow proactive maintenance plans. Further, depending on the failure mode, different proactive control strategies can be applied to converters in order to improve the reliability and availability of power electronic systems. Many efforts have been devoted to the reduction of power electronic system wear-out failure induced by accumulated degradation and catastrophic failure triggered by single-event overstress. The wear-out failure under field operation could be mitigated by scheduled maintenances based on lifetime prediction and condition monitoring.

7.2 Inside the Elements

Electronic packaging refers to enclosures for integrated circuits, passive devices, and circuit cards. The effectiveness of an electronic system, as well as its reliability and cost, is strongly determined by the packaging materials used. This is of fundamental importance for signal and power transmission, heat dissipation, electromagnetic interference shielding, and protection from environmental factors such as moisture, contamination, chemicals, and radiation. Random and wear-out failures have a close relation to the construction process and packaging of the devices. In this section, the inside structures of the devices in power electronic systems are presented.

Resilient Power Electronic Systems, First Edition. Shahriyar Kaboli, Saeed Peyghami, and Frede Blaabjerg.
© 2022 John Wiley & Sons Ltd. Published 2022 by John Wiley & Sons Ltd.
Companion website: www.wiley.com/go/kaboli/resilientpower

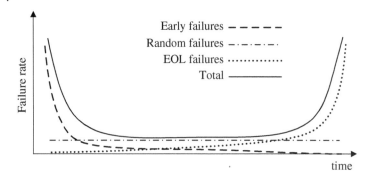

Figure 7.1 Weibull distribution of the failure rate – bathtub shape.

7.2.1 Active Devices

Power modules are power switching/control circuit elements integrated into convenient isolated-base packages that offer a broad spectrum of commonly used switching circuit configurations and ratings. Power modules are often power semiconductor devices, and also provide an easy way to cool the device as well as connect them to the outer circuit. Power modules are mechanically and thermally optimized for ease of assembly, long life, and reliable operation. There are several common structures that power modules are available in, such as an insulated gate bipolar transistor (IGBT) or MOSFET. A power electronic module or power module acts as a physical container for the storage of several power components, usually power semiconductor devices. Besides die attach soldering and aluminum heavy wire bonding, new packaging concepts are pursued to build power electronics systems that will provide improved thermal performance and higher reliability. Besides modules that contain a single power electronic switch (such as MOSFET, IGBT, BJT, Thyristor, GTO, or JFET) or diode, classical power modules contain multiple semiconductor dies that are connected to form an electrical circuit of a certain structure, called topology. Modules also contain other components such as ceramic capacitors to minimize switching voltage overshoots and negative temperature coefficient (NTC) thermistors to monitor the module's substrate temperature. Examples of broadly available topologies implemented in modules are:

- switch (MOSFET, IGBT), with an antiparallel diode
- bridge rectifier containing four (one-phase) or six (three-phase) diodes
- half bridge (inverter leg, with two switches and their corresponding antiparallel diodes)
- H-Bridge (four switches and the corresponding antiparallel diodes)
- boost or power factor correction (one (or two) switches with one (or two) high-frequency rectifying diodes)
- ANPFC (active neutral-point factor correction) (power factor correction leg with two switches and their corresponding antiparallel diodes and four high-frequency rectifying diodes)
- three level NPC (neutral-point-clamped) (I-Type) (multilevel inverter leg with four switches and their corresponding antiparallel diodes)
- three level MNPC (modified neutral-point-clamped) (T-Type) (multilevel inverter leg with four switches and their corresponding antiparallel diodes)
- three level ANPC (active neutral-point-clamped) (multilevel inverter leg with six switches and their corresponding antiparallel diodes)
- three level H6.5 (consisting of six switches (four fast IGBTs/two slower IGBTs) and five fast diodes)
- three-phase inverter (six switches and the corresponding antiparallel diodes)
- Power Interface Module (PIM) (consisting of the input rectifier, power factor correction, and inverter stages)

- Intelligent Power Module (IPM) (consisting of the power stages with their dedicated gate drive protection circuits and may also be integrated with the input rectifier and power factor correction stages)

Figure 7.2 shows the inside view of a power module consisting of several IGBTs and diodes. Generally, a power electronic module can be divided into die structures and the requirements for packaging.

- Die structure
  A die, in the context of integrated circuits, is a small block of semiconducting material on which a given functional circuit is fabricated. Typically, integrated circuits are produced in large batches on a single wafer of electronic-grade silicon or other semiconductors through processes such as photolithography. The wafer is cut into many pieces, each containing one copy of the circuit. Each of these pieces is called a die. Semiconductor multi-die structures have intermediate vertical side chips, as shown in Figure 7.3, and packages housing such semiconductor multi-die structures are described. In an example, a multi-die semiconductor structure includes a first main stacked die structure having a first substantially horizontal arrangement of semiconductor

Figure 7.2   Inside view of a power module.

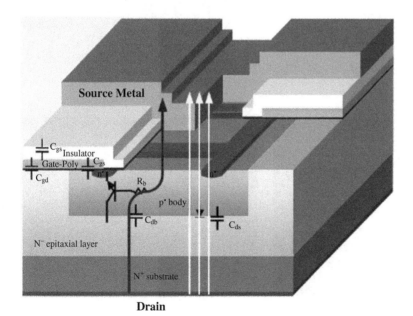

Figure 7.3   Vertical view of a MOSFET die. *Source:* Used with permission from SCILLC dba ansemi.

dies. A second structure having a second substantially horizontal arrangement of semiconductor dies is also included. An intermediate vertical side chip is disposed between and electrically coupled to the first and second structures.

- Package structure

In a power module, a pad is mounted on the die and the die is attached to the direct bonded copper (DBC) substrate by the solder. A DBC substrate has good electrical insulation and allows for heat dissipation. The structure is bonded to the copper plate with the solder for better heat dissipation. The wire is then bonded to the pad surface through ultrasonic wedge bonding technology, as shown in Figures 7.4–7.7.

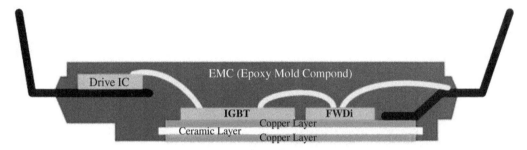

Figure 7.4 Vertical view of a multi-die power module. *Source:* Used with permission from SCILLC dba onsemi.

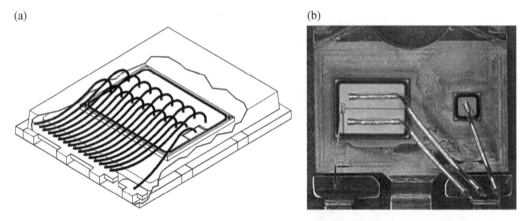

Figure 7.5 Wire bonds in power modules: (a) MicroFET, (b) TO-247 IGBT. *Source:* Used with permission from SCILLC dba onsemi.

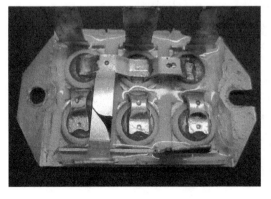

Figure 7.6 High current contacts in a three-phase rectifier.

In addition to the traditional screw contacts, the electrical connection between the module and other parts of the power electronic system can also be achieved by pin contacts, press-fit contacts, spring contacts that inherently press on contact areas, or by a pure pressure contact where corrosion-proof surface areas are directly pressed together. Press-fit pins achieve a very high reliability and ease the mounting process without the need for soldering, as shown in Figure 7.8. Compared to press-fit connections, spring contacts have the benefit of allowing easy and non-destructive removal of the connection several times, as for inspection or replacement of a module, for instance. Both contact types have a rather limited current-carrying capability due to their comparatively low cross-sectional area and small contact surface. Therefore, modules often contain multiple pins or springs for each of the electrical power connections.

7.2.2 Passive Devices

The passive devices have simpler structure than the active devices. However, their structure contains some weak points.

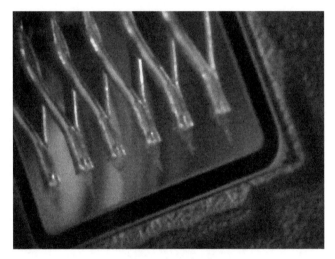

Figure 7.7   Wire bond contacts in a power module.

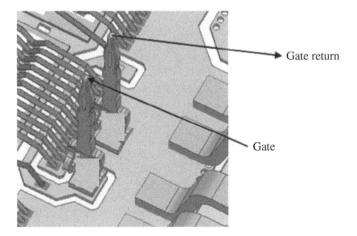

Figure 7.8   Press-fit contacts of a power module. *Source:* Used with permission from SCILLC dba onsemi.

- Capacitors

  The basic structure of a capacitor consists of two parallel metal foils (very thin sheets of metal). The metal foils act as an electrode. Dielectric material is used as an insulator to separate the metal foils. The insulating materials, such as glass, rubber, ceramic, plastic, and paper, are used as a dielectric material. The capacitor nominal working voltage indicates how much voltage the capacitor can withstand in the long term. In capacitors, the thickness of the electrolyte element and the insulating material determine the working voltage of the capacitor. Figure 7.9 shows the conceptual structure of a capacitor. Various capacitor types have different structures. Figure 7.10 shows the structure of an electrolytic capacitor and Figure 7.11 shows a ceramic capacitor. An overvoltage will damage the dielectric of a capacitor and may affect the performance and life of a capacitor or cause catastrophic failure.

- Magnetic devices

  An inductor usually consists of a coil of conducting material, typically insulated copper wire, wrapped around a core either of plastic (to create an air-core inductor) or of a ferromagnetic material, as shown in Figure 7.12a. Small inductors can be etched directly onto a printed circuit

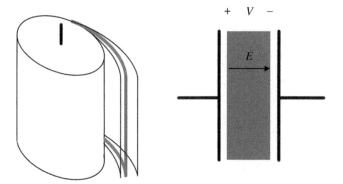

Figure 7.9 Schematic view and simplified model of a capacitor.

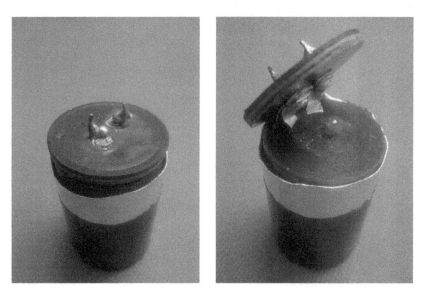

Figure 7.10 Physical structure of a real electrolyte capacitor.

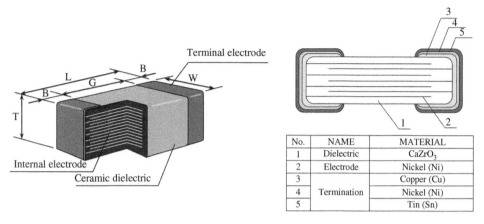

Figure 7.11 Physical structure of a real ceramic capacitor. *Source:* TDK (with permission).

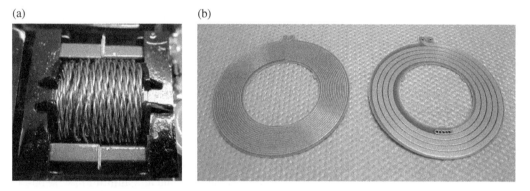

Figure 7.12 Structure of an inductor: (a) with ferrite core, (b) planar inductors.

board by laying out the trace in a spiral pattern, as shown in Figure 7.12b. Some planar inductors use a planar core. Small value inductors can also be built on integrated circuits using the same processes that are used to make interconnects.

7.3 Faults in Active Devices

Power electronic applications come with very high requirements in terms of reliability. Active devices, i.e. solid-state switches, are the key elements of the power electronic systems, but show a high failure rate in a power electronic system.

7.3.1 Semiconductor-Related Faults

Semiconductor-related faults or die-related faults form mainly random faults. These faults happen due to single-event failure in the die, as shown in Figure 7.13. Much effort has been devoted to the reduction of device wear-out failure induced by accumulated degradation and catastrophic failure

Figure 7.13 A damaged die of an IGBT module.

triggered by single-event overstress. The wear-out failure under field operation can be mitigated by scheduled maintenances based on lifetime prediction and condition monitoring. However, catastrophic failure is difficult to predict and thus may lead to serious consequences in power electronic converters.

7.3.2 Package-Related Faults

Faults in the package level are mainly wear-out type faults. They are usually related to the connections in the package. Various metallurgical failure modes of gold and aluminum wire bonds are reported. Examples are taken from both low and high-power devices. Whenever possible, known methods of avoiding these failure modes are given. Wire bond failure modes can be divided into two categories. The first is comprised of those failure modes that are caused by poorly controlled or poorly designed manufacturing processes that result in a lower product yield or higher per unit bonding cost, as well as those processes that predispose the device to early field failure. The second category is comprised of failure modes of adequately made bonds that are caused to fail by environmental stresses during the operating life of the device. These include non-optimized bonding schedules, cracks in the heels of ultrasonic bonds, intermetallic formation, poor metallization, and inadequate glassivation removal. Assuming that the package is hermetic, or for plastic devices that humidity and other corrosion producers are not present, the primary wire bond failure modes in the second category result from environmental temperature exposure and the number of power or thermal cycles experienced by the device during operation. Both of these can induce intermetallic formation. The latter can cause metallurgical flexure fatigue at the bond heel or in the wire.

7.3.2.1 Wire Bond Lift-Off

Lift-off of Al bonding wires is one common failure mode of power modules during long-time operation. In reference [1], the failure mechanism of lift-off of Al wires was investigated and the major factors were discussed based on both experiments and finite element simulations. This indicates that lift-off of Al wires is mainly determined by the interfacial thermal stress at the bonding interface.

Thermal expansion of Al wires and thermal mismatch of Al-Si interfaces contribute to the interfacial thermal stress which is affected by the resistance heat of Al wires and power loss of Si chips. Accordingly, new lifetime models for the Al wires lift-off failure mode of power modules are proposed and verified through power cycling tests. The conduction current, the heating time, and the junction temperature swing are three major factors for the fatigue life model of power modules under Al wires lift-off failure mode. Figure 7.14 shows the contact of a wire bond before and after lift-off. It is seen that the contact temperature increases, which accelerates the wear-out process of the module.

7.3.2.2 Heel Cracking

Heel crack is one of the most complicated reliability problems of wire bonding in the power electronic package. Figure 7.15 shows the heel point of a wire bond. In reference [2], the effects of solder reflow to the heel crack for the 2–5 mil aluminum wire is investigated. The results show that the plastic strain at the heel region induced by the wire bonding process, influence of the molding process, and the coefficient of thermal expansion (CTE) mismatches between different

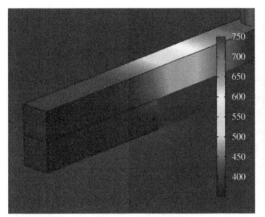

Figure 7.14 Effect of wire bond lift-off on the temperature of the contact.

Figure 7.15 The heel point of a wire bond.

components in package are the main causes of heel crack that happened in the reflow. With respect to the trend for the process to be lead free, the simulation is also processed under the three temperature hierarchies with different peak reflow temperatures and wetting times. From von Mises stress and related plastic strain distributions, it can been seen that, during the reflow, the heel region of the wire endures larger stress and plastic strain than other areas, and with the peak reflow temperature and wetting time increasing, the plastic strain also increases by about 20%, which is very critical for material fatigue. Reference [3] presents a comprehensive long-term reliability analysis of commercially available SiC MOSFETs under high-temperature operation and high-temperature swing, degradation related key precursors, and possible causes behind them. For this purpose, discrete SiC devices are power cycled and all datasheet parameters are recorded at certain intervals with the aid of the curve tracer. Variation of electrical parameters throughout the tests is presented in order to assess their correlation with the aging/degradation state of the switch. Among them, gate oxide charge trapping related threshold voltage drift and corresponding state resistance variation have been observed for all samples. For some samples, bond wire heel cracking is found to be the root cause of sudden state resistance and body diode voltage increases. The discussions regarding aging precursors are supported by failure analysis obtained through the decapsulation of failed devices. Finally, the findings are evaluated in order to define the suitability of electrical parameters as an aging precursor parameter in the light of practical implementation related issues.

7.3.2.3 Solder Fatigue

Since solder joints provide the mechanical and electrical interconnect between the package and the die, they are susceptible to failures during thermal cycling, as shown in Figure 7.16 [4]. For this reason, ensuring solder joint reliability is one of the most critical design aspects of electronic assemblies. Solder joints of electronic assemblies are complex elements and therefore building accuracy in a life prediction model is not an easy task. The acknowledged complexity arises from the following: the three-dimensional package structures with solder joints are subjected to multiaxial nonlinear material behaviors, complex joint shapes, and multiaxial loading. Failure of solder joints is a complex sequence of possible failure mechanisms involving grain/phase coarsening, grain boundary sliding, matrix creep, microvoid formation, and linking, resulting in crack initiation and crack propagation. In the case of SnAgCu solder joints, the damage accumulation process

Figure 7.16 Effect of solder fatigue on a power module. *Source:* Used with permission from SCILLC dba onsemi.

leads to much less coarsening of the microstructure. The effects of substrate finishes and component finishes on the reliability are considerable [4]. Soft solders react with metallizations to form interfacial intermetallics. Intermetallics grow with time and temperature. Metallization consumption by intermetallic growth and intermetallics within the solder can be observed. Often the soldering process itself induces anomalies (e.g. changes in the reflow temperature profile, different cooling rates, excessive voids inside the solder, brittle phase formation, and concentration gradients of elemental or metallurgical composition when the joint is formed). Smaller solder joints may impose increased interfacial effects and decreased fatigue-creep phenomena (joint volume versus interfacial effects) [4]. In reference [5], plastic strain versus fatigue life data are presented for tests. It was found that these data could be correlated by the Coffin-Manson fatigue law, with an exponent of approximately 0.5 for the tests run at −35 to 125 °C. At 150 °C, the exponent was reduced to 0.37. These results were obtained for plastic strain limited tests. Different results are obtained when total strain limits are employed. This difference is discussed as well as the influence of cycling frequency and temperature changes. A model is presented that describes the influence of plastic strain and cycling frequency. Corrections to the model predicted fatigue life, which account for temperature changes, cycling waveshape, and joint geometries, are also considered.

7.3.2.4 Crack Fatigue in Ceramics

Thermal loadings stress whole electronic packages while powering the semiconductor and by environmental temperature changes. Direct copper bonding (DCB) substrates are widely used in electronic applications due to their high thermal conductivity performance. They consist of copper layers sintered onto a ceramic sheet. Because of the very high process temperature, thermal induced stresses arise along the substrate while cooling to ambient temperatures. The DCB substrates usually carry residual stresses, which are concentrated along the edges of copper structures [6]. Under service conditions, or even during the electronic manufacturing process, the stress concentration might increase and in consequence copper structures rip off the substrate by cracking and conchoidal fractures in the ceramic sheet. To avoid the critical cracking situation, knowledge about the residual stress condition is required. The target will be a stress mapping across the DCB substrate. Maximum tensile stress in the ceramics during the thermal cycle test (from −40 to 250 °C) of active metal brazing (AMB) substrate is estimated using the finite element method analysis, because such a tensile stress is the driving force of Cu plate delamination from the ceramic plate. In order to accelerate thermal fatigue of the AMB substrate, a tensile stress 1.5–2.1 times larger than the maximum thermal stress at −40 °C is applied to the ceramic plate by four-point bending the AMB substrate at 250 °C repeatedly at a frequency of 1 Hz. The time to failure by repeated bending of the substrate is less than 1/40 of the time to delamination of the Cu plate by thermal cycling [7].

7.3.2.5 Corrosion of Bond Wires

One of the failure mechanisms is corrosion of the contact between the copper (Cu) parts and the aluminum (Al) bond-pad, consisting of various intermetallic compounds (IMCs), which are more sensitive to corrosion compared to gold (Au) Al IMCs. The studies elaborate on three corrosion mechanisms present in the Cu–Al system: interfacial Cu–Al IMC corrosion, bulk Cu–Al IMC corrosion, and Al bond pad corrosion [8]. For the first mechanism, which is dominant, an empirical corrosion model is introduced. To gather data for this model, a recently developed method for analyzing the IMC contact area to study the dynamics of the dominant mechanism is used. Data is collected from various devices, which are exposed to accelerated aging conditions. The focus of this reference is on unbiased conditions using a wide temperature and humidity range. In total, four epoxy molding compound types have been investigated and are compared to each other, using the empirical model proposed in this reference. Finally, it is shown that the model allows the prediction of the lifetime of Cu–Al ball contacts for different application conditions and also allows the selection of an appropriate compound type. A detailed description of the Cu–Al wire bond interface is presented in reference [9], which can possibly explain the often-observed corrosion failures in humidity reliability tests. Using microstructural analysis techniques, it is shown that the unstressed interface contains up to three intermetallic phases, where the Cu-rich phases are located at the Cu–ball interface. Using the humidity stress test, only the high-Cu containing intermetallic layers close to the Cu wire ball bond undergo a corrosion process, whilst the Cu-lean layers are stable in all environment stress tests. The failing layers consist out of an amorphous Al-based oxide matrix with embedded Cu precipitates. The failure process can be explained as galvanic corrosion of the intermetallic phases. The Cu-rich phases corrode faster compared to Al-rich phases, since they have an electrochemical potential lower than the Cu cathode and form, hypothetically, a less stable self-passivation oxide. The observed time-to-failure is then determined by the composition, thickness, and volume of the intermetallic layers. This was verified by reliability test results performed on open wire bonds and on plastic encapsulated products.

7.3.2.6 Delamination of Bond Wires

Stitch crack has been a recurring failure mechanism for many years in semiconductor packaging. Currently, the combination of highly filled mold compounds and copper wire increases the risk for stitch cracks after temperature cycling (TC). Firstly, highly filled mold compounds generally have a CTE that is much lower than that of copper, and a much higher elastic modulus compared to traditional compounds, which results in a higher stress on the bond wire compared to traditional compounds having a CTE that matches the CTE of copper. Secondly, the copper wire experiences large plastic deformation while forming the stitch during wire bonding at elevated temperatures, which leads to local "embrittlement." The combination of both effects, i.e. increased stress and reduced resistance to fatigue, leads to a higher risk of cracking. However, in general, the stitch is sufficiently strong to withstand the increased stress. In reference [10], it is shown that delamination is a prerequisite for stitch cracking to occur because it has a pronounced effect on the stress in the stitch as it reduces the support by the lead, resulting in more force being exerted on the stitch. Simulations reveal increased stress for a stitch where the mold compound is delaminated compared to a fully adherent compound. The results of the simulations match the results of TC reliability tests and scanning acoustic tomography to identify delaminated regions within the package.

7.4 Thermal Cycling

The temperature of power semiconductor devices is one of the main issues affecting the performance, availability, and reliability of power converters. In reference [11], the effect of junction temperature swing duration on the lifetime of transfer molded power IGBT modules is studied and a relevant lifetime factor is modeled. This study is based on 39 accelerated power cycling test results under six different conditions by an advanced power cycling test setup, which allows tested modules to be operated under more realistic electrical conditions during the power cycling test. The analysis of the test results and the temperature swing duration-dependent lifetime factor under different definitions and confidence levels are presented. This study enables the temperature rise time effect on a lifetime model of IGBT modules to be included for its lifetime estimation and may result in an improved lifetime prediction of IGBT modules under given mission profiles of converters. A post-failure analysis of the tested IGBT modules is also performed. The main reason for the temperature swing is the thermal impedance of the die and module. Figure 7.17 shows that all

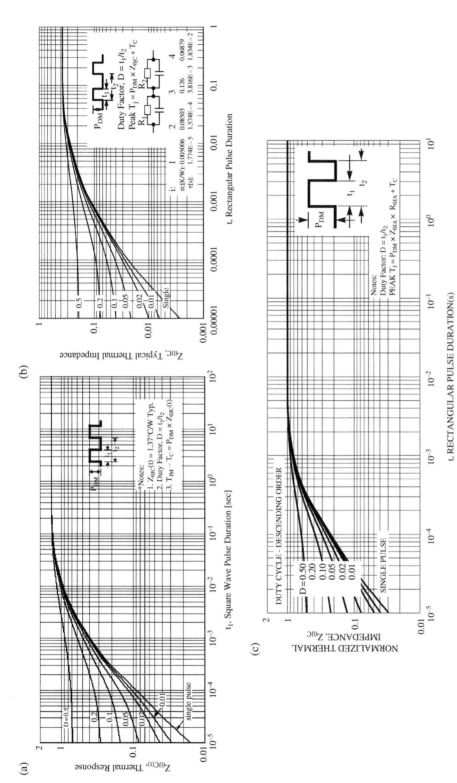

Figure 7.17 The thermal impedance of various devices: (a) diode, (b) IGBT, (c) MOSFET. *Source:* Used with permission from SCILLC dba onsemi.

devices have this thermal impedance. Thus, the junction temperature can increase immediately after the power flow starts while the other parts still remain at the standing temperature. Figure 7.18 shows the temperature profile of a die for 100 μs power cycling. It is seen that the die temperature increases rapidly while the contact temperature is constant.

A power cycling test is the most important reliability test for power modules. This test consists in periodically applying a current to a device mounted onto a heat sink, as shown in Figure 7.19. This leads to a power loss in the entire module and results in a rise in the semiconductor temperature. In reference [12], the different kinds of semiconductors and power modules used for traction applications are described. Experimental and simulation methods employed for power cycling tests are presented. The module weak points and fatigue processes are pointed out. It gives a clear overview of all studies dealing with power cycling that were carried out.

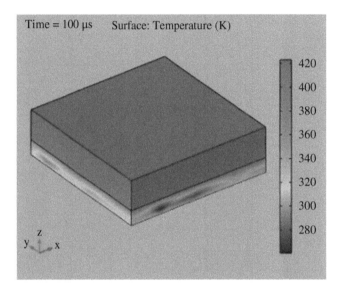

Figure 7.18 The temperature profile of a die for 100 μs power cycling.

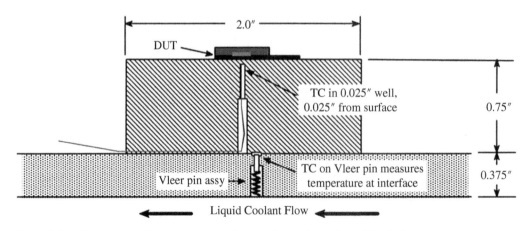

Figure 7.19 The power cycling setup. *Source:* Used with permission from SCILLC dba onsemi.

The numerous advantages of power modules and their ongoing development for higher voltage and current ratings make them interesting for traction applications. These applications imply high reliability requirements. One important requirement is the ability to withstand power cycles. Power cycles cause temperature changes, which lead to a mechanical stress that can result in a failure. Lifting of bond wires is thereby the predominant failure mechanism. A fast power cycling test method activating the main failure mechanism has been developed that allows the reproduction of millions of temperature changes in a short time. The applicability of fast testing is supported by a mechanical analysis. Test results show the number of cycles to failure as a function of temperature changes for an IGBT single switch [13].

Knowledge of the instantaneous junction temperature is essential for effective health management of power converters, enabling safe operation of the power semiconductors under all operating conditions. Methods based on fixed thermal models are typically unable to compensate for degradation of the thermal path resulting from aging and the effect of variable cooling conditions. Thermosensitive electrical parameters (TSEPs), on the other hand, can give an estimate of junction temperature, but measurement inaccuracies and the masking effect of varying operating conditions can corrupt the estimate. Reference [14] presents a robust and non-invasive real-time estimate of junction temperature that can provide enhanced accuracy under all operating and cooling conditions when compared to model-based or TSEP-based methods alone. The proposed method uses a Kalman filter to fuse the advantages of model-based estimates and an online measurement of TSEPs. Junction temperature measurements are obtained from an online measurement of the on-state voltage, at high current and processed by a Kalman filter, which implements a predict-correct mechanism to generate an adaptive estimate of T_j. It is shown that the residual signal from the Kalman filter may be used to detect changes in thermal model parameters, thus allowing the assessment of thermal path degradation. The algorithm is implemented on a full-bridge inverter and the results are verified with an IR camera.

As a key component in the wind turbine system, the power electronic converter and its power semiconductors suffer from complicated power loadings related to climate conditions, and are proven to have high failure rates. Therefore, a correct lifetime estimation of a wind power converter is crucial for reliability improvement and also for cost reduction of wind power technology. A relative more advanced approach is proposed in reference [15], which is based on the loading and strength analysis of devices and takes into account different time constants of the thermal behaviors in a power converter. With the established methods for loading and lifetime estimation for power devices, more detailed information of the lifetime-related performance in a wind power converter can be obtained. Some experimental results are also included to validate the thermal behavior of the power device under different mission profiles [15].

7.5 Faults in Passive Devices

Passive devices are also subjected to wear-out failures. In this category, capacitors show a higher failure rate than resistors and magnetic devices.

7.5.1 Faults in Resistors

The resistor damage always appears on the open circuit. Figure 7.20 shows two types of damaged resistors.

Figure 7.20 Two damaged resistors.

7.5.2 Faults in Capacitors

Electrolytic filter capacitors are frequently responsible for static converter breakdowns. Figure 7.21 shows an open-circuit contact fault in a capacitor while Figure 7.22 shows a damaged capacitor due to a high-temperature failure. To predict these faults, a method to set a predictive maintenance is applied. The best indicator of fault of the output filter capacitors is an increase of ESR (equivalent series resistance). The output-voltage ripple of the converter increases with respect to ESR. In order to avoid errors due to load variations, the output-voltage ripple is filtered at the switching frequency of the converter. The problem is that this filtered component is not only dependent on the aging of the capacitors but also on the ambient temperature, output current, and input voltage of the converter. Thus, to predict the failure of the capacitors, this component is processed with these parameters and the remaining time before failure is deduced [16]. A tool has been developed to establish the predictive maintenance of the converter. First, a reference system including all the converter parameters is built for the converter in its sound state, i.e. using sound electrolytic filter capacitors. Then, all these parameters are processed and compared on line with the reference system, thereby computing the lifetime of these capacitors.

Operating at high temperatures decreases the lifetime of the capacitor. Figure 7.23 shows how the impact of temperature increases the lifetime of a capacitor. In these studies, a hot spot of the capacitor is considered, as shown in Figure 7.24.

Corrosion of the metallic film in cylindrical metallized film capacitors generally progresses from the end caps toward the center of the film [17]. Promoted by the penetration of moisture from outside, the corrosion is usually more advanced in the outer layers where the windings are less compressed. This progression can result in increasing lengths of metallization being disconnected from direct connection to the zinc-sprayed end caps, resulting in a long serial connection back to some point deeper in the spiral winding of the capacitor. In addition, vestigial remnants of metal, more resistant to corrosion, may remain at various points. Theoretical analysis indicates that heavy

Figure 7.21 The open-circuit contact fault in a capacitor.

Figure 7.22 A damaged capacitor due to the high-temperature failure.

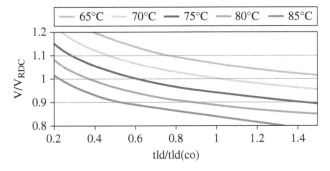

Figure 7.23 Effect of temperature rise on the lifetime of the capacitors. *Source:* TDK (with permission).

Thermal map, cross section

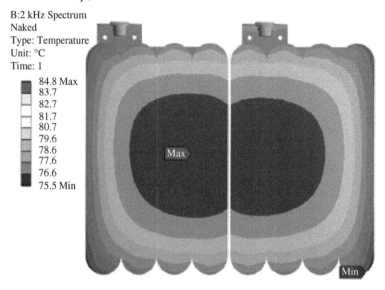

Current spectrum considered

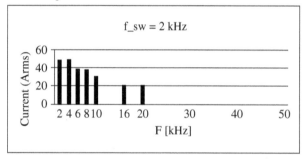

Figure 7.24 Hotspot of a capacitor in power cycling studies. *Source:* TDK (with permission).

series current at the still-connected end of the metal film and excessive current density in the vestigial links can result in localized heating that exceeds the tolerance of the dielectric, resulting in catastrophic failure. Such heating and probable fusing of remnant links are prime suspects in the catastrophic failure of metallized polypropylene capacitors.

Vibration and mechanical shocks usually lead to the mechanical damage of the capacitors, especially at the lead points, as shown in Figures 7.25–7.27.

The normal voltage failures in multilayer ceramic capacitors (MLCs) using the temperature–humidity–bias test (THB) are described in reference [18]. The cumulative failure data showed that the MLC failures occurred in several stages during the THB test, with the extent of failures depending on the quality of the capacitor lot and the bias voltage. THB failures increased after some of the MLCs had undergone a barrel-plating operation, indicating moisture penetration and ionic contaminants as the likely cause of accelerating the failure rate. In the cross-section of failed MLCs, the authors observed large holes caused by high-temperature explosive events occurring inside the MLC. There were also internal cracks connecting electrodes of opposite polarity, with silver inclusions along the length of the cracks. These observations strongly suggested that silver migration was the cause of short-circuit paths, leading to subsequent failure of the MLC.

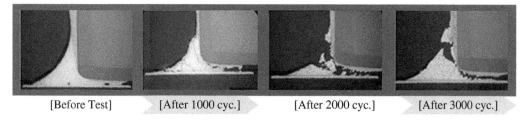

[Before Test]   [After 1000 cyc.]   [After 2000 cyc.]   [After 3000 cyc.]

Figure 7.25 Damaging of the capacitor contacts due to mechanical vibration. *Source:* TDK (with permission).

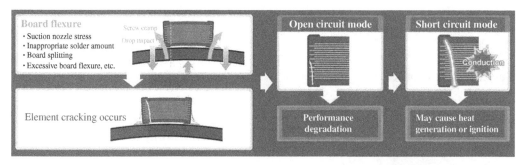

Figure 7.26 Cracks appearing in the capacitor due to the mechanical shocks. *Source:* TDK (with permission).

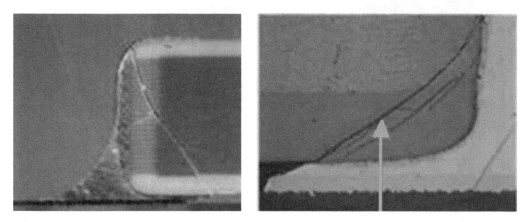

Figure 7.27 Two crack photos at the ceramic capacitor contacts. *Source:* TDK (with permission).

7.5.3 Faults in Magnetic Devices

The main long-term failure factor of the magnetic elements is partial discharge (PD). The PDs occur due to the uneven voltage distributions of the insulator. This unbalanced voltage distribution is due to unavoidable existing voids through the insulator. For more clarity, an arbitrary insulator with its void is presented in Figure 7.28. This void usually consists of air and has a different permittivity from the insulator. Owing to this difference and the unequal dimensions, a simple model of the insulator and its void is depicted in Figure 7.28. This model consists of two different capacitors modeling the capacitance of void and the insulator. Considering this model after applying the voltage, the voltage across the void exceeds the breakdown voltage and a PD occurs. This

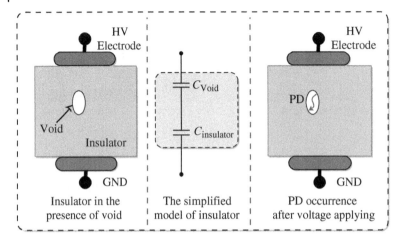

Figure 7.28 The simplified model of an insulator in the presence of voids.

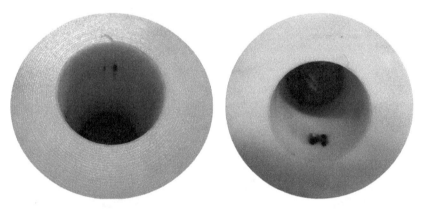

Figure 7.29 Two damaged insulators due to the PD power loss in the long term.

process occurs whenever the total applied voltage is more than a predetermined level known as the inception voltage. In this condition, one or several PDs occur and in each PD a level of heat is released. The released energy through the insulator in PDs is the main factor deteriorating the insulators, as shown in Figure 7.29.

7.6 Summary and Conclusions

In this chapter, the element level failure causes and mechanisms in power electronic systems were described. The summary of the study is as follows:

1) The wear-out phase begins at the system end of life. Wear-out studies gives some benefits. First, it allows proactive maintenance plans. Further, depending on the failure mode, different proactive control strategies can be applied to converters in order to improve the reliability and availability of power electronic systems. Many efforts have been devoted to the reduction of

power electronic system wear-out failure induced by accumulated degradation and catastrophic failure triggered by single-event overstress. The wear-out failure under field operation could be mitigated by scheduled maintenances based on lifetime prediction and condition monitoring.

2) A power electronic module or power module acts as a physical container for the storage of several power components, usually power semiconductor devices. The classical power modules contain multiple semiconductor dies that are connected to form an electrical circuit of a certain structure, called topology. Modules also contain other components such as ceramic capacitors to minimize switching voltage overshoots and NTC thermistors to monitor the module's substrate temperature. Generally, a power electronic module can be divided into die structures and the requirements for packaging.

3) Semiconductor-related faults or die-related faults form mainly random faults. These faults happen due to single event failure in the die. Many efforts have been devoted to the reduction of device wear-out failure induced by accumulated degradation and catastrophic failure triggered by single-event overstress. The wear-out failure under field operation could be mitigated by scheduled maintenances based on lifetime prediction and condition monitoring. However, the catastrophic failure is difficult to predict and thus may lead to serious consequences of power electronic converters.

4) The main wear-out failures of the power modules are: wire bond lift-off, solder fatigue, ceramic cracking, heel cracking, corrosion, and delamination.

5) The capacitors are the main part of the passive devices subjected to a high failure rate. High temperature, opening of the leads, and cracks in ceramic are the main failure factors of the capacitors.

References

1 Huang, Y., Jia, Y., Luo, Y. et al. (2020). Lifting-off of Al bonding wires in IGBT modules under power cycling: Failure mechanism and lifetime model. *IEEE Journal of Emerging and Selected Topics in Power Electronics* 8 (3): 3162–3173.

2 He, L., Pan, S., Wang, L., and Zhang, D.W. (2006). Heel crack and lead-free soldering issues affecting power electronics packages. In: *TENCON 2006–2006 IEEE Region 10 Conference*, Hong Kong, China, 1.

3 Ugur, E., Yang, F., Pu, S. et al. (2019). Degradation assessment and precursor identification for SiC MOSFETs under high temp cycling. *IEEE Transactions on Industry Applications* 55 (3): 2858–2867.

4 Schubert, A., Dudek, R., Auerswald, E. et al. (2003). Fatigue life models for SnAgCu and SnPb solder joints evaluated by experiments and simulation. In: *In Proceedings of the 53rd Electronic Components and Technology Conference, 2003*, New Orleans, LA, USA, 603–610.

5 Solomon, H. (1986). Fatigue of 60/40 solder. *IEEE Transactions on Components, Hybrids, and Manufacturing Technology* 9 (4): 423–432.

6 Muench, S., Roellig, M., Cikalova, U. et al. (2017). A laser speckle photometry based non-destructive method for measuring stress conditions in direct-copper-bonded ceramics for power electronic application. In: *2017 18th International Conference on Thermal, Mechanical and Multi-Physics Simulation and Experiments in Microelectronics and Microsystems (EuroSimE)*, Dresden, Germany, 1–8.

7 Miyazaki, H., Hyuga, H., Hirao, K. et al. (2018). Accelerated thermal fatigue test of metallized ceramic substrates for SiC power modules by repeated four-point bending. In: *2018 IEEE 30th International Symposium on Power Semiconductor Devices and ICs (ISPSD)*, Chicago, IL, USA, 264–267.

8 Rongen, R., O'Halloran, G.M., Mavinkurve, A. et al. (2014). Lifetime prediction of Cu–Al wire bonded contacts for different mould compounds. In: *2014 IEEE 64th Electronic Components and Technology Conference (ECTC)*, Orlando, FL, USA, 411–418.

9 Boettcher, T., Rother, M., Liedtke, S. et al. (2010). On the intermetallic corrosion of Cu–Al wire bonds. In: *2010 12th Electronics Packaging Technology Conference*, Singapore, 585–590.

10 van Soestbergen, M., Mavinkurve, A., Shantaram, S., and Zaal, J.J.M. (2017). Delamination-induced stitch crack of copper wires. In: *2017 18th International Conference on Thermal, Mechanical and Multi-Physics Simulation and Experiments in Microelectronics and Microsystems (EuroSimE)*, Dresden, Germany, 1–4.

11 Choi, U., Blaabjerg, F., and Jørgensen, S. (2017). Study on effect of junction temperature swing duration on lifetime of transfer molded power IGBT modules. *IEEE Transactions on Power Electronics* 32 (8): 6434–6443.

12 Durand, C., Klingler, M., Coutellier, D., and Naceur, H. (2016). Power cycling reliability of power module: a survey. *IEEE Transactions on Device and Materials Reliability* 16 (1): 80–97.

13 Held, M., Jacob, P., Nicoletti, G. et al. (1997). Fast power cycling test of IGBT modules in traction application. In: *Proceedings of Second International Conference on Power Electronics and Drive Systems*, Singapore, vol. 1, 425–430.

14 Eleffendi, M.A. and Johnson, C.M. (2016). Application of Kalman filter to estimate junction temperature in IGBT power modules. *IEEE Transactions on Power Electronics* 31 (2): 1576–1587.

15 Ma, K., Liserre, M., Blaabjerg, F., and Kerekes, T. (2015). Thermal loading and lifetime estimation for power device considering mission profiles in wind power converter. *IEEE Transactions on Power Electronics* 30 (2): 590–602.

16 Lahyani, A., Venet, P., Grellet, G., and Viverge, P. (1998). Failure prediction of electrolytic capacitors during operation of a switchmode power supply. *IEEE Transactions on Power Electronics* 13 (6): 1199–1207.

17 Brown, R.W. (2006). Linking corrosion and catastrophic failure in low-power metallized polypropylene capacitors. *IEEE Transactions on Device and Materials Reliability* 6 (2): 326–333.

18 Ling, H.C. and Jackson, A.M. (1989). Correlation of silver migration with temperature-humidity-bias (THB) failures in multilayer ceramic capacitors. *IEEE Transactions on Components, Hybrids, and Manufacturing Technology* 12 (1): 130–137.

8

External Faults

8.1 Origins of the External Faults

In the previous chapters, we studied the internal faults that cause non-resilient operation of the power electronic system. The service of a power electronic system may also be interrupted because of other factors. One of these factors is the external faults. External faults occur beyond the borders of the converter: this means that the load supporting is interrupted due to external factors. For example, a transient phenomenon in the load or in the power source of the converter causes a reaction from the converter protection. Thus, the converter may be shut down while the load of the converter power source returns to a normal state. Protection systems act when a fault occurs in the system. Their performance is very important; isolation of the converter is not always the best choice because this strategy has a bad effect on the availability of the converter. Two main types of external faults are those that occur outside the input and output ports of the power electronic converters, as shown in Figure 8.1. The faults at the input port come from the power source. For example, a lightning strike in the power grid is a sample of external faults at the input port. The faults at the output port of the converter are due to load faults. For example, a short circuit in the load is an example of external faults at the output port. Other external faults are faults affecting the converter from the other electric parasitic ports, such as radiation effects or those non-electrical paths, such as environmental effects.

8.1.1 Faults in the Power Source

Various faulty conditions occur at the power source port of the converter. In a polyphase system, a fault may affect all phases equally, which is a symmetrical fault. If only some phases are affected, the resulting asymmetrical fault becomes more complicated to analyze due to the simplifying assumption that an equal current magnitude in all phases is no longer applicable. The analysis of this type of fault is often simplified by using methods such as symmetrical components. The design of systems to detect and interrupt power system faults is the main objective of power system protection. Lightning is another failure factor of the systems from the power source. Lightning has the ability to destroy the electronic equipment. Electrical engineers have spent over 130 years devising ways to control lightning and power surges in the electrical distribution grid. Both lightning surges and power line switching surges are represented as double exponential waveforms during testing, as shown in Figure 8.2. This means that a lightning surge rises to its peak value in 1.2 μs and decays to one-half of this in 50 μs. A switching surge waveform is much longer; it rises to its peak in roughly 200 μs and decays to half its value in a few thousand microseconds. In reference [1],

Resilient Power Electronic Systems, First Edition. Shahriyar Kaboli, Saeed Peyghami, and Frede Blaabjerg.
© 2022 John Wiley & Sons Ltd. Published 2022 by John Wiley & Sons Ltd.
Companion website: www.wiley.com/go/kaboli/resilientpower

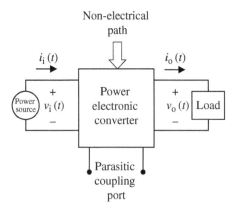

Non-electrical path

$i_i(t)$ $i_o(t)$

Power source $v_i(t)$ +   −

Power electronic converter

$v_o(t)$ +   −   Load

Parasitic coupling port

Figure 8.1   The paths of external faults in a power electronic converter.

(a)

(b)

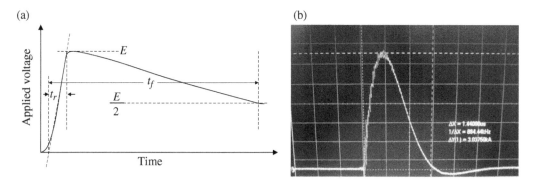

Figure 8.2   Lightning waveforms: (a) standard waveform used in the tests, (b) a real lightning test current.

a computer model for transient analysis of a network of buried and above ground conductors is presented. The model is based on the electromagnetic field theory approach and the modified image theory. Validation of the model is achieved by comparison with field measurements. The model is applied for computation of transient voltages to remote ground of large grounding grid conductors. Also computation of longitudinal and leakage currents, transient impedance, electromagnetic fields, and transient induced voltages is possible. This model is aimed to help in electromagnetic compatibility (EMC) and lightning protection studies that involve electrical and electronic systems connected to grounding systems.

The power quality issues are the other failure factors. Reference [2] summarizes the main problems and solutions of the power quality in microgrids, distributed-energy-storage systems, and ac/dc hybrid microgrids. First, the power quality enhancement of grid-interactive microgrids is presented. Then, the cooperative control to enhance voltage harmonics and unbalances in microgrids is reviewed. Afterwards, the use of a static synchronous compensator in grid-connected microgrids is introduced in order to improve voltage sags and unbalances. Finally, the coordinated control of distributed storage systems and ac/dc hybrid microgrids is explained. The cause of voltage sags in industrial plants, their impact on equipment operation, and possible solutions are described in reference [3]. The definition proposed focuses on system faults as the major cause of voltage sags. The sensitivity of different types of industrial equipment, including adjustable speed drive controls, programmable logic controllers, and motor contactors, is analyzed. Available methods of power conditioning for this sensitive equipment are also described. Figure 8.3 shows a voltage sag in the DC power system.

   In the distributed power sources, there are other types of problem. Nowadays, the majority of the photovoltaic (PV) power sources are connected to the public grid. One of the main connection problems occurs when voltage sags appear in the grid due to short circuits, lightning, etc. International standards regulate the grid connection of PV systems, forcing the source to remain connected during short-time grid-voltage faults. As a consequence, during the voltage sag, the source should operate with increasing converter currents to maintain the injection of the generated power. This abnormal operation may result in non-desired system disconnections due to over-current. Reference [4] proposes a controller for a PV three-phase inverter that ensures minimum peak values in the grid-injected currents, as compared with conventional controllers. From the system analysis, a design method is presented in order to set the parameters of the control scheme. Current-voltage and power-voltage characteristics of large PV arrays under partially shaded conditions are characterized by multiple steps and peaks, as shown in Figure 8.4. This makes tracking of the actual maximum power point (MPP) a difficult task. In addition, most of the existing schemes are unable to extract maximum power from the PV array under these conditions. Reference [5] proposes a novel algorithm to track the global power peak under partially shaded conditions.

Figure 8.3 A voltage sag in the DC power system.

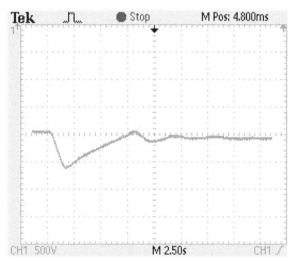

(a)  (b)

Figure 8.4 Problems associated with distributed power sources: (a) partially shaded solar plant, (b) snow cover on the solar plant.

The formulation of the algorithm is based on several critical observations made out of an extensive study of the PV characteristics and the behavior of the global and local peaks under partially shaded conditions. In reference [5], the proposed algorithm works in conjunction with a dc–dc converter. In order to accelerate the tracking speed, a feedforward control scheme for operating the dc–dc converter is also proposed, which uses the reference voltage information from the tracking algorithm to shift the operation toward the MPP. The tracking time with this controller is about one-tenth as compared with a conventional controller.

8.1.2 Faults in the Load

In an electric power system, overcurrent or excess current is a situation where a larger than intended electric current exists through a conductor, leading to excessive generation of heat, and the risk of fire or damage to equipment. Possible causes for overcurrent include short circuits, excessive load, and incorrect design [6]. Figure 8.5 shows the output voltage and current of a high voltage power supply (HVPS). A fault occurs in the load and the load current increases. In this figure, the protection system responds after 300 micro second and shuts down the converter output. It saves the converter and even the load but the converter is not available after the fault.

These faults force the power electronic converter to have a proper protection system. Fault detection (FD) in power electronic converters is necessary in embedded and safety critical applications to prevent further damage. Fast FD is a mandatory step in order to make a suitable response to a fault in one of the semiconductor devices. Reference [7] presents a fast and robust method for fault diagnosis in non-isolated dc–dc converters. FD is based on time and current criteria that observe the slope of the inductor current over time. By using a hybrid structure via coordinated operation of two FD subsystems it has been found that they work in parallel. No additional sensors, which increase the system cost and reduce reliability, are required for this detection method. The effects of input disturbances and closed-loop control are also considered. In the experimental setup, a field programmable gate array digital target is used for the implementation of the proposed method, in order to perform a very fast switch FD. Results show that, with the presented method, FD is robust and can be done in a few microseconds.

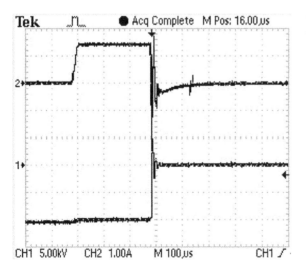

Figure 8.5 The output voltage and current of a high voltage power supply in an overcurrent fault.

The loads may have other faulty conditions that increase stress on the converter. A broken rotor bar is one of the commonly encountered induction motor faults that may cause serious motor damage to the motor if not detected early. Rotor windings in squirrel cage induction motors are usually manufactured from aluminum alloy. Larger motors generally have rotors and end-rings fabricated out of these, whereas motors with ratings less than a few hundred horsepower generally have die-cast aluminum alloy rotor cages. Replacement of the rotor core in larger motors is costly; therefore, by detecting broken rotor bars early, such secondary deterioration can be avoided. The rotor can be repaired at a fraction of the cost of rotor replacement, not to mention averting production revenue losses due to unplanned downtime. Figure 8.6 shows a squirrel cage rotor.

Some of the more common secondary effects of broken rotor bars are:

- Broken bars can cause sparking, a serious concern in hazardous areas.
- If one or more rotor bars are broken, the healthy bars are forced to carry additional current, leading to rotor core damage from persistent elevated temperatures in the vicinity of the broken bars and current passing through the core from broken to healthy bars.
- Broken bars cause torque and speed oscillations in the rotor, provoking premature wear of bearings and other driven components.
- As the rotor rotates at a high radial speed, broken rotor bars can lift out of the slot due to centrifugal force and strike against the stator winding, causing a catastrophic motor failure.

In addition, the motor short-circuit fault and rotor misalignment leads to overcurrent faults in the converters, as shown in Figure 8.7.

8.1.3 Faults in the Parasitic Ports

Some of the external faults of the converters act via non-deterministic ports of the converter. This means that they have electrical effects on the converter elements that are not applied through the input and output ports of the converter. The faults caused by radiation is one of them. Radiation may be either ionizing or non-ionizing. Non-ionizing radiation, which is common near the Earth's surface, consists of lower frequencies of the electromagnetic spectrum and lacks sufficient energy to remove electrons or create ions from molecules or atoms. Sources of non-ionizing radiation are

Figure 8.6 Rotor of an induction motor with top: end ring, right: bearing, left: rotor.

(a)

(b)

(c)

Figure 8.7 Problems in the motors that lead to external faults for the converters: (a) rotor misalignment, (b) stator overcurrent, (c) rotor overcurrent.

cellular phones, microwave devices, and power lines. Ionizing radiation involves high energy and includes alpha particles, beta particles, gamma rays, and galactic cosmic rays, as shown in Figure 8.8, which are some of the primary types of space radiation.

Gamma radiation is penetrating and can affect most electrical equipment. Simple equipment (like motors, switches, incandescent lights, wiring, and solenoids) is very radiation resistant and may never show any radiation effects, even after a very large radiation exposure. Diodes and computer chips (electronics) are much more sensitive to gamma radiation. Some electronic devices exhibit a recovery after being exposed to gamma radiation, after the radiation has stopped, but the recovery is hardly ever back to 100% functionality. However, if the electronics are exposed to gamma radiation while unpowered, the gamma radiation effects are less. Radiation has two main effects on electronics: single-event effects and total ionizing dose effects. Single-event effects are caused by highly energetic particles, which can cause bit flips in digital circuits or voltage spikes in analog circuits. The total ionizing dose has an accumulating effect, depending on the dose rate and the total time a circuit is exposed to radiation. Ionizing radiation creates electron-hole pairs in the

Figure 8.8 The trajectories of subatomic particles in the cloud chamber. *Source:* Cosmic Rays Laboratory, Sharif University of Technology.

electronics, changing the transistor parameters and eventually destroying them. It can also cause leakage currents between circuits.

8.1.4 Faults in the Non-electrical Paths

Environmental factors affect the power electronic systems via the non-electrical paths. The most important environmental failure factors are:

- Humidity
  Humidity is the amount of water vapor in the air. Water vapor is the gaseous state of water and is invisible. There are three main measurements of humidity: absolute, relative, and specific. Absolute humidity is the water content of air. Relative humidity, expressed as a percent, measures the current absolute humidity relative to the maximum for that temperature. Specific humidity is a ratio of the water vapor content of the mixture to the total air content on a mass. Many electronic devices have humidity specifications, for example, 5–95%. At the top end of the range, moisture may increase the conductivity of permeable insulators, leading to malfunction. Too low humidity may make materials brittle or cause electrostatic discharge. A particular danger to electronic items, regardless of the stated operating humidity range, is condensation. When an electronic item is moved from a cold place to a warm humid place, condensation may coat circuit boards and other insulators, leading to a short circuit inside the equipment. Figure 8.9 shows the increasing water vapor density in the air with increasing ambient temperature.
- Mechanical shock and vibration
  Vibration fatigue is one of the main mechanisms that will cause the failure of electronic devices. If the natural frequency of a printed circuit board (PCB) and its case do not obey the octave rule, the vibration of the PCB and the case will couple with each other, and stress applied to the PCB will be amplified, resulting in early failure. Electronic equipment can be subjected to many different forms of vibration over a wide range of frequencies and acceleration levels. All electronic equipment will be subjected to some type of vibration during its lifetime. If the vibration is not due to an active association with a machine or a moving vehicle, then it may be due to transportation of equipment from a manufacturer to a customer. Vibration is usually considered to be an undesirable condition because it can produce many different types of failure in electronic

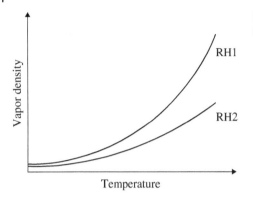

Figure 8.9 Relation between vapor density and relative humidity and temperature.

Figure 8.10 A high-power inverter mounted under a train.

equipment. Mechanical vibrations can have different sources. In vehicles such as automobiles, trucks, and trains, as shown in Figure 8.10, most of the vibration is due to the rough surfaces over which these vehicles travel. Portable electronic devices such as pagers, palm-top organizers, and compactly designed cell phones are also vulnerable to damage from mechanical shock and vibration. Over a period of time, the post-shock ringing vibration can fatigue boards and connectors, which creates hard-to-spot electrical problems. Therefore, the testing of electronic assemblies should include shock and vibration analysis. Vibration can cause damage to power electronic converters in several ways. First, it can accelerate failure by causing indentations. Second, it can loosen windings and cause mechanical damage to insulation by fracturing, flaking, or eroding of the material. Third, the excessive movement it causes can result in lead wires becoming brittle. As a result of these problems, whenever vibration is located in a power electronic converter, its source should be located quickly and corrected.

- Dust

 Dust/dirt can accumulate on a board surface, reducing dielectric strength, as shown in Figure 8.11. Dust may absorb the water in the air and can lead to short circuits on the boards.

- Corrosion

 Corrosion is a natural process that converts a refined metal into a more chemically stable form, such as oxide, hydroxide, carbonate, or sulfide. It is the gradual destruction of materials (usually a metal) by chemical and/or electrochemical reaction with their environment. The corrosion phenomenon affects the element leads and terminals of the converters. From general

Figure 8.11 The accumulated dust on a power supply board.

Figure 8.12 Corrosion on the iron packet of a converter cooling fan.

experiences of life, we probably have some idea about what corrosion is and have experienced the higher levels of corrosion that occur in the presence of moisture and the hostile gas species that are often present in the atmosphere. The phenomenon of corrosion involves reactions that lead to the creation of ionic species, by either loss or gain of electrons. Take the case of the rusting of iron, where metallic iron is converted into various oxides or hydroxides when exposed to moist air, as shown in Figure 8.12.

8.2 Resilience During External Faults

External faults cause interruption to the load support. There are two reasons for the non-resilient operation of the power electronic systems during external faults. One is the role of the external fault as a stressor of converter elements. Another reason is interaction between the external fault and the protection system.

8.2.1 External Faults as the Stressors

External faults cause an increase in the stress of the converter elements. In a normal operation, the power electronic system works based on its mission profile. Mission profiles such as environmental and operational conditions along with the system structure, including energy resources, grid and converter topologies, induce stress on different converters and thereby play a significant role in power electronic systems reliability [8]. An external fault such as a short circuit at the converter output increases the defined levels of the system mission profile. Thus, the temperature swing and maximum temperature of the converter elements increase. Temperature swing and maximum temperature are two of the critical stressors on the most failure-prone components of converters, i.e. capacitors and power semiconductors. Temperature-related stressors generate electrothermal stress on these components, ultimately triggering high potential failure mechanisms. Failure of any component may cause converter outage and system shutdown. Thermal cycling is a critical stressor on semiconductor elements as they are the main functional components of power electronic converters. Load variation of a converter causes temperature swing, which intensively affects the lifetime of semiconductor switches. In reference [9], an active thermal controlling method is proposed in order to enhance the overall system reliability. The proposed strategy applies power sharing among the converters by taking into account the lifetime of the power switches in the paralleled converters under different loading conditions. Hence, the lifetime of the converters is equally consumed in terms of load variations, and the overall system reliability is improved. Other external faults act in a similar way. Figures 8.13–8.16 show some samples of the faults in the elements of power electronic converters due to the external stressors. Figure 8.17 shows that the die area of insulated gate

Figure 8.13 Protecting devices against lightning at the input of the power converters. *Source:* Littlefuse Co. (with permission).

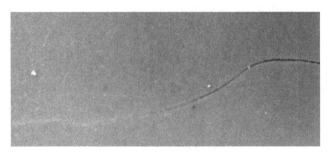

Figure 8.14 Crack propagation in a ceramic layer of a power module due to mechanical shock.

Figure 8.15 Crack on a power IC package due to the pressure of the absorbed humidity.

Figure 8.16 Corrosion signs on the various electronic elements.

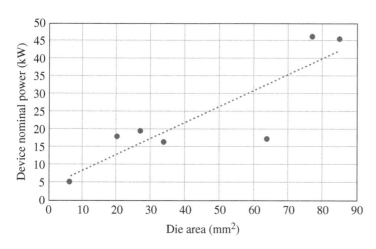

Figure 8.17 Increasing the die area of various IGBTs versus their nominal power.

bipolar transistors (IGBTs) increases with increasing nominal power. This means that they are more subjected to radiation failures.

8.2.2 Influence of the Protection System

From the load point of view, it is not important that either the converter is damaged or the converter protection system isolates the load/power source in faulty condition. In both cases, the load supplying is interrupted. It means that the protection system operates properly from the protection point of view. But it is not the suitable scenario of the protection from the resilience point of view. As an example, the protection system of a HVPS is described. The power supply drives a vacuum tube and uses a crowbar branch for protecting the tube against the vacuum arc as shown in Figure 8.18. The conventional protecting crowbar structures such as thyratron [10], ignitron [11], spark gap [12], and series-connected thyristors do not have the turn-off capability through their command signal after triggering. This means that after detecting the fault and turning on the crowbar, the crowbar remains in an on-state until all the energy-storage components of the HVPS are completely discharged and the current of the crowbar falls to zero. When the fault is cleared, all of the energy storage components must be fully charged in order to return the HVPS to normal operation. Since the HVPS restoration time after the fault can reach several tenths of a second depending on the specifications of the crowbar and HVPS, this restoration time is not an issue in the case of permanent faults because the repairing time of the faulty system is much longer than the restoration time of the HVPS. However, the HVPS is faced with some temporary faults in the tube. When the dc breakdown occurs inside the vacuum tube, an arc on the negative electrode of the tube is generated. This arc is normally known as a vacuum arc. The vacuum arc causes the material of the negative electrode to vaporize. These vaporized materials form a conducting path between the electrodes of the tube. The vacuum arc is a temporary fault [13, 14]. To remove this fault in the vacuum tube, the HVPS should be cut off for a very brief time interval. This time interval is named the arc clearing time interval and is usually on the order of a few

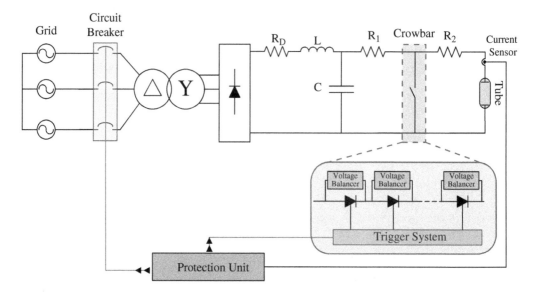

Figure 8.18 A conventional high-voltage power supply with the series-connected thyristor crowbar.

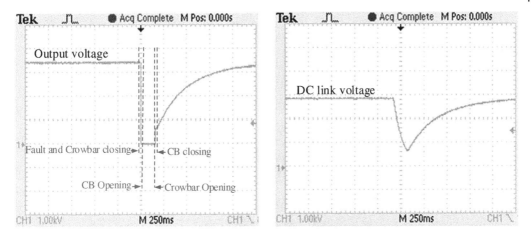

Figure 8.19 Waveforms of voltage variations under a transient fault with the thyristor crowbar.

microseconds [15]. It is much smaller than the restoration time of the HVPS. This disadvantage reduces power supply availability and resilience against the vacuum arc fault in the tube, as shown in Figure 8.19.

There is a very important point in the above-mentioned example: the external fault (vacuum arc) was temporary. All of the efforts for resilient operation of the HVPS are realized because the fault is cleared after a moment and the HVPS could then continue the load support. Therefore, this type of external fault is an important point in resilience.

8.3 Types of External Faults

According to the above-mentioned note, the type of external fault affects the operation of the power converter from a resilience point of view. The faults may be a single event, repetitive transient, or permanent.

8.3.1 Single Events

A single event is a change of state caused by one single phenomenon affecting a sensitive node in a system. In power electronic systems, there are various examples of single event faults:

- High energy radiation and cosmic ray
  High energy and cosmic rays are some of the destroyer single events, as shown in Figure 8.20. In reference [16], high-voltage diodes with active areas between 1 and 32 cm^2 were irradiated either with carbon ions having energies between 17 and 252 MeV or particles of 98 MeV or neutrons with energies up to 800 MeV. As the voltage across the devices was raised all of them failed eventually, even if there was a 1 MΩ resistor in series. With the high-energy carbon ions it could be shown that the failure can be triggered at locations that are hundreds of microns away from the pn-junction. The neutron experiment indicates that there may be a steep fall-off in failure rate at the lowest voltages. A phenomenological expression for predicting atmospheric neutron-induced failure rates in silicon carbide (SiC) power devices is presented in reference [17]. This expression relates the local electric field to a terrestrial neutron-induced failure rate and is

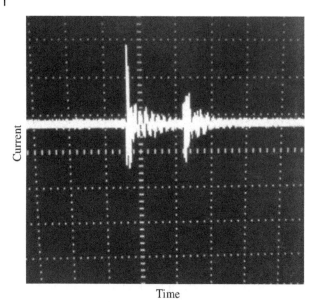

Time

Figure 8.20 Single-event property of the cosmic ray. *Source*: Cosmic Rays Laboratory, Sharif University of Technology.

derived using empirical data that show commonalities between failures of different SiC power devices under this type of radiation. This reference also presents a physics-based approach that provides a similar functional form for predicting these failure rates. The proposed closed-form expression is then used to demonstrate the use of this model in predicting failure rates in a hypothetical silicon carbide power device modeled using a technology computer-aided design tool. In reference [18], a method for the calculation of failure rates due to cosmic rays is presented. The method is based on the output of standard device simulation tools and is applied to IGBTs and free-wheeling diodes. Different models for the failure rate density are compared with respect to their consistency with experimental data. The method is applied both to the active area and to the edge termination of IGBTs. Furthermore, the influence of gate voltage on the failure rate of IGBTs is investigated. The method can be used to predict failure rates over a wide range of voltage classes and to detect weak points of device design. The applied voltage (Vcc) dependence of single effect burnout (SEB) characteristics of the failure-in-time (FIT) is generally estimated by the accelerated test, because it takes a long time to cause SEB under natural conditions. It is therefore meaningful to confirm the relationship between the field failure rate (FFR), SEB, FIT, and cold bias stability (CBS) characteristics. Through a physical analysis in reference [18], the destruction point is confirmed to be located around the electric field peak position during the SEB experiment using neutron irradiation. After both SEB curve fitting and sufficient numbers of analyses for the destruction points, the first major factor to characterize the SEB curve is confirmed to be the electric field strength.

- Vacuum arc

  The high-voltage power supplies are used in various applications such as medical instruments, material processing, water treatment, insulation testing, and high-power vacuum tubes [15]. In vacuum tube applications, the HVPS serves as the cathode power supply. Vacuum tubes tend to be the costliest element of their systems. However, they are prone to vacuum arc faults as a consequence of their special nature. According to the manufacturers' recommendations, the fault current of the vacuum tubes should be interrupted within a specified time, and, in the meantime, the injected fault energy into the tube (arc energy) must not exceed a specified limit. For the majority of vacuum tubes, the fault current interruption time limit and arc energy limit are in the regions of 10 μs and 10 J, respectively. It is worth noting that the energy that is stored in the bulk dc-link of the output stage of HVPS can be several kilojoules. Hence, the existence of a fast and reliable protection mechanism is essential. The arc fault in the tube can be modeled by the arc voltage (V_f) and arc resistance (R_f) according to Figure 8.21, depicting the arc voltage versus its current for different electrode materials. As can be seen, the value of R_f is negligible and the average voltage for most commonly used metals is about 20 V. When the crowbar current reaches its peak value – or, equivalently, the crowbar diverts the whole of the fault current through its path – the current flowing through the tube will not be equal to zero. This is because the voltage drop across the thyristors can lead to a current flowing through the tube. This current is named the residual current. The vacuum arc is a temporary fault, as shown in Figure 8.22.

- Electrostatic discharge
  Electrostatic discharge (ESD) is the release of static electricity when two objects come into contact. Many electronic devices are susceptible to low voltage ESD events. For example, hard-drive components are sensitive to only 10 V. For this reason, manufacturers of electronic devices

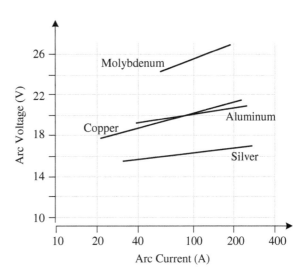

Figure 8.21 The dc volt-ampere characteristics of vacuum arcs.

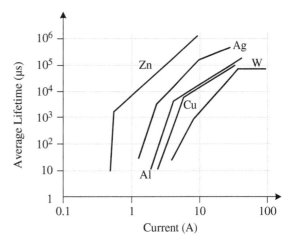

Figure 8.22 The average lifetime of the vacuum arc fault for different electrode materials.

incorporate measures to prevent ESD events throughout the manufacturing, testing, shipping, and handling processes. In reference [19], several aspects of ESD are described from the point of view of the test, design, product, and reliability engineering. A review of the ESD phenomena along with the test methods, the appropriate on-chip protection techniques, and the impact of process technology advances from CMOS to BiCMOS on the ESD sensitivity of IC protection circuits are presented. The status of understanding in the field of ESD failure physics and the current approaches for modeling are discussed.

8.3.2 Transient Repetitive Faults

Some of the temporary faults are repetitive. They are repeated in every start-up of the system or other change in the system operating point.

- Transformer inrush current
  In a transformer, when power is applied, the residual field will cause a high inrush current until the effect of the permanent magnetism is reduced. Transformer inrush currents are high-magnitude, harmonic-rich currents generated when transformer cores are driven into saturation during energization, as shown in Figure 8.23. These currents have undesirable effects, including potential damage or loss-of-life to the transformer, protective relay mis-operation, and reduced power quality on the system. Controlled transformer switching can potentially eliminate these transients if residual core and core flux transients are taken into account in the closing algorithm. Reference [20] explores the theoretical considerations of core flux transients, and based on these studies algorithms were developed that allow controlled energization of most transformers without inrush current. The wavelet transform is a powerful tool in the analysis of the power transformer transient phenomena because of its ability to extract information from the transient signals simultaneously in both the time and frequency domains. Reference [21] presents a novel technique for accurate discrimination between an internal fault and a magnetizing inrush current in the power transformer by combining wavelet transforms with neural networks. The wavelet transform is firstly applied to decompose the differential current signals of the power transformer into a series of detailed wavelet components. The spectral energies of the wavelet components are calculated and then employed to train a neural network to discriminate an internal fault from the magnetizing inrush current. The simulated results presented

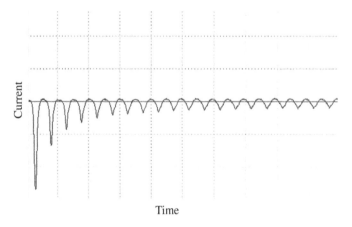

Figure 8.23 The transformer inrush current versus the time.

clearly show that the proposed technique can accurately discriminate between an internal fault and a magnetizing inrush current in power transformer protection.

• Transients in the converter control system

Some of the temporary faults are due to the problems in the control system of the converters, as shown in Figure 8.24. The designers usually consider these effects but they are a potential source of external faults. Both passive and active control methods are used to prevent such problems. Pulse width modulation (PWM) voltage source converters are becoming a popular interface to the power grid for many applications. Hence, issues related to the reduction of PWM harmonics injection in the power grid are becoming more relevant. The use of high-order filters like *LCL* filters is a standard solution to provide the proper attenuation of a PWM carrier and sideband voltage harmonics. However, those grid filters introduce potentially unstable dynamics that should be properly damped either passively or actively. The second solution suffers from control and system complexity (a high number of sensors and a high-order controller), even if it is more attractive due to the absence of losses in the damping resistors and due to its flexibility. An interesting and straightforward active damping solution consists in plugging in, in a cascade to the main controller, a filter that should damp the unstable dynamics. No more sensors are needed, but there are open issues such as preserving the bandwidth, robustness, and limited complexity. Reference [22] provides a systematic approach to the design of filter-based active damping methods. The tuning procedures, performance, robustness, and limitations of the different solutions are discussed using theoretical analysis, selected simulation, and experimental results. Reference [23] presents a detailed description of the finite control set model predictive control (FCS-MPC) applied to power converters. Several key aspects related to this methodology are, in depth, presented and compared with traditional power converter control techniques, such as linear controllers with PWM-based methods. The basic concepts, operating principles, control diagrams, and results are used to provide a comparison between the different control strategies. The analysis is performed on a traditional three-phase voltage source inverter, used as a simple and comprehensive reference frame. However, additional topologies and power systems are addressed to highlight differences, potentialities, and challenges of FCS-MPC. The possibility to address different or additional

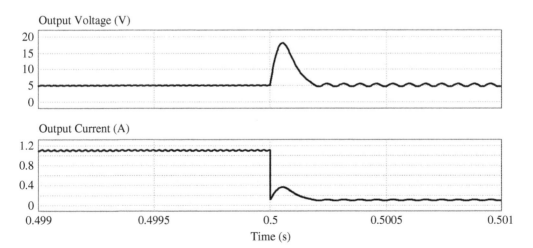

Figure 8.24 Overvoltage fault at the output of a buck converter due to the load change.

control objectives easily in a single cost function enables a simple, flexible, and improved performance controller for power-conversion systems.

8.3.3 Permanent Faults

A permanent short circuit is an example of permanent faults. In reference [24], cable faults in voltage source converter-based dc networks are analyzed in detail with the identification and definition of the most serious stages of the fault that need to be avoided. A fault location method is proposed because this is a prerequisite for an effective design of a fault protection scheme. It is demonstrated that it is relatively easy to evaluate the distance to a short-circuit fault using a voltage reference comparison. For the more difficult challenge of locating ground faults, a method of estimating both the ground resistance and the distance to the fault is proposed by analyzing the initial stage of the fault transient. Analysis of the proposed method is provided and is based on simulation results, with a range of fault resistances, distances, and operational conditions considered.

8.4 Fault Clearance

The simplest configuration of a power electronic system is formed by a source, a converter, and a load, as shown in Figure 8.25. In this configuration, the power electronic converter supplies a single load via a single source. If a permanent fault occurs in the load, resilient operation of the converter is meaningless because the load is permanently damaged. However, if the fault in the load is temporary, resilient operation of the converter is important because the converter can continue to supply the load after the fault clearance.

Another configuration of the power electronic system is a power source that supplies several loads via one power electronic converter, as shown in Figure 8.26. In this case, interesting situations occur during load faults. In this case, the converter should be resilient, even against permanent faults in the load, because other loads should be supplied. Therefore, isolating mechanisms of the faulty loads is mandatory for resilient operation. Other types of converter configurations were discussed in Chapter 3.

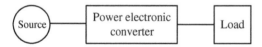

Figure 8.25 A power electronic system with a single source–single converter–single load configuration.

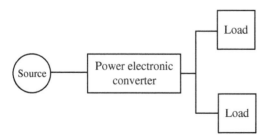

Figure 8.26 A power electronic system with a single source–single converter–multiload configuration.

8.5   Summary and Conclusions

In this chapter, the external faults that lead to unavailability of the power electronic converters were described. The summary of the study is as follows:

1) Two main types of external faults are the faults that occur outside the input and output ports of the power electronic converters. Other external faults are those that affect the converter from other electric ports, such as radiation effects, or on electrical paths, such as environmental effects.
2) External faults cause the increase in stress of the converter elements. In normal operation, the power electronic system works based on its mission profile. Mission profiles such as environmental and operational conditions along with the system structure, including energy resources, grid and converter topologies, induce stress on different converters and thereby play a significant role in power electronic system reliability. An external fault such as a short circuit at the converter output increases the defined levels of the system mission profile. Thus, the temperature swing and other stressors of the converter elements increase.
3) From the load point of view, it is not important that either the converter is damaged or the converter protection system isolates the load/power source in a faulty condition. In both cases, the load supply is interrupted. This means that the protection system operates properly from a protection point of view. However, it is not a suitable scenario for the protection from a resilience point of view.
4) Any type of external faults, either temporary or permanent, affect the operation of the power converter from a resilience point of view. The faults may be a single event, repetitive transient, or permanent.
5) Fault clearance plays an important role in the resilient operation of the converters. Resilience against temporary faults and isolating the faulty loads are two important factors for resilience.

References

1 Grcev, L.D. (1996). Computer analysis of transient voltages in large grounding systems. *IEEE Transactions on Power Delivery* 11 (2): 815–823.
2 Guerrero, J.M., Loh, P.C., Lee, T., and Chandorkar, M. (2013). Advanced control architectures for intelligent microgrids—Part II: Power quality, energy storage, and AC/DC microgrids. *IEEE Transactions on Industrial Electronics* 60 (4): 1263–1270.
3 McGranaghan, M.F., Mueller, D.R., and Samotyj, M.J. (1993). Voltage sags in industrial systems. *IEEE Transactions on Industry Applications* 29 (2): 397–403.
4 Miret, J., Castilla, M., Camacho, A. et al. (2012). Control scheme for photovoltaic three-phase inverters to minimize peak currents during unbalanced grid-voltage sags. *IEEE Transactions on Power Electronics* 27 (10): 4262–4271.
5 Patel, H. and Agarwal, V. (2008). Maximum power point tracking scheme for PV systems operating under partially shaded conditions. *IEEE Transactions on Industrial Electronics* 55 (4): 1689–1698.
6 Choi, U., Blaabjerg, F., and Lee, K. (2015). Study and handling methods of power IGBT module failures in power electronic converter systems. *IEEE Transactions on Power Electronics* 30 (5): 2517–2533.
7 Shahbazi, M., Jamshidpour, E., Poure, P. et al. (2013). Open- and short-circuit switch fault diagnosis for nonisolated dc–dc converters using field programmable gate Array. *IEEE Transactions on Industrial Electronics* 60 (9): 4136–4146.

8 Peyghami, S., Wang, H., Davari, P., and Blaabjerg, F. (2019). Mission-profile-based system-level reliability analysis in dc microgrids. *IEEE Transactions on Industry Applications* 55 (5): 5055–5067.

9 Peyghami, S., Davari, P., and Blaabjerg, F. (2018). System-level lifetime-oriented power sharing control of paralleled dc/dc converters. In: *2018 IEEE Applied Power Electronics Conference and Exposition (APEC)*, San Antonio, TX, USA, 1890–1895.

10 Brunner, O., Ravida, G., and Valuch, D. (2012). Performance of the crowbar of the LHC high power RF system. *Conference Proceedings* C1205201: 3641–3643.

11 Bettini, P. and Lorenzi, A.D. (1996). Misfiring protection and monitoring of the RFX toroidal circuit ignitron crowbar. In: *Proceedings of 1996 International Power Modulator Symposium*, 137–140.

12 Clark, G. and Thio, Y. (1984). Design and operation of a self activating crowbar switch. *IEEE Transactions on Magnetics* 20 (2): 364–365.

13 Smeets, R.P.P. (1987). Low-current behavior and current chopping of vacuum arcs. PhD thesis, Technische Universiteit Eindhoven.

14 Cobine, J.D. and Farrall, G.A. (1963). Recovery characteristics of vacuum arcs. *Transactions of the American Institute of Electrical Engineers. Part I: Communication and Electronics* 82 (2): 246–253.

15 Gilmour, A.S. Jr. (2011). *Klystrons, Traveling Wave Tubes, Magnetrons, Crossed-field Amplifiers, and Gyrotrons*, 1e. Artech House Publishers.

16 Voss, P., Maier, K., Męczyński, W. et al. (1997). Irradiation experiments with high-voltage power devices as a possible means to predict failure rates due to cosmic rays. In: *Proceedings of 9th International Symposium on Power Semiconductor Devices and IC's*, Weimar, Germany, 169–172.

17 Akturk, A., McGarrity, J.M., Goldsman, N. et al. (2019). Predicting cosmic ray-induced failures in silicon carbide power devices. *IEEE Transactions on Nuclear Science* 66 (7): 1828–1832.

18 Pfirsch, F. and Soelkner, G. (2010). Simulation of cosmic ray failures rates using semiempirical models. In: *2010 22nd International Symposium on Power Semiconductor Devices & IC's (ISPSD)*, Hiroshima, Japan, 125–128.

19 Duvvury, C. and Amerasekera, A. (1993). ESD: A pervasive reliability concern for IC technologies. *Proceedings of the IEEE* 81 (5): 690–702.

20 Brunke, J.H. and Frohlich, K.J. (2001). Elimination of transformer inrush currents by controlled switching. I. Theoretical considerations. *IEEE Transactions on Power Delivery* 16 (2): 276–280.

21 Mao, P.L. and Aggarwal, R.K. (2001). A novel approach to the classification of the transient phenomena in power transformers using combined wavelet transform and neural network. *IEEE Transactions on Power Delivery* 16 (4): 654–660.

22 Dannehl, J., Liserre, M., and Fuchs, F.W. (2011). Filter-based active damping of voltage source converters with LCL filter. *IEEE Transactions on Industrial Electronics* 58 (8): 3623–3633.

23 Kouro, S., Cortes, P., Vargas, R. et al. (2009). Model predictive control—A simple and powerful method to control power converters. *IEEE Transactions on Industrial Electronics* 56 (6): 1826–1838.

24 Yang, J., Fletcher, J.E., and O'Reilly, J. (2012). Short-circuit and ground fault analyses and location in VSC-based DC network cables. *IEEE Transactions on Industrial Electronics* 59 (10): 3827–3837.

9

Malfunctioning
Influence of Noise and Disturbance

9.1 Electromagnetic Pollution

The role of electrical energy in our everyday lives has grown by leaps and bounds during the twentieth century. In the early decades of the 1900s, experts in most fields took a direct approach to achieving their goals, without much regard to or understanding of the negative side effects of their technological innovations. Thus, the deleterious consequences of rapid development came into the limelight only later. In recent decades, some of those consequences have reached international proportions, making it necessary to study them, with the ultimate aim of reducing or eliminating the disagreeable effects.

One such problem is environmental electromagnetic pollution. Unacceptably high levels of electromagnetic disturbances can prevent electrical and electronic devices, apparatus, and systems from operating properly in a common electromagnetic environment. A device is considered to be electromagnetically compatible only if its impacts are tolerated by all other devices operating in the same environment. To ensure that this compatibility exists, a relatively new engineering discipline, electromagnetic compatibility (EMC), has evolved. EMC is the field of electrical engineering that studies, analyzes, and solves electromagnetic interaction problems.

Achieving EMC requires us to view disturbances from two standpoints: electromagnetic emissions and electromagnetic susceptibility. Because electromagnetic noise propagates by conduction and radiation, the scope of problems outlined above continues to broaden, as shown in Figure 9.1 [1].

In the first half of the twentieth century, electromagnetic disturbance sources, for the most part, were limited to motor-driven machinery and switching apparatus. However, with the rapid spread of power semiconductors and power electronic systems, interference levels on power mains have increased significantly in intensity and frequency of occurrence. At the same time, the world is becoming more densely populated with devices that are increasingly sensitive to electromagnetic disturbances. In industrial spheres, electronic control systems, data processing equipment, and other sensitive devices play an increasingly important role. The developments have produced quite a serious situation.

9.2 Description of Electromagnetic Disturbances

The electromagnetic emissions of electrical equipment are not easy to precisely specify and classify, but we can attempt to do so if we know some of the characteristics of the offending signals. To a degree, classifications are arbitrary, but they can help us to understand the electromagnetic

Resilient Power Electronic Systems, First Edition. Shahriyar Kaboli, Saeed Peyghami, and Frede Blaabjerg.
© 2022 John Wiley & Sons Ltd. Published 2022 by John Wiley & Sons Ltd.
Companion website: www.wiley.com/go/kaboli/resilientpower

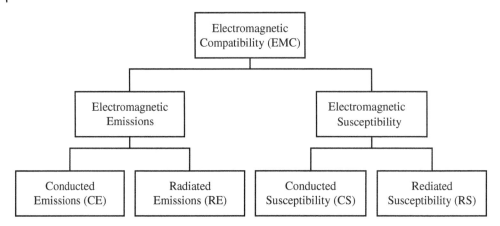

Figure 9.1 Areas of electromagnetic compatibility.

emission of power electronic equipment. Generally, the frequency content, character, and transmission mode provide the basis for classifying man-made electromagnetic disturbance, but it is not unusual to categorize them in terms of energy content, waveform, and other factors.

9.2.1 Classifying Disturbances by Frequency Content

Figure 9.2 illustrates the assorted high-frequency (HF) disturbances by frequency content. Electromagnetic disturbances with an upper limit of 0–1250 Hz or 2000 Hz increase losses on the mains and distort the voltage waveform. Therefore, examination and elimination of this type of electromagnetic noise is a sphere independent from the issue of HF disturbances. The frequency range of 1.25–150 kHz is not often examined by EMC engineers, although electromagnetic disturbances in this frequency range are coming to light more and more. The acceptable level of harmonics in this frequency range is specified in many national standards.

The range of radio frequency disturbances starts at 150 kHz. This range is generally divided into the band of 0.15–30 MHz and 30–300 MHz. The reason for this division is that different transmission modes and measurement methods are applied to HF disturbances.

However, one cannot sufficiently classify HF disturbances in terms of the frequency content only; the character must also be examined. A major distinction is between narrowband

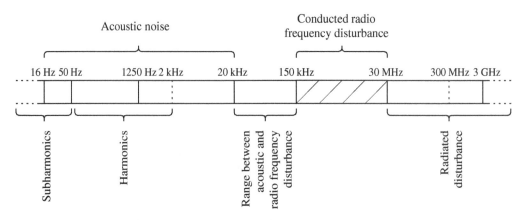

Figure 9.2 Frequency range of electromagnetic disturbances.

and broadband signals, and the difference can be a determining factor in the appropriate application of emission level standards and in solving electromagnetic interference (EMI) problems.

9.2.2 Classifying Disturbances by Character

Electromagnetic disturbances will often affect the mains voltage. These disturbances can be of long or short duration. Changes of long duration usually are not included in the domain of EMC, as they mainly cause alterations in the rms value of the mains voltage. The duration of short changes runs from a few seconds down to less than a microsecond. Short-duration electromagnetic disturbances appear as distortions on the main voltage.

Electromagnetic disturbances of short duration can be divided into three groups:

- Noise and ripple, which is more or less a permanent alteration of the voltage curve. Figure 9.3 shows some types of noise and ripple.

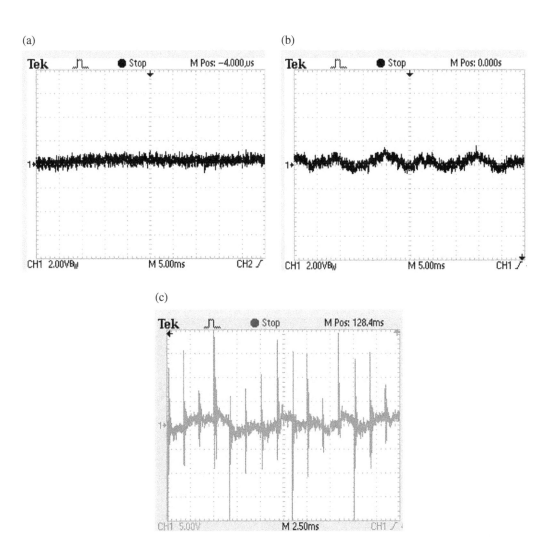

Figure 9.3   Some types of the noise and ripple: (a) the ground noise, (b) voltage ripple, (c) high voltage transients.

- Impulse and single event disturbance, which are positive and negative peaks superimposed on the mains voltage. In many cases, they are not periodic. Figure 9.4 shows some types of impulse and single events.
- Transients and other transitional processes. Figure 9.5 shows some samples of the transients.

9.2.3 Classifying Disturbances by Transmission Mode

Electromagnetic disturbances travel by conduction on wiring and by radiation in space. Electromagnetic disturbances below approximately 10 MHz spread primarily by conduction, while at higher frequencies, radiation becomes dominant. Also, disturbances may get into circuits that are closely spaced via inductive and capacitive couplings. The different characteristics pertaining to the transmission mode are explained by the fact that in the frequency range of 0.15–30 MHz, usually only the conducted EMI must be measured and suppressed. In this frequency range, measurement of radiated EMI is only required by certain standards and recommendations. Up the

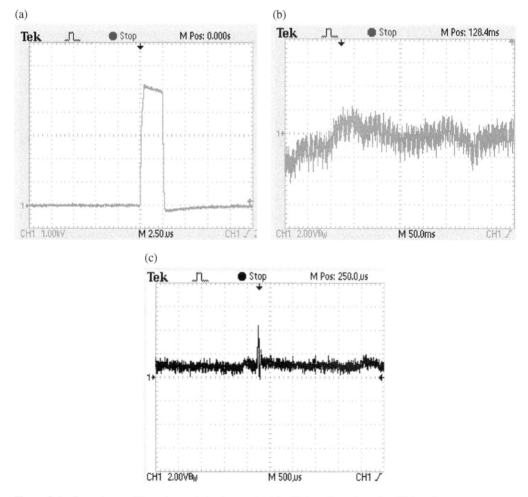

Figure 9.4 Some types of impulse and single events: (a) a high-voltage impulse, (b) low-frequency fluctuations, (c) single event disturbance.

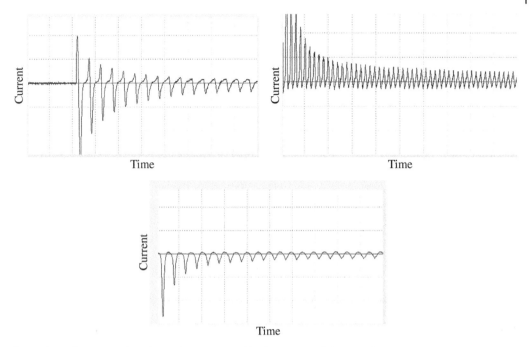

Figure 9.5 Three samples of transients occurred in the startup of the transformers.

frequency content of the semiconductor-generated EMI reaches some specified number of megahertz, only the conducted EMI of these equipment generally should be measured.

9.3 EMI in Power Electronic Equipment

A dramatic increase in the number of line-powered electronic equipment (computers and other office equipment, electronic ballasts, variable-speed drives, and consumer electronics, e.g. color televisions) has taken place. These items of equipment draw distorted, and often fluctuating, line current; they also generate high-frequency conducted and radiated noise due to the sharp edges of the waveforms characteristic of the switching power processors employed in them. As a result of the finite grid impedance, the distorted line current increases the distribution losses and causes voltage distortion; also, the fluctuation leads to visible flicker of the emitted light of lamps. The conducted and radiated high-frequency noise interferes with radio and TV reception, communication via cellular telephones, and data transmission. The result is a gradually deteriorating electromagnetic environment. Reference [1] presents the mechanisms that generate low-frequency and high-frequency electrical noise, lists the problems caused by the various noise types, provides an overview of the standards that establish noise limits, discusses and evaluates the various mitigation techniques, and raises concerns regarding the impact of the circuitry that has to be added to the equipment to meet the harmonic limits and the pitfalls and deficiencies of the line-harmonics regulation standards.

9.3.1 Application of Noise and Ripple in Power Electronic Equipment

The noise and associated phenomena such as ripple and fluctuations are used in power electronics for controlling the converters or fault diagnosis. Switching regulators with ripple-based control are conceptually simple, have fast transient responses to both line and load perturbations, and some versions operate with a switching frequency that is proportional to the load current under the discontinuous conduction mode. These characteristics make the ripple regulators especially well-suited for power management applications in computers and portable electronic devices. Reference [2] presents an overview of the various ripple-based control techniques, discusses their merits and limitations, and introduces techniques for reducing the noise sensitivity and the sensitivity to capacitor parameters, improving the frequency stability and the dc regulation, and avoiding fast-scale instability. The main advantages of noise measurements are that the tests are less destructive, faster, and more sensitive than DC measurements after accelerated life tests. The following topics are addressed: (i) the kind of noise spectra in view of reliability diagnostics, such as thermal noise, shot noise, typical poor-device indicators like burst noise and generation-recombination noise, (ii) why conduction noise is a quality indicator; (iii) the quality of electrical contacts and vias; (iv) electromigration damage; (v) the reliability in diode-type devices like solar cells, laser diodes, and bipolar transistors; and (vi) the series resistance in modern short channel MESFET, MODFET, and MOST devices [3].

9.3.2 EMI Issues Caused by Power Electronic Equipment

Most electrical equipment that draws power from the mains generates HF signals that could negatively affect the operation of other connected equipment. To minimize the effects the high-level disturbances, there has been considerable interest in examining and understanding how industrial, commercial, and residential products generate noise. In order to find the most effective and economic solutions for EMI control in this rugged environment, we must understand how semiconductor devices and circuits generate and are affected by electromagnetic noise.

As an instance, a rectifier can be considered. The simplest rectifier (diode type) can be regarded as a switching element. This element acts as a short circuit for forward bias and as an open circuit for reverse bias. In practice, rectifier switching from one state to the other does not occur instantaneously. The switch-on operation of a power rectifier is shown in Figure 9.6. The voltage on the rectifier changes from the off-state to the on-state in time t_0. The on-state current increases quickly

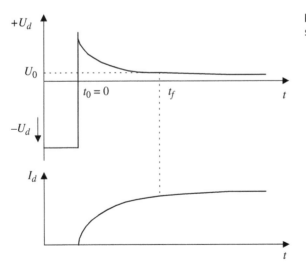

Figure 9.6 The rectifier switch-on waveforms.

but, in the first moment, a relatively high on-state voltage appears on the rectifier. This voltage falls back to its nominal value in the time interval t_f. This time is needed for charge carriers to get into the depletion (p–n) region. The voltage spike is actually a broadband emission. The rectifier produces even a higher emission during its switch-off operation, by the same token [1].

Other power semiconductor devices generate HF disturbance just as diodes. For example, silicon-controlled rectifiers (SCR) and power transistors, like diode rectifiers, generate HF distur-bances during both switch-on and switch-off operations. However, in contrast to rectifiers, the high-frequency noise levels are much higher at switch-on than at switch-off. Another type of disturbance caused by power electronic equipment is the distortion in their voltages and currents. For example, the main disadvantages of conventional ac/dc converters are injection of low-frequency harmonics into the ac power supply and also their poor power factor. The most com-mon technique to eliminate low-frequency harmonics and to improve the power factor is the pulse width modulation (PWM) current mode control. In this method, the converter input cur-rent is compared to a sinusoidal reference signal synchronized with converter input voltage and the error signal is used to command the converter switches after passing through a PI controller and PWM modulator, as shown in Figure 9.7. In reference [4], three-phase power factor correc-tion (PFC) rectifier topologies with sinusoidal input currents and a controlled output voltage are derived from known single-phase PFC rectifier systems and/or passive three-phase diode rectifi-ers. The systems are classified into hybrid and fully active PWM boost-type or buck-type rectifiers, and their functionality and basic control concepts are briefly described. This facilitates the under-standing of the operating principle of three-phase PFC rectifiers starting from single-phase sys-tems, and organizes and completes the knowledge base with a new hybrid three-phase buck-type PFC rectifier topology denominated as a Swiss rectifier. Finally, core topics of future research on three-phase PFC rectifier systems are discussed, such as the analysis of novel hybrid buck-type PFC rectifier topologies, the direct input current control of buck-type systems, and the multi-objective optimization of PFC rectifier systems. The second part of this reference is dedicated to a comparative evaluation of four rectifier systems offering a high potential for industrial

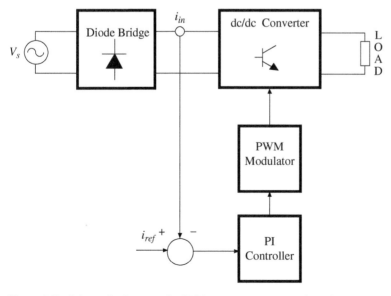

Figure 9.7 Schematic diagram of a PWM-current mode control ac/dc converter.

applications based on simple and demonstrative performance metrics concerning the semiconductor stresses, the loading and volume of the main passive components, the differential mode and the common mode EMI noise levels, and ultimately the achievable converter efficiency and power density.

Application of this modulation technique shifts harmonics to higher frequencies and improves the power factor, but leads to another problem, which is generation of high-frequency harmonics at the PWM switching frequency and its multiples. These harmonics are generated in the RF range and constitute the main cause of high EMI noise levels in these types of converters [5]. A PWM-controlled active filter generates harmonics at the switching frequency and its multiples in the RF range, and the concentrated power spectrum may produce EMI problems. In reference [5], the random PWM (RPWM) technique is applied in order to spread the noise spectrum over a wide range, thus considerably reducing the amplitudes of these harmonics and the consequent EMI problems. To study the operation of active filters, the case of an ac/dc converter along with a power-factor corrector is considered as a nonlinear load and a series active filter, respectively. A line impedance stabilization network is used to study the RF noise emanating from the converter. A noise model to study the EMI emission is presented and used in reference [5]. A theoretical analysis of the RF noise power spectrum is carried out in order to demonstrate the advantages of the RPWM technique over conventional PWM. Experimental results confirm the validity of the theoretical calculations and simulation results, and demonstrate the effectiveness of applying the RPWM technique in reducing the RF noise level.

9.3.3 Effect of EMI on the Resilient Operation

The importance of EMI in resilient behavior of power electronic systems is very high. EMI causes unwanted shutdowns of the systems and malfunctioning of the control systems. These effects lead to non-resilient operation of the system. Figure 9.8 shows the effect of EMI on the measured voltage signal of a power electronic converter. Figure 9.8a is the ac part of the signal without EMI and Figure 9.8b shows the same point with EMI effect. The shown spikes and induced ripples lead to malfunctioning of the protection and control system and reduces the availability of the converter.

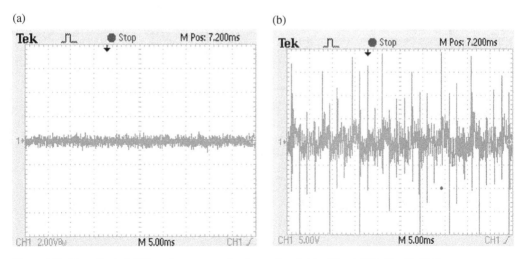

Figure 9.8 The effect of EMI on the measured voltage signal: (a) without EMI, (b) with EMI.

• **Case study**

The high-voltage dc power supplies are essential components in various industries and medical applications such as testing equipment, water purification, and vacuum tubes. Among these applications, some applications are inherently susceptible to the short-circuit fault (SCF) due to the possibility of an arc occurrence in the load. When the SCF occurs in the load, the output current increases drastically. Hence in the high-voltage dc power supply (HVPS), the correct operation of the protection and the fault detection (FD) system is essential to prevent serious damage both in the converter and the load. The structure of the HVPS with the protection system and the FD system is depicted in Figure 9.9.

Therefore, SCF detection is mandatory in an HVPS to prevent fatal damage. The majority of converters employ a single sensor to detect the SCF. This attribute increases the interference vulnerability of the FD system in the presence of noise. Therefore, miss detections and false alarms are possible to occur. Miss detections and false alarms are harmful catastrophes in most applications. A commonly used method to suppress the noise impacts is using a low-bandwidth low-pass filter. However, the use of the low-bandwidth low-pass filter reduces the speed of FD due to filter delay. The conventional FD system consists of a load current feedback that is measured by the output current sensor and the protection system is made up of a circuit breaker in the input and a crowbar. When the SCF occurs in the HVPS, the output current increases sharply. Thus, the FD system can diagnose the SCF. Subsequently, the protection system is activated and the crowbar is triggered to divert the stored energy in the output capacitor from the faulty load and the circuit breaker interrupts the input current flow.

Many proposals have addressed the FD in different topologies of power electronics converters [6–11]. In references [6] and [7], the SCF detection is based on monitoring the gate-emitter voltage of IGBT. Due to the sharp increase of the output current in the SCF condition, the current changing rate is an appropriate parameter to employ in the FD system. Hence in reference [8], the voltage at a parasitic inductor is used for SCF detection. Several papers have addressed the observer-based [9] and the model-based [10] FD methods. Another diagnosis method based on the duty cycle and the inductor current sloop is presented in reference [11]. Most FD methods use a single FD system. The structure of a conventional FD system is depicted in Figure 9.10, where a single variable of the converter, x, is used to process the FD.

Generally, the FD system may declare two different incorrect states about the converter situation; firstly, miss detections during the faulty conditions and, secondly, false alarms during normal

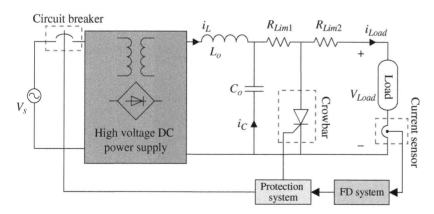

Figure 9.9 Structure of the HVPS with the protection and the FD system.

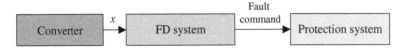

Figure 9.10 Conventional FD system.

operation of the HVPS [12]. Miss detections may stem from sensors and FD system component failures [13, 14]. In the case of the faulty condition, miss detections can be excessively hazardous due to the possibility of harmful damage in both the converter and the load. False alarms can be the result of noise impacts in the FD system [15]. Disturbing the correct function of the converter without a rational reason is one of the major consequences of false alarms. Moreover, frequent occurrence of false alarms reduces the availability of the converter. In noise-free environments, all methods can detect the SCF correctly and do not report any false alarm. However, in the presence of noise, the FD system is exposed to interference due to the use of a single sensor. As a result, false alarms are repeatedly reported and the availability of the converter decreases considerably.

The structure of the conventional FD system is illustrated in Figure 9.11. This FD system consists of a current sensor that measures the load current, a buffer for isolating the sample signal, a low-pass filter for suppressing the noise impacts, a comparator for detecting the current increment, and a latch. In normal conditions, the output of the comparator is "-1" and the protection system is inactive because the current is less than *Threshold*. When the SCF occurs, the load current increases severely and exceeds *Threshold*. Thus, the output of the comparator becomes "+1" and the SCF is detected and the protection system is activated. The conventional FD system operates accurately in noise-free environments. By contrast, the FD system may encounter interference from noise in noisy environments, as shown in Figure 9.12. As shown in Figure 9.11, in order to prevent false alarms problems, the low-pass filter is used in the FD system to suppress noise impacts. As the bandwidth of the low-pass filter decreases, the probability of the noise impacts

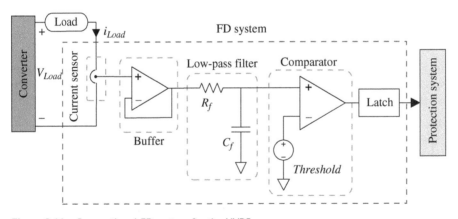

Figure 9.11 Conventional FD system for the HVPS.

Figure 9.12 Effect of noise on the protection command of the converter.

and false alarms reduces, while the speed of the FD system decreases due to the low-pass filter delay. Quick FD is extremely vital in high-voltage applications. Hence a fast novel FD scheme is mandatory in the presence of noise.

9.4 Conducted EMI Measurement

Electromagnetic emissions produced by power electronic equipment are usually broadband and coherent, in the range from the operating frequency up to a number of megahertz. Conducted EMI usually should be measured within this frequency range. Electromagnetic disturbances can appear in the form of *common-mode* (often called *asymmetrical*) and *differential-mode* (*symmetrical*) voltage and current components. The definition of common-mode and differential-mode components is illustrated in Figure 9.13a [1]. The common-mode and differential-mode components are defined by voltages and currents, measured on the mains terminals, as follows:

$$U_d = U_1 - U_2, I_d = \frac{I_1 - I_2}{2} \tag{9.1}$$

$$U_c = \frac{U_1 + U_2}{2}, I_c = I_1 - I_2 \tag{9.2}$$

(a)

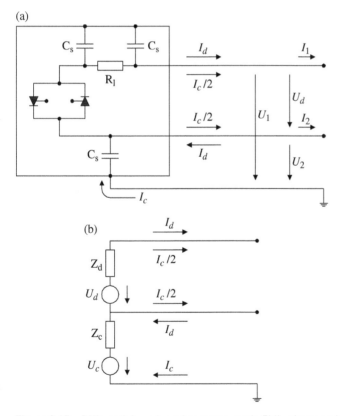

(b)

Figure 9.13 Differential-mode and common-mode EMI voltage and current components: (a) typical EMI source and (b) the HF substitution circuit of the EMI source.

where U_d is the differential-mode voltage component, I_d is the differential-mode current component, U_c is the common-mode voltage component, and I_c is the common-mode current component. Furthermore, I_1, I_2, U_1, and U_2 are currents and voltages defined in Figure 9.13.

In Figure 9.13, the HF equivalent circuit of an EMI source is depicted. The differential-mode component flows in the supply wires (including the neutral wire). The differential-mode voltage component can be measured between phase conductors as well. The common-mode current component flows from the phase and neutral conductor toward the Earth. The circuit for the common-mode component is closed by the impedance Z_c, representing the stray capacitances between the earthed parts and the circuit.

To carry out compliance measurements, standard regulations require a standard network, namely, "Line Impedance Stabilization Network" or LISN. LISN is an industrial network that must have the following characteristics in order to satisfy measurement conditions:

– Transferring power from source to converter,
– Providing a low impedance path for noise measurement,
– Reduction of the effect of source impedance in noise measurement.

The analysis of an industrial LISN would be too complicated. Therefore, usually a simple model with sufficient accuracy is used. A commonly used LISN is shown in Figure 9.14.

When only the differential mode noise is of concern, the common ground of the LISN, shown in Figure 9.14, is no longer considered and only a series circuit between phases exists. With this structure and assuming that the power supply is a short circuit for a dynamic analysis [2], we can represent the equivalent noise circuit as in Figure 9.15, in which I_{ex} denotes the converter input current [2].

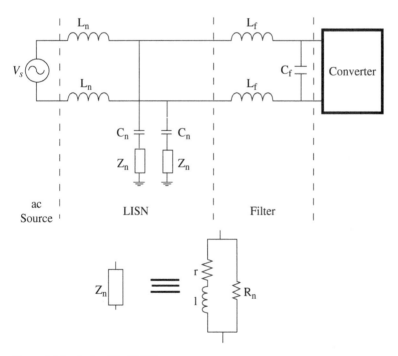

Figure 9.14 A simplified LISN.

Figure 9.15 Noise equivalent circuit.

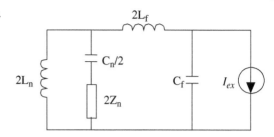

There are two ways to calculate noise, the direct method and the indirect method. In the direct method the LISN is added to the circuit under study and the frequency spectrum of the LISN output voltage, which is the voltage across Z_n, is calculated. In the indirect method, the system function is extracted from the noise equivalent circuit, as the one shown in Figure 9.15. According to the noise equivalent linear circuit, the system function $H\,(j\omega)$ can be calculated as:

$$H(j\omega) = \frac{V_{Z_n}(j\omega)}{I_{ex}(j\omega)} \tag{9.3}$$

where $V_{Zn}(j\omega)$ and $I_{ex}(j\omega)$ are the LISN output voltage and converter input current, respectively. As a result, the system function $H(s)$ is calculated as a function of the noise equivalent circuit parameters as:

$$H(s) = \frac{n_0 + n_1 s + n_2 s^2 + n_3 s^3 + n_4 s^4}{d_0 + d_1 s + d_2 s^2 + d_3 s^3 + d_4 s^4 + d_5 s^5} \tag{9.4}$$

where the coefficients n_i and d_j correspond to the parameters of the noise equivalent circuit.

In this method, $I_{ex}(t)$ can be computed by computer simulation of the converter. Then by applying FFT to $I_{ex}(t)$, $I_{ex}(j\omega)$ can also be calculated. Therefore, $V_{Zn}(j\omega)$ will be computed using equation (9.3). Simulation results have shown that there is no significant difference between the results obtained from these two methods, although the method applying the system function is more convenient for calculation of the noise level.

9.5 Noise Suppression

For most electrical equipments employing semiconductors, including the power electronic equipment, EMI must not exceed specific acceptable levels. For economic reasons, EMC should be considered early in the equipment design stage. As the product development progresses from design to test and production, the range of available noise suppression techniques decreases steadily. As a first step, one should analyze the equipment's noise generation characteristics to determine the required level of EMI suppression. Because EMI filters are not cheap, it is best to avoid the practice of automatically designing and building an EMI filter into a system without any previous attention to its EMI characteristics.

Semiconductor circuits usually produce differential-mode noise components. Common-mode EMI can be very effectively controlled by reducing stray capacitances between the circuit and the grounded parts. First, reduce the stray capacitances between heat sinks and the ground. In addition, EMI generated by switching processes can be reduced by proper design and also via use of RC snubbers. In many cases, suitable noise suppression can be achieved by applying EMI filter capacitors. The filter capacitor effectively short circuits the differential- and common-mode noise components close to their generation points.

It is important to note that conducted noise can be decreased effectively by applying some techniques in control units of a power electronic circuit. For example, in the PWM current control mode ac/dc converter, which is presented in the previous section, a solution for reducing EMI noise is to apply an RPWM control method [2]. This technique is suitable for distributing the power spectrum of EMI noise and reducing the amplitudes of high-frequency harmonics. Thus, by taking advantage of the RPWM technique, EMI problems caused by high frequency harmonics in the PWM technique can be overcome.

The widespread use of semiconductor modules in power electronic equipment has required a more detailed EMS analysis. Because these adjacent high-power circuits and sensitive electronic components interact so profoundly, the design engineer must consider not only external environmental disturbances but also the EMS of semiconductor circuits to internally grounded noise.

EMI suppression techniques should be considered in the early design stages rather than waiting until problems appear during testing or field operations. It is much more economical to solve EMS shortcomings early on. As the design process progresses, available EMS reduction techniques become fewer while possible source-victim paths become more numerous and more difficult to analyze. A system designed without regard to EMS will almost surely have some noise problems, some of which may be difficult and expensive to solve.

The first step in analyzing an existing EMS problem is to determine how the noise source (culprit) and the receiver (victim) are coupled. From an EMS perspective, noise generated by the culprit can no longer be suppressed, and the victim cannot be made less sensitive. Therefore, we must modify the transmission channel. Coupling occurs by conduction, radiation, or both. Conductive noise coupling can be decreased via the use of proper circuit connections, whereas radiated EMI is reduced through shielding techniques.

9.5.1 Grounding

In electrical engineering, ground or earth can refer to the reference point in an electrical circuit from which voltages are measured, a common return path for electric current, or a direct physical connection to the Earth.

Electrical circuits may be connected to ground (earth) for several reasons. In mains-powered equipment, exposed metal parts are connected to ground to prevent user contact with dangerous voltage if electrical insulation fails. Connections to ground limit the build-up of static electricity when handling flammable products or electrostatic-sensitive devices. In some telegraph and power transmission circuits, the earth itself can be used as one conductor of the circuit, saving the cost of installing a separate return conductor (see single-wire earth return).

For measurement purposes, the Earth serves as a (reasonably) constant potential reference against which other potentials can be measured. An electrical ground system should have an appropriate current-carrying capability to serve as an adequate zero-voltage reference level. In electronic circuit theory, a "ground" is usually idealized as an infinite source or sink for charge, which can absorb an unlimited amount of current without changing its potential. Where a real ground connection has a significant resistance, the approximation of zero potential is no longer valid. Stray voltages or earth potential rise effects will occur, which may create noise in signals or if large enough will produce an electric shock hazard.

There are two primary reasons for grounding devices, cables, equipment, and systems. The first reason is to prevent shock and fire hazards in the event that an equipment frame or housing develops a high voltage due to lightning or an accidental breakdown of wiring or components. The second reason is to reduce EMI effects resulting from electromagnetic fields, common impedance, or other forms of interference coupling.

9.5.2 Shielding

Electromagnetic shielding is the practice of reducing the electromagnetic field in a space by blocking the field with barriers made of conductive or magnetic materials. Shielding is typically applied to enclosures to isolate electrical devices from the "outside world," and to cables to isolate wires from the environment through which the cable runs. Electromagnetic shielding that blocks radio frequency electromagnetic radiation is also known as RF shielding.

The shielding can reduce the coupling of radio waves, electromagnetic fields, and electrostatic fields. A conductive enclosure used to block electrostatic fields is also known as a Faraday cage. The amount of reduction depends very much upon the material used, its thickness, the size of the shielded volume, and the frequency of the fields of interest and the size, shape, and orientation of apertures in a shield to an incident electromagnetic field.

9.5.3 Control Techniques for EMC

Operation in a polluted environment has difficulties that should be considered in the control and operation of the systems. Reference [16] presents a new synchronization method that employs an enhanced phase-locked loop (EPLL) system. The operational concept of the EPLL is novel and based on a nonlinear dynamical system. As compared with the existing synchronization methods, the introduced EPLL-based synchronization method provides a higher degree of immunity and insensitivity to noise, harmonics, and other types of pollution that exist in the signal used as the basis of synchronization. The salient feature of the EPLL-based synchronization method over conventional synchronization methods is its frequency adaptivity, which permits satisfactory operation when the centre frequency of the base signal varies. The proposed EPLL-based method of synchronization is also capable of coping with unbalanced system scenarios. Structural simplicity of the EPLL-based method greatly simplifies its implementation in digital software and/or hardware environments as an integral part of a digital control platform for power electronic converters. The primary application of the proposed synchronization method is for the distributed generation units, e.g. wind generation systems, which utilize power electronic converters as an integral part of their systems.

9.6 EMC Standards

Efforts to establish acceptable HF disturbance emission levels, however, have run up against several difficulties. Many years ago, experts examined HF pollution of the mains. Initial efforts to achieve EMC in this realm focused on limiting the HF emissions of equipment connected to the mains. Preliminary EMI measurement methods were developed at that time, and several limits for HF emission of electrical equipment were determined. A more comprehensive study of EMS came into being only in the second half of the twentieth century, simulated by more widespread use of electronic devices. CISPR (Comité International Spécial des Perturbations Radioélectriques or International Committee for Radio Interference) was the first international organization authorized to promulgate international recommendations on the subject of radio interference.

After World War II, CISPR became a special committee of the International Electrotechnical Commissions or IEC. There are some subcommittees in CISPR as follows [1]:

- Subcommittee A: Interference measuring devices, measurement methods.
- Subcommittee B: EMI from industrial, scientific, and medical apparatus (ISM).
- Subcommittee C: Noise caused by high power cables, high voltage equipment, and electrical traction.
- Subcommittee D: Ignition interference from motor vehicles, combustion engines, and related subjects.
- Subcommittee E: EMS for radio and television receivers.
- Subcommittee F: EMI in domestic appliances, fluorescent tubes, and similar devices.

Limits for EMI emission of electrical equipment were first established only for the frequency range 0.15–30 MHz, thus meeting broadcast requirements. These limits were later extended downward to the frequency range of 10–150 kHz [1]. CISPR requires the measurement and attenuation of HF emissions in the frequency range of 30–300 MHz. Thereafter, several countries turned to the IEC with a request to take up the problem of EMI with regard to semiconductor apparatus connected to mains. For surveying this special field, the IEC established a new subcommittee under the mark TC77. The subcommittee TC77 works in cooperation with several nations and CISPR.

IEC joined the study of EMS requirements in the 1960s. This subject was first addressed by IEC subcommittee TC65 and later subcommittee TC77 joined in the effort.

In the early 1990s, several subcommittees of the IEC continued to study EMI phenomena. The subcommittee TC110 and the organization ACEC (Advisor Committee on Electromagnetic Compatibility) strive jointly for harmony in that comprehensive work.

9.7 Summary and Conclusions

In the course of their daily routine, experts in the field of power electronics more and more often encounter the problem of high-frequency interference. In practice, EMC issues are usually ignored until a problem is revealed by testing or in normal operation. As a result, EMC fixes tend to be applied at the test or even production stages of product development, which can lead to solutions that are unsatisfactory, unnecessarily expensive, or both.

To avoid this situation, those who are involved in design, development, production, and operation of semiconductor equipment must be able to identify and solve EMI problems as early as possible. Although a great deal of written material on EMC has appeared in technical journals and conference records, these sources constitute a collection of miscellaneous subjects that do not always interrelate and are difficult to use in engineering practice. The main topics of this chapter are summarized as follows:

1) Some failure factors may not damage the system but interfere with its proper operation.
2) Noise is a common interfering factor in the systems. Interference mitigation and hence EMC is achieved by addressing both emission and susceptibility issues, i.e. quieting the sources of interference and hardening the potential victims. The coupling path between source and victim may also be separately addressed to increase its attenuation. There are two methods of noise coupling: conducted and radiated.
3) Grounding has a key role in controlling the effect of noise and interference in circuits. The most important law in grounding is that all electric references must have the same potential in a power network.

4) An EMI filter is an electronic passive device that is used in order to suppress conducted interference that is present on a signal or power line. EMI filters can be used to suppress interference that is generated by the device or by other equipment in order to make a device more immune to electromagnetic interference signals present in the environment.
5) Some fault alarms are not effective for the immediate shutdown of systems.

References

1 Redl, R. (2001). Electromagnetic environmental impact of power electronics equipment. *Proceedings of the IEEE* 89 (6): 926–938.

2 Redl, R. and Sun, J. (2009). Ripple-based control of switching regulators – An overview. *IEEE Transactions on Power Electronics* 24 (12): 2669–2680.

3 Vandamme, L.K.J. (1994). Noise as a diagnostic tool for quality and reliability of electronic devices. *IEEE Transactions on Electron Devices* 41 (11): 2176–2187.

4 Kolar, J.W. and Friedli, T. (2013). The essence of three-phase PFC rectifier systems – Part I. *IEEE Transactions on Power Electronics* 28 (1): 176–198.

5 Kaboli, S., Mahdavi, J., and Agah, A. (2007). Application of random PWM technique for reducing the conducted electromagnetic emissions in active filters. *IEEE Transactions on Industrial Electronics* 54 (4): 2333–2343.

6 Rodríguez, M.A., Claudio, A., Theilliol, D., and Vela, L.G. (2007). A new fault detection technique for IGBT based on gate voltage monitoring. In: *2007 IEEE Power Electronics Specialists Conference*, 1001–1005.

7 Rodríguez-Blanco, M.A., Claudio-Sanchez, A., Theilliol, D. et al. (2011). A failure-detection strategy for IGBT based on gate-voltage behavior applied to a motor drive system. *IEEE Transactions on Industrial Electronics* 58 (5): 1625–1633.

8 Wang, Z., Shi, X., Tolbert, L.M. et al. (2014). A di/dt feedback-based active gate driver for smart switching and fast overcurrent protection of IGBT modules. *IEEE Transactions on Power Electronics* 29 (7): 3720–3732.

9 Deng, F., Chen, Z., Khan, M.R., and Zhu, R. (2015). Fault detection and localization method for modular multilevel converters. *IEEE Transactions on Power Electronics* 30 (5): 2721–2732.

10 Izadian, A. and Khayyer, P. (2010). Application of Kalman filters in model-based fault diagnosis of a dc–dc boost converter. In: *IECON 2010 – 36th Annual Conference on IEEE Industrial Electronics Society*, 369–372.

11 Shahbazi, M., Jamshidpour, E., Poure, P. et al. (2013). Open- and short-circuit switch fault diagnosis for nonisolated dc–dc converters using field programmable gate Array. *IEEE Transactions on Industrial Electronics* 60 (9): 4136–4146.

12 Fowler, K.R. and Land, H.B. (2004). System design that minimizes both missed detections and false alarms: A case study in arc fault detection. In: *Proceedings of the 21st IEEE Instrumentation and Measurement Technology Conference (IEEE Cat. No.04CH37510)*, vol. 3, 2213–2216.

13 Hussain, S., Mokhtar, M., and Howe, J.M. (2015). Sensor failure detection, identification, and accommodation using fully connected cascade neural network. *IEEE Transactions on Industrial Electronics* 62 (3): 1683–1692.

14 Liu, Y., Stettenbenz, M., and Bazzi, A.M. (2019). Smooth fault-tolerant control of induction motor drives with sensor failures. *IEEE Transactions on Power Electronics* 34 (4): 3544–3552.

15 Vu, V.T., Pettersson, M.I., Machado, R. et al. (2017). False alarm reduction in wavelength-resolution SAR change detection using adaptive noise canceler. *IEEE Transactions on Geoscience and Remote Sensing* 55 (1): 591–599.

16 Karimi-Ghartemani, M. and Iravani, M.R. (2004). A method for synchronization of power electronic converters in polluted and variable-frequency environments. *IEEE Transactions on Power Systems* 19 (3): 1263–1270.

Part III

Health Estimation of the Power Electronic Systems

10

Condition Monitoring

10.1 Reasons for Condition Monitoring

Achieving the resilience capability needs an important tool: condition monitoring (CM) of the power converter. Condition monitoring has already been proven to be a cost-effective means of enhancing reliability and improving customer service in power equipment, such as transformers and rotating electrical machinery. CM for power semiconductor devices in power electronic converters is at a more embryonic stage, but as progress is made in understanding semiconductor device failure modes, appropriate sensor technologies, and signal processing techniques, this situation will rapidly improve. Figure 10.1 generally shows the variation of a system health versus the time. In the sample system, there is a plan to work until time t_f. However, a fault occurrence causes the lifetime of the system to be shortened to time t_d. Condition monitoring is the tool used to predict the fault in an earlier time or to detect the fault in a short time to save the system. Reference [1] describes the current state of the art in CM research for power electronics. Reliability models for power electronics, including dominant failure mechanisms of devices, are described first. This is followed by a description of recently proposed CM techniques. The benefits and limitations of these techniques are then discussed. It is intended that this review will provide the basis for future developments in power electronics CM. The increasing importance of energy conversion devices and their widespread use in uncountable applications have motivated significant research efforts. Reference [2] presents an analysis of the state of the art in this field. The analyzed contributions were published in most relevant journals and magazines or presented in either specific conferences in the area or more broadly scoped events.

In this chapter, the basic approaches for condition monitoring of electric power converters are presented. There are two general goals for monitoring the state of an electric power converter:

- Monitoring for controlling a variable (usually output voltage of a power converter)
- Condition monitoring for preventing a catastrophic failure

Monitoring is a key function during implementation of any control process. All closed-loop control schemes work based on monitoring the output variable of the process. However, converter condition monitoring methods are also used for informing about the state of the converter from a failure point of view. Many reliability improvement techniques need to have a view about the state of the electric power converter. In this chapter, commonly used methods for condition monitoring of power converters are described. Condition monitoring is the technique used to monitor a parameter in a power converter in order to identify a considerable change, which is an indication of a developing fault. Conditional monitoring of power converters has many benefits for the

Resilient Power Electronic Systems, First Edition. Shahriyar Kaboli, Saeed Peyghami, and Frede Blaabjerg.
© 2022 John Wiley & Sons Ltd. Published 2022 by John Wiley & Sons Ltd.
Companion website: www.wiley.com/go/kaboli/resilientpower

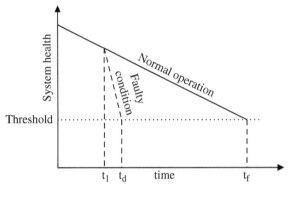

Figure 10.1 The system health diagram versus the time for normal and faulty conditions.

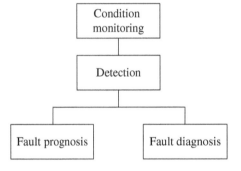

Figure 10.2 Flowchart of condition monitoring areas.

converter. Condition monitoring is important in certain conditions that would shorten a normal lifespan and can give information before those conditions lead to a major failure. Condition monitoring allows scheduling the maintenance to prevent failure and avoid its consequences. Figure 10.2 shows the condition monitoring blocks and goals. The condition monitoring is done based on the detection of the system parameters and variables. Fault prognosis and fault diagnosis are two goals of the CM.

10.1.1 Fault Prognosis

System fault prognostic techniques have been the subjects of considerable research into condition-based maintenance systems in recent times due to the potential advantages that can be gained from reducing downtime, decreasing maintenance costs, and increasing converter availability. Engineering systems, such as aircraft, industrial processes, manufacturing systems, transportation systems, electrical and electronic systems, etc., are becoming more complex and are subjected to failure modes that impact adversely on their reliability, availability, safety, and maintainability. Such critical assets are required to be available when needed, and maintained on the basis of their current condition rather than on the basis of scheduled or breakdown maintenance practices. Moreover, on-line, real-time fault diagnosis and prognosis can assist the operator in avoiding catastrophic events. Recent advances in condition-based maintenance and prognostics and health management have prompted the development of new and innovative algorithms for fault, or incipient failure, diagnosis, and failure prognosis aimed at improving the performance of critical systems. Reference [3] introduces an integrated systems-based framework for diagnosis and prognosis that is generic and applicable to a variety of engineering systems. The enabling technologies are based on suitable health monitoring hardware and software, data processing methods that focus on extracting features, or condition indicators from raw data via data mining and sensor fusion tools, accurate diagnostic and prognostic algorithms that borrow from Bayesian estimation theory, and specifically particle filtering, fatigue, or degradation modeling, and real-time measurements to declare a fault with prescribed confidence and given false alarm rate while predicting accurately and precisely the remaining useful life of the failing component/system. Potential benefits to industry include reduced maintenance costs, improved equipment uptime, and safety. Figure 10.3 shows the variation of an IGBT collector-emitter voltage in the long term. Increasing the studied voltage is a sign of a wire bond lift-off. Thus, it is concluded that the IGBT is at its end of life and should be replaced before failure. Note that no fault occurs in this process and the fault prognosis algorithm "predicts" the upcoming fault.

10.1.2 Fault Diagnosis

Fault diagnosis in industrial processes is a challenging task that demands effective and timely decision-making procedures under extreme conditions of noisy measurements, highly interrelated data, large number of inputs, and complex interaction between the symptoms and faults. Fault diagnosis determines which fault occurred, in other words, determines the root(s) of the out-of-control status. Process fault diagnosis involves interpreting the current status of the plant when given sensor readings and process knowledge. Early diagnosis of process faults, while the plant is still operating in a controllable region, can help to avoid events progressing and reduce the amount of productivity losses during an abnormal event. Figure 10.4 shows the detected current of a high voltage power supply and the reaction of its protection system in the output shutdown. Fault diagnosis algorithms exist in many industrial processes. The fault diagnosis is well-known knowledge in the field of electrical machines. Reference [4] presents a review of the developments in the field of diagnosis of electrical machines and drives based on artificial intelligence (AI). It covers the application of expert systems, artificial neural networks (ANNs), and fuzzy logic systems that can be integrated into each other and also with more traditional techniques. The application of genetic algorithms is also considered. In general, a diagnostic

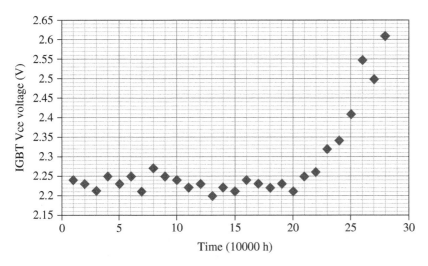

Figure 10.3 Variation of an IGBT collector–emitter voltage in the long term.

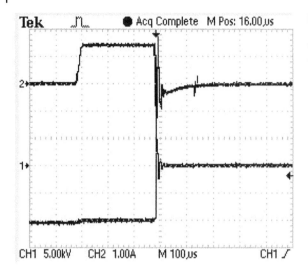

Figure 10.4 The detected current of a high-voltage power supply and the reaction of its protection system in the output shutdown.

procedure starts from a fault tree developed on the basis of the physical behavior of the electrical system under consideration. In this phase, the knowledge of well-tested models able to simulate the electrical machine in different fault conditions is fundamental to obtain the patterns characterizing the faults. The fault-tree navigation performed by an expert system inference engine leads to the choice of suitable diagnostic indexes, referring to a particular fault and relevant to build an input data set for specific AI (NNs, fuzzy logic, or neuro-fuzzy) systems. The discussed methodologies, which play a general role in the diagnostic field, are applied to an induction machine, utilizing as input signals the instantaneous voltages and currents. In addition, the supply converter is also considered to incorporate in the diagnostic procedure the most typical failures of power electronic components.

In the field of power electronics, fault diagnosis is an important topic to prevent catastrophic failures. As the power electronic system is expanded, fault diagnosis becomes complicated. Reference [5] is related to faults that can appear in multilevel (ML) inverters, which have a high number of components. This is a subject of increasing importance in high-power inverters. First, methods to identify a fault are classified and briefly described for each topology. In addition, a number of strategies and hardware modifications that allow for operation in faulty conditions are presented. As a result of the analyzed works, it can be concluded that ML inverters can significantly increase their availability and are able to operate even with some faulty components.

10.2 Aims of Condition Monitoring

Condition monitoring provides information about the growth of failure in power converters. A decision about this failure is related to the data that are obtained from the condition monitoring system. For example, monitoring a long-term high-output current beyond nominal specifications of a power electronic converter usually causes the protection system to operate and take it out of service. As another alternative, monitoring of a high-temperature hotspot in this converter may lead to application of a derating scenario that holds the converter in service.

10.2.1 Maintenance Scheduling

A failure in industrial equipment results in not only the loss of productivity but also timely services to customers, and may even lead to safety and environmental problems. This emphasizes the need for maintenance in manufacturing operations of organizations. Maintenance is of great importance in keeping availability and reliability levels of production facilities, product quality, etc. Condition monitoring helps to detect the preliminary signs of the faults. Thus, a timely maintenance program prevents a shutdown of the system, as shown in Figure 10.5. A predictive-maintenance structure for a gradually deteriorating single-unit system (continuous time/continuous state) is presented in reference [6]. The proposed decision model enables an optimal inspection and replacement decision in order to balance the cost engaged by failure and unavailability on an infinite horizon. Two maintenance decision variables are considered: the preventive replacement threshold and the inspection schedule based on the system state. In order to assess the performance of the proposed maintenance structure, a mathematical model for the maintained system cost is developed using regenerative and semi-regenerative process theories. Numerical experiments show that the expected maintenance cost rate on an infinite horizon can be minimized by a joint optimization of the replacement threshold and the periodic inspection times. The proposed maintenance structure performs better than classical preventive maintenance policies, which can be treated as particular cases. Using the proposed maintenance structure, a well-adapted strategy can automatically be selected for the maintenance decision-maker, depending on the characteristics of the wear process and on the different unit costs. Even limit cases can be reached: for example, in the case of expensive inspection and costly preventive replacement, the optimal policy comes close to a systematic periodic replacement policy. Most of the classical maintenance strategies (periodic inspection/replacement policy, systematic periodic replacement, corrective policy) can be emulated by adopting some specific inspection scheduling rules and replacement thresholds. In a more general way, the proposed maintenance structure shows its adaptability to different possible characteristics of the maintained single-unit system. Maintenance scheduling is important in power systems because of the necessity of resilient operation of the power systems. Reference [7] addresses generation maintenance scheduling in a competitive electric energy environment. In a centralized setting, the system operator derives a maintenance scheduling plan that achieves the desired reliability while minimizing cost and imposes it on all producers. In a competitive environment, this is not possible because the operator is still in charge of maintaining an adequate level of reliability, but the target of each producer is to maximize its own profits, which conflicts in general with the reliability objective of the operator. This reference proposes a technically sound coordinating mechanism based on incentives/disincentives among producers and the

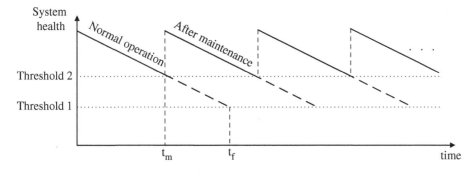

Figure 10.5 Effect of timely maintenance on the system health.

operator, which allows producers to maximize their respective profits while the operator ensures an appropriate level of reliability. Increasing the penetration of power electronic-based power sources in the modern power systems leads to increasing the importance of proper maintenance scheduling of the power electronic systems. Reference [8] develops models and the associated solution tools for devising optimal maintenance strategies, helping to reduce operation costs and enhancing the marketability of wind power. It is considered a multistate deteriorating wind turbine subject to several failure modes. We also examine a number of critical factors, affecting the feasibility of maintenance, especially the dynamic weather conditions, which makes the subsequent modeling and the resulting strategy season-dependent. Reference [8] formulates the problem as a partially observed Markov decision process with heterogeneous parameters. The model is solved using a backward dynamic programming method, producing a dynamic strategy. Reference [8] also highlights the benefits of the resulting strategy through a case study using data from the wind industry. The case study shows that the optimal policy can be adapted to the operating conditions, choosing the most cost-effective action. Compared with fixed, scheduled maintenances and a static strategy, the dynamic strategy can achieve considerable improvements in both reliability and costs.

10.2.2 Derating

Components are usually rated at high voltages and temperatures. However, applying these conditions simultaneously can compromise the components and lead to earlier failure. Derating is when a system or component is operated below its normal operating limit. This reduces the deterioration rate of the component and minimizes failures attributed to extreme operating conditions, as shown in Figure 10.6. Reference [9] examines the proper application of induction machines when supplied with unbalanced voltages in the presence of over- and undervoltages. Differences in the definition of voltage unbalance are also examined. The approach adopted is to use NEMA derating for unbalanced voltages as a basis to include the effects of undervoltages and overvoltages, through motor loss calculations. One of the most important causes of failure in wind power systems is due to the failures of the power converter and to one of its most critical components, the power semiconductor devices. Reference [10] proposes a novel derating strategy for the wind turbine system based on the reliability performance of the converter and the total energy production throughout its entire lifetime. An advanced reliability design tool is first established and demonstrated, in which the wind power system together with the thermal cycling of the power semiconductor devices are modeled and characterized under a typical wind turbine system mission profile. Based

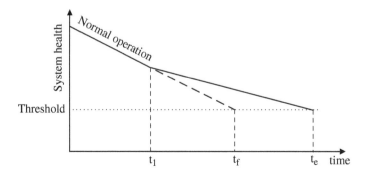

Figure 10.6 Effect of derating on the system health.

on the reliability design tools, the expected lifetime of the converter for a given mission profile can be quantified under different output power levels, and an optimization algorithm can be applied to extract the starting point and the amount of converter power derating that is necessary in order to obtain the target lifetime requirement with a maximum energy production capability. A nonlinear optimization algorithm has been implemented and various case studies of lifetime requirements have been analyzed. Finally, an optimized derating strategy for the wind turbine system has been designed and its impact has been highlighted.

10.2.3 Operation Management

The environmental and economical benefits of the microgrid, and consequently its acceptability and degree of proliferation in the utility power industry, are primarily determined by the envisioned controller capabilities and the operational features. Depending on the type and depth of penetration of distributed energy resource units, load characteristics and power quality constraints, and market participation strategies, the required control and operational strategies of a microgrid can be significantly, and even conceptually, different from those of the conventional power systems [11]. Condition monitoring helps to develop the managing algorithms for the power system with regards to the health index on the units.

10.3 Methods of Condition Monitoring

Condition monitoring techniques are very important for reducing in-service failures and unscheduled down-time and allow the implementation of intelligent maintenance. Therefore, the main goal of all approaches for condition monitoring is monitoring of parameters and variables of a power converter. The mechanism of this monitoring is different in various techniques of monitoring. There are two main approaches for condition monitoring: sensor-less and sensor-based methods. In sensor-based methods, a physical sensor is used to measure and monitor the desired parameter or variable. In the sensor-less method, the job is done based on calculations.

10.3.1 Sensor-Based Methods

One technique uses an individual sensor for each parameter that should be monitored. As an example, temperature sensors give valuable data about the temperature profile of a converter. These techniques are reliant on a range of sensors based principally on piezoelectric, electrodynamic, eddy current, inductive, magnetic, and thermal technologies. The sensor-based method was historically the first method used for condition monitoring. It has some features and advantages that enables it still to be used in a power converter. It provides faster access to data derived from online measurements and gives a more efficient response should a remedial action prove necessary. Although the benefits and even the necessity of using online sensors are recognized, several factors need to be considered prior to their installation in industrial applications. Typical industrial environments often involve fluids, temperatures, pressures, and flows that could be harmful to the delicate electronic sensor components. It is important to note that as the number of monitored parameters increases, the number of necessary sensors also increases. Thus, there is a limitation to the number of sensors used because the sensor response signals need to be collected and transmitted with minimal loss or interference from the surrounding equipment. The challenge then lies in constructing sensors rugged enough to withstand the rigors of the application and cost-effective enough to allow for their broad deployment throughout industrial enterprises. Therefore, a minimum number of sensors for sufficient data for converter monitoring is very important. Some information can help to reduce the sensor number. The main approach of this information is recognizing the points of the converter that are under considerable stress or are more sensitive to stress. Some examples of this information are listed as follows:

- Recognizing the most important point of a converter from an overtemperature point of view. It is important to note that it is usual to consider the hotspot of the converter as this point. Although this is a correct assumption in many cases, it is not always the most important point. For example, stator winding of a motor in an inverter-driven ASD may be the hotspot of ASD. However, the most important point of ASD is as a junction of inverter switches, because they are more sensitive to overtemperature than motor winding.
- Recognizing the most important point of a converter from an overvoltage point of view. Similar to temperature, the most important point of a converter from an overvoltage point of view is a sensitive point to overvoltage not a point with maximum overvoltage in the converter. For example, semiconductor switches are more sensitive to overvoltage than a transformer in a switch mode dc power supply.
- Analyzing the mechanical forces and finding the weakness of the converter.

As an example, overtemperature is a common reason for failure in power systems. Hence, temperature monitoring is a useful tool to prevent thermal damage of converters. Figure 10.7 shows the temperature sensors that are implemented on the heat sink and inside the module to detect an overtemperature fault.

10.3.2 Sensor-Less Methods

Sensor-less methodology is a solution for eliminating the physical sensors needed for the control process. It is historically applied to servomotor systems for controlling goals. However, sensor-less methods are also used for normal condition monitoring. Monitoring the state of mechanical machine tool components is of growing importance for increasing machine tool availability and reducing inspection efforts and costs. Open numerical controls offer a new opportunity for getting access to signals of drives as well as of the control itself and to integrate the end-user's individual

(a)　　　　　　　　　　　　　　　　　　　(b)

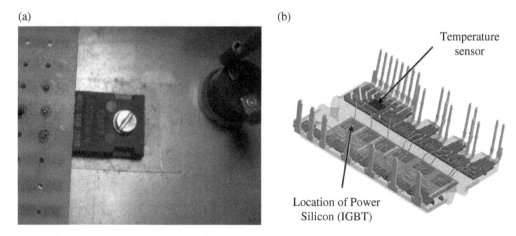

Figure 10.7 Application of temperature sensor for condition monitoring, (a) on the heat sink, (b) inside the package. *Source:* Used with permission from SCILLC dba onsemi.

applications. Because the actual drive signals contain information on the mechanical behavior of the drive chain components, their use for condition monitoring mechanical drive components is a reliable and cost-effective approach. Controlled induction motor drives without mechanical speed sensors at the motor shaft have the attractions of low cost and high reliability. To replace the sensor, the information on the rotor speed is extracted from measured stator voltages and currents at the motor terminals. Vector-controlled drives require an estimation of the magnitude and spatial orientation of the fundamental magnetic flux waves in the stator or in the rotor. Open-loop estimators or closed-loop observers are used for this purpose. They differ with respect to accuracy, robustness, and sensitivity against model parameter variations. Dynamic performance and steady-state speed accuracy in the low-speed range can be achieved by exploiting parasitic effects of the machine. The overview in reference [12] uses signal flow graphs of complex space vector quantities to provide an insightful description of the systems used in sensor-less control of induction motors. Reference [13] introduces a new direct torque and flux control based on space-vector modulation for induction motor sensor-less drives. It is able to reduce the acoustical noise, torque, flux, current, and speed pulsations during the steady state. The direct torque-controlled transient merits are preserved, while better quality steady-state performance is produced in sensor-less implementation for a wide speed range. The flux and torque estimator is presented and an improved voltage-current model speed observer is introduced. It is concluded that the proposed control topology produces better results for steady-state operation than the classical control. Reference [14] proposes a method for sensor-less on-line vibration monitoring of induction machines based on the relationship between the current harmonics in the machine and their related vibration harmonics. Initially, the vibration monitoring system records two baseline measurements of current and vibration with the machine operating under normal conditions. The baseline data is then evaluated to determine the critical frequencies to monitor online. Once these frequencies are determined, the baseline vibration measurement is simply used to scale the current harmonic signal to an estimated vibration level. Based on theoretical analysis, simulation results, and the experimental results shown here, a linear relationship between the current harmonics and vibration level can be assumed. The results of two experiments on a three-phase 230 V, 10 hp induction motor operating under no load are discussed and show the feasibility of this method for sensor-less on-line vibration monitoring. In reference [15], a new preventive maintenance strategy is proposed for the sensor-less condition monitoring of inverter dc-link aluminum electrolytic capacitors based on equivalent series resistance (ESR) and capacitance C estimation. The main concept of the proposed method is to estimate the ESR and C of the capacitor, using the inverter whenever the motor is stopped. The parameters are estimated from the dc-link voltage and stator current measurements available in the inverter with the switched dc-link voltage applied to the motor stator winding. The temperature of the capacitor can also be estimated under thermal equilibrium based on the stator resistance estimate to take the influence of temperature variation into account. An experimental study on a 250-W permanent magnet synchronous motor performed under accelerated capacitor degradation is presented to verify the proposed technique. It is shown that the proposed method provides a reliable and sensitive indication of capacitor aging without additional hardware requirements for reliable, efficient, and safe operation of the inverter and driven process.

10.4 Detection System

The detection block in Figure 10.2 is the first step in condition monitoring. The detection system consists of both hardware and software tools.

10.4.1 Data Acquisition Systems

Data acquisition is the process of sampling signals that measure real-world physical conditions and converting the resulting samples into digital numeric values that can be manipulated by a computer. Data acquisition applications are controlled by software programs developed using various general purpose programming languages such as LabVIEW.

10.4.1.1 Data Loggers

A data logger is an electronic device that records data over time or in relation to the location either with a built-in instrument or sensor or via external instruments and sensors. Increasingly, but not entirely, they are based on a digital processor (or computer). They generally are small, battery powered, portable, and equipped with a microprocessor, internal memory for data storage, and sensors. Some data loggers interface with a personal computer and utilize software to activate the data logger and view and analyze the collected data, while others have a local interface device (keypad, LCD) and can be used as a standalone device. Data loggers vary between general purpose types for a range of measurement applications to very specific devices for measuring in one environment or application type only. It is common for general purpose types to be programmable; however, many remain as static machines with only a limited number of or no changeable parameters. Electronic data loggers have replaced chart recorders in many applications. One of the primary benefits of using data loggers is the ability to automatically collect data on a 24-hour basis. Upon activation, data loggers are typically deployed and left unattended to measure and record information for the duration of the monitoring period. Figure 10.8 shows a typical data logger board. It contains some analog input channels. This data is obtained from various sensors which are distributed inside the converter. The output resistance of these sensors is usually high and need to be buffered at the first step. Therefore, there are some input buffers on this board that are implemented by operational amplifiers. These buffered analog inputs are analyzed by a microprocessor. The result of this analysis activates some isolated digital switch or sends comments via a LAN connection.

Figure 10.8 A data logger board including input analog inputs, high-impedance input buffer, I/O LAN connector, and isolated digital switch.

10.4.1.2 Event Recording

A sequence of events recorder is an intelligent standalone microprocessor-based system, which monitors external inputs and records the time and sequence of the changes. Sequence of events recorders usually have an external time source. When wired inputs change state, the time and state of each change is recorded. Figure 10.9 shows a user page of a data logger software used for event recording in an industrial site. This software is associated with LABview software with a data logger board as its hardware.

10.4.2 Signal Processing Tools

With advancements in digital electronics and reduced component costs in recent years, monitoring instruments for use in condition-based maintenance programs have become more cost-effective and dependable. Machinery does not need to be taken out of service as many tests are done online, and in many cases very little expertise is required for testing and data interpretation. This enables the user to make well-informed decisions for planning maintenance and repairs, which ultimately leads to increased productivity. There are two categories in signal processing: digital and analog signal processing. A novel fast-diagnostic method for open-switch faults in inverters without sensors is proposed to improve the reliability of the power electronic system in reference [16]. The presented method is achieved by analysis of the switching function model of the inverter under both healthy and faulty conditions. Due to the different voltages endured by each power switch under healthy conditions from open-switch faults in some cases, open-circuit faults can be detected by sensing the collector-emitter voltages of the lower power switches in each leg. The diagnostic scheme employs the simple hardware circuit to obtain indirectly the voltages of lower power switches and to get rid of high cost and complexity as a result of sensors. Also, this method minimizes the detection time and is available for the inverter faults in the systems with open-loop or closed-loop current control strategies. To ensure the effectiveness and reliability of the proposed diagnostic method, the rising-edge delays for switching signals are used to avoid wrong diagnostic results because of the ON–OFF processes of power switches, and the delay durations are determined by considering various factors. Simulation and experimental results validate the proposed scheme for detecting switch and leg open-circuit faults. With an increasing number of wind

Figure 10.9 User manual of a data logger software: history record (1), start to next record (2). *Source*: NATIONAL INSTRUMENTS CORP.

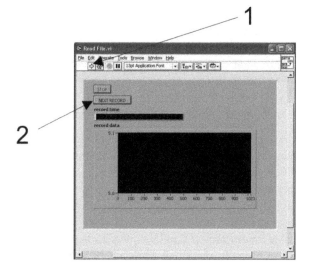

turbines being erected offshore, there is a need for cost-effective, predictive, and proactive mainte-nance. A large fraction of wind turbine downtime is due to bearing failures, particularly in the generator and gearbox. One way of assessing impending problems is to install vibration sensors in key positions on these subassemblies. Such equipment can be costly and requires sophisticated software for analysis of the data. An alternative approach, which does not require extra sensors, is investigated in reference [17]. This involves monitoring the power output of a variable-speed wind turbine generator and processing the data using a wavelet in order to extract the strength of par-ticular frequency components, characteristic of faults. This has been done for doubly-fed induction generators (DFIGs), commonly used in modern variable-speed wind turbines. The technique is first validated on a test rig under controlled fault conditions and is then applied to two operational wind turbine DFIGs where generator shaft misalignment was detected. For one of these turbines, the technique detected a problem three months before a bearing failure was recorded.

10.4.3 Measurement Tools

The above-mentioned methods need to measure devices in order for them to operate. With regards to fast and comprehensive application of sensor-less methods, the most important measuring devices are voltage and current sensors. In addition, thanks to the critical role of overtemperature in the failure process, temperature measurements are preferred to perform sensor-based tests. A sensor is a device that converts a physical property into a corresponding electrical signal. An acquisition system to measure different properties depends on the sensors that are suited to detect those properties. Signal conditioning may be necessary if the signal from the transducer is not suitable for hardware. The signal may need to be filtered or amplified in most cases.

10.4.3.1 Thermal Measurement
There are many techniques for temperature measurement. In the case of condition monitoring, small size as well as acceptable accuracy are important factors. Based on this view, some of the most famous temperature sensors are listed as follows:

- A thermocouple is a temperature-measuring device consisting of two dissimilar conductors that contact each other at one or more spots, where a temperature differential is experienced by the different conductors (or semiconductors). It produces a voltage when the temperature of one of the spots differs from the reference temperature at other parts of the circuit.
- Thermistors are thermally sensitive resistors whose prime function is to exhibit a large, predict-able, and precise change in electrical resistance when subjected to a corresponding change in body temperature. Negative Temperature Coefficient (NTC) thermistors exhibit a decrease in electrical resistance when subjected to an increase in body temperature, as shown in Figure 10.10, and Positive Temperature Coefficient (PTC) thermistors exhibit an increase in electrical resist-ance when subjected to an increase in body temperature.

One of the most important approaches to temperature measurement is the method based on the prediction. Reference [18] presents a technique to predict the die temperature of a MOSFET based on an empirical model derived following an offline thermal characterization. First, a method is presented for the near-simultaneous measurement of the die temperature during controlled power dissipation. The method uses a linear arbitrary waveform power controller which is momentarily disconnected at regular intervals to allow the forward voltage drop of the MOSFET's antiparallel diode to be measured. Careful timing ensures that the power dissipation is not significantly affected by the repeated disconnection of the power controller. Second, a pseudorandom binary sequence-based

Resistance versus temperature

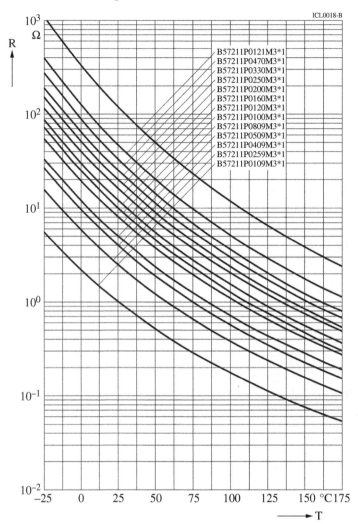

Figure 10.10 Variation of resistance of an NTC versus temperature. *Source:* TDK Co. (with permission).

system identification approach is used to determine the thermal transfer impedance, or cross-coupling between the dice of two devices on shared cooling using the near-simultaneous measurement and control method. A set of infinite impulse response digital filters are fitted to the cross-coupling characteristics and are used to form a temperature predictor. Experimental verification shows excellent agreement between measured and predicted temperature responses to power dissipation. Results confirm the usefulness of the technique for predicting die temperatures in real time without the need for on-die sensors.

10.4.3.2 Current Measurement

A current transformer (CT) is used for measurement of alternating electric currents. CTs, together with voltage transformers (VTs) (potential transformers [PTs]), are known as instrument transformers. When the current in a circuit is too high to apply directly to measuring instruments, a CT produces a reduced current accurately proportional to the current in the circuit, which can be conveniently connected to measuring and recording instruments. A CT isolates the measuring instruments from what may be very high voltage in the monitored circuit. CTs are commonly used in metering and protective relays in the electrical power industry. Figure 10.11 shows a CT mounted on a high-voltage circuit breaker. This CT has a high-voltage isolation between a high-voltage line and the measured current signal.

Figure 10.12 shows an application of a high-frequency CT for monitoring a high-frequency current of a pulsed power supply. Other sensors are the hall effect type of CTs, which can be used for a high-frequency current. This is an important characteristic of this type of CT. However, hall effect CTs need an auxiliary power supply to work. Passive CTs usually have a lower frequency

Figure 10.11 Application of CT for current sensing in a switching power supply.

Figure 10.12 Application of high-frequency CT for current sensing in a pulsed power supply.

bandwidth than hall effects types of CTs. Figure 10.13 shows a sample of an IGBT current measured using a Rogowski coil as an air-core CT. Figure 10.14 shows a Hall-effect current sensor used for measuring a dc-link current of an inverter. The dc-link current is a direct current but conventional passive CTs based on transformers cannot be used.

10.4.3.3 Voltage Measurement

If the isolation requirement is not considered, the voltage can be simply measured using a divider, as shown in Figure 10.15. For the isolation requirement, isolated methods of voltage measurement are needed. Instrument transformers are high accuracy class electrical devices used to isolate or transform voltage or current levels. The most common usage of instrument transformers is to operate instruments or metering from high-voltage or high-current circuits, safely isolating secondary control circuitry from the high voltages or currents. The primary winding of the transformer is connected to the high-voltage or high-current circuit, and the meter or relay is connected to the secondary circuit. Instrument transformers may also be used as an isolation transformer so that secondary quantities may be used in phase shifting without affecting other primary connected devices.

Figure 10.13 A sample of IGBT current measured by a Rogowski coil.

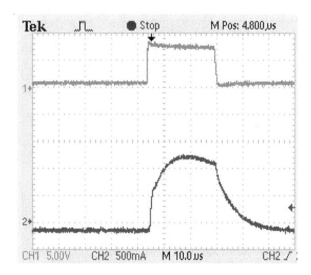

Figure 10.14 A Hall-effect current sensor, which is used for current measuring in an inverter dc bus.

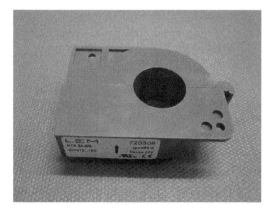

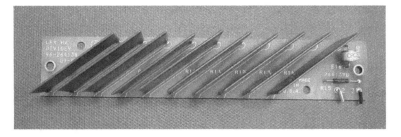

Figure 10.15 A high-voltage divider.

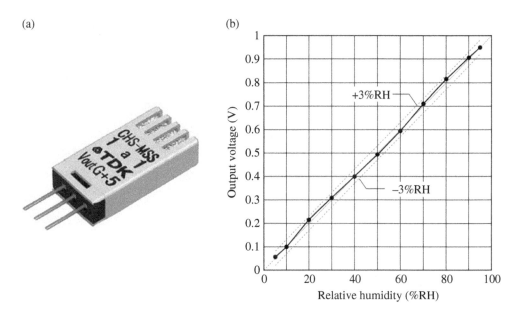

Figure 10.16 A humidity sensor (a), and its characteristic (b). *Source:* TDK (with permission).

10.4.3.4 Environmental Measurement

Figure 10.16 shows a typical environmental sensor (temperature and humidity). In this case, the environmental parameters, such as ambient temperature and humidity, are measured and sent to the control unit. The control unit applies proper commands to maintain the ambient parameters in a specific range based on the data resulting from this sensor.

10.4.4 Isolation

Measuring devices must be reliably isolated because of safety and EMC problems. One of the most safe methods is application of optical isolators. An optical fiber is a flexible, transparent fiber made of high-quality extruded glass or plastic. It can function as a waveguide to transmit light between the two ends of the fiber. Power over fiber optic cables can also work to deliver an electric current for low-power electric devices. The field of applied science and engineering concerned with the design and application of optical fibers is known as fiber optics.

Optical fibers are widely used in fiber-optic communications, where they permit transmission over longer distances and at higher bandwidths (data rates) than wire cables. Fibers are used instead of metal wires because signals travel along them with less loss and are also immune to electromagnetic interference. Fibers are also used for illumination and are wrapped in bundles so that they may be used to carry images, thus allowing viewing in confined spaces. Specially designed fibers are used for a variety of other applications, including sensors and fiber lasers.

Optical fiber cables carry the information over light waves, which travel in the fibers due to the properties of the fiber materials, similar to light traveling in free space. The light waves (one form of electromagnetic radiation) are unaffected by other electromagnetic radiation nearby. The optical fiber is electrically non-conductive, so it does not act as an antenna to pick up electromagnetic signals that may be present nearby. Therefore, the information traveling inside the optical fiber cables is immune to electromagnetic interference, e.g. radio transmitters, power cables adjacent to the fiber cables, or even electromagnetic pulses generated by nuclear devices.

Optical fibers are made and drawn from silica glass, which is a non-conductor of electricity and so there are no ground loops and leakage of any type of current. Optical fibers are thus laid down along with high-voltage cables on electricity poles due to its electrical insulator behavior. Figure 10.17 shows an application of a fiber optic link for measuring high-voltage dc voltage. In this converter, there is a floating dc voltage, which is biased to a high-voltage power supply. To measure this floating dc voltage, it is converted to a digital code by a microprocessor in the high-voltage side and sent to the low-voltage side via a fiber optic link. It is converted again to an analog signal in the low-voltage side by a digital-to-analog converter.

10.5 Summary and Conclusions

Condition monitoring is an important part of the methods for reliability improvement. The output of this part is real information about the present condition of the converter. In this chapter, various

Figure 10.17 Application of a fiber optic for isolated condition monitoring.

techniques for condition monitoring of electric power converters were described. The main topics of this chapter are summarized as follows:

1) There are two main approaches for condition monitoring: sensor-based and sensor-less methods.
2) Sensor-based methods usually give a precise value of certain parameters but they suffer from some drawbacks, such as a need for some space for mounting, sensitivity to noise, and are physically damaging.
3) Sensor-less methods give the desired parameters based on some calculations that start from a minimum number of measured parameters. They have no need for a physical space and can be quite immune to noise. However, they are dependent on the initial measured parameters. Accuracy is always a concern in these methods.
4) Nowadays, sensor-less methods are popular in controlling electric machines. They are used not only for condition monitoring but also for controlling a desired parameter, such as electromagnetic torque.

References

1 Yang, S., Xiang, D., Bryant, A. et al. (2010). Condition monitoring for device reliability in power electronic converters: A review. *IEEE Transactions on Power Electronics* 25 (11): 2734–2752.

2 Riera-Guasp, M., Antonino-Daviu, J.A., and Capolino, G. (2015). Advances in electrical machine, power electronic, and drive condition monitoring and fault detection: State of the art. *IEEE Transactions on Industrial Electronics* 62 (3): 1746–1759.

3 Ly, C., Tom, K., Byington, C.S. et al. (2009). Fault diagnosis and failure prognosis for engineering systems: A global perspective. In: *In 2009 IEEE International Conference on Automation Science and Engineering*, 108–115. Bangalore, India.

4 Filippetti, F., Franceschini, G., Tassoni, C., and Vas, P. (2000). Recent developments of induction motor drives fault diagnosis using AI techniques. *IEEE Transactions on Industrial Electronics* 47 (5): 994–1004.

5 Lezana, P., Pou, J., Meynard, T.A. et al. (2010). Survey on fault operation on multilevel inverters. *IEEE Transactions on Industrial Electronics* 57 (7): 2207–2218.

6 Grall, A., Dieulle, L., Berenguer, C., and Roussignol, M. (2002). Continuous-time predictive-maintenance scheduling for a deteriorating system. *IEEE Transactions on Reliability* 51 (2): 141–150.

7 Conejo, A.J., Garcia-Bertrand, R., and Diaz-Salazar, M. (2005). Generation maintenance scheduling in restructured power systems. *IEEE Transactions on Power Systems* 20 (2): 984–992.

8 Byon, E. and Ding, Y. (2010). Season-dependent condition-based maintenance for a wind turbine using a partially observed Markov decision process. *IEEE Transactions on Power Systems* 25 (4): 1823–1834.

9 Pillay, P., Hofmann, P., and Manyage, M. (2002). Derating of induction motors operating with a combination of unbalanced voltages and over or undervoltages. *IEEE Transactions on Energy Conversion* 17 (4): 485–491.

10 Vernica, I., Ma, K., and Blaabjerg, F. (2018). Optimal derating strategy of power electronics converter for maximum wind energy production with lifetime information of power devices. *IEEE Journal of Emerging and Selected Topics in Power Electronics* 6 (1): 267–276.

11 Katiraei, F., Iravani, R., Hatziargyriou, N., and Dimeas, A. (2008). Microgrids management. *IEEE Power and Energy Magazine* 6 (3): 54–65.

12 Holtz, J. (2002). Sensorless control of induction motor drives. *Proceedings of the IEEE* 90 (8): 1359–1394.

13 Lascu, C., Boldea, I., and Blaabjerg, F. (2000). A modified direct torque control for induction motor sensorless drive. *IEEE Transactions on Industry Applications* 36 (1): 122–130.

14 Riley, C.M., Lin, B.K., Habetler, T.G., and Schoen, R.R. (1998). A method for sensorless on-line vibration monitoring of induction machines. *IEEE Transactions on Industry Applications* 34 (6): 1240–1245.

15 Lee, K., Kim, M., Yoon, J. et al. (2008). Condition monitoring of DC-link electrolytic capacitors in adjustable-speed drives. *IEEE Transactions on Industry Applications* 44 (5): 1606–1613.

16 An, Q., Sun, L., Zhao, K., and Sun, L. (2011). Switching function model-based fast-diagnostic method of open-switch faults in inverters without sensors. *IEEE Transactions on Power Electronics* 26 (1): 119–126.

17 Watson, S.J., Xiang, B.J., Yang, W. et al. (2010). Condition monitoring of the power output of wind turbine generators using wavelets. *IEEE Transactions on Energy Conversion* 25 (3): 715–721.

18 Davidson, J.N., Stone, D.A., Foster, M.P., and Gladwin, D.T. (2016). Measurement and characterization technique for real-time die temperature prediction of MOSFET-based power electronics. *IEEE Transactions on Power Electronics* 31 (6): 4378–4388.

11

Fault Prognosis

11.1 Importance of the Fault Prognosis

System fault prognostic techniques have been the considerable subject of the condition-based maintenance system in recent times due to the potential advantages that could be gained from reducing downtime, decreasing maintenance costs, and increasing machine availability. In the maintenance of engineering systems, condition monitoring, fault diagnosis, and fault prognosis constitute some of the principal tasks, as shown in Figure 11.1. In a resilient power electronic system, fault prognosis is an effective tool for preventing the unavailability of the system.

With the increase in the number of machines within processing plants and their operational complexities, many engineers and researchers have started to look for automated solutions for these tasks. In most of the proposed solutions, these dynamic systems are modeled using tools to diagnose and predict faults in those systems. In modern industries, the increasing cost pressure forces utilities to operate their tools as efficiently as possible. At the same time, other requirements force them to maintain high quality levels. In this delicate area of conflict, comprehensive and detailed asset management methods, considering both costs and delivered quality, are essential for long-term success. Such comprehensive asset management methods comprise probabilistic reliability calculations, both for the current network state and also for expected future scenarios considering the long-term effects of decisions. The prognosis of component reliability performance in dependency of the chosen asset management strategies is a vital necessity [1]. However, suitable models and data for such a prognosis are not easily available. Reference [1] presents a comprehensive asset management approach with a special focus on the relevance of a component reliability prognosis.

A fault prognosis is an essential task in the power electronic systems. This task is very important in systems with mission critical. In addition, it prevents the unscheduled shutdowns in the system with difficult availability. Accurate remaining useful life prognosis of bearings in wind turbines can effectively help to schedule maintenance strategy and reduce operational costs at wind farms. An unscented particle filter is good at state tracking in nonlinear problems. In reference [2], a robust model-based approach based on an improved unscented particle filter is presented to deal with a bearing life prognosis in wind turbines, which involves: (i) the mean of sigma points after an unscented Kalman transform is regarded as the particles in the particle filter to guarantee the particles aggregation; (ii) several past measurements are utilized to estimate the likelihood function of the current step; (iii) a uniform distribution is adopted for resampling particles to make them diversify. The presented remaining useful life prognosis approach depends more on the measurement, rather than the initial parameters of the degradation model, which makes it

Resilient Power Electronic Systems, First Edition. Shahriyar Kaboli, Saeed Peyghami, and Frede Blaabjerg.
© 2022 John Wiley & Sons Ltd. Published 2022 by John Wiley & Sons Ltd.
Companion website: www.wiley.com/go/kaboli/resilientpower

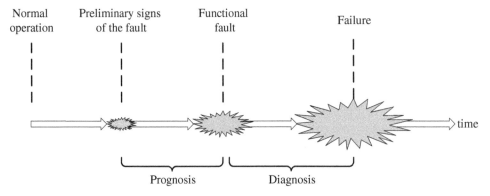

Figure 11.1 Fault prognosis and diagnosis intervals in a faulty system.

practicable for on-site wind turbines. Three life-cycle bearings from wind turbine high-speed shafts demonstrate the effectiveness of the proposed approach.

Distributed generation, dc transmission and distribution, and power electronics for ac/dc/ac conversion bring the advantages of increased control and flexibility in electric power management. These advantages also introduce challenges that must be managed. An electrical environment where voltage is no longer strictly sinusoidal implies that a new approach for the design and maintenance of electrical insulation systems has to be devised. A non-sinusoidal voltage supply often causes increased failure probability and, thus, reduced reliability and life of an electrical asset. Reference [3] addresses the implications of the transient and steady waveform distortion introduced by power electronic systems and a dc supply, for which the electrical insulation was designed and tested through consolidated criteria based on a sinusoidal voltage supply. Based on condition monitoring, a health index definition is proposed, which allows the condition of the insulation of an electrical apparatus to be assessed as a function of operation time, and, based on the aging and life models, the evaluation of maintenance actions and of the feasibility and extent of life extension plans to be carried out.

11.2 Methods of Fault Prognosis

Condition monitoring and prognosis are effective methodologies to reduce the downtime and maintenance cost and improve the reliability and lifespan of power electronic converter systems. As stated in the literature, current fault prognosis methods can be generally classified as three types: model-based methods, knowledge-based methods, and data-based methods, as shown in Figure 11.2.

11.2.1 Model-Based Fault Prognosis

Model-based fault diagnosis, using statistical techniques, residual generation, and parameter estimation, has been an active area of research [4]. Model-based methods are approaches that build accurate mathematical descriptions of systems using physics or first-principle. The difference between the outputs of the model and the real data of the system is used for predicting the fault, as shown in Figure 11.3. The identification and update of the model parameters generally require

Figure 11.2 Various fault prognosis methods.

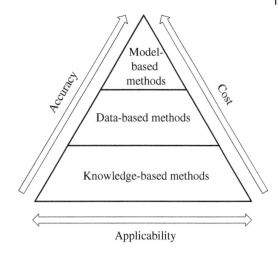

Figure 11.3 Block diagram of the model-based fault diagnosis.

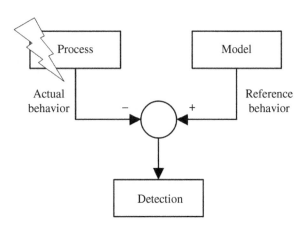

specifically designed experiments and statistical methods, respectively. As a result, they tend to outperform other kinds of models when sufficient knowledge of physical mechanisms are available. However, as industrial systems become more and more complex, exact mathematical knowledge about the system is generally unavailable. Additionally, model-based methods are often built case by case. Reference [5] presents a model-based method to implement the condition estimation and a data-driven method to conduct the prognosis for a boost converter. First, the equivalent circuit for a non-ideal boost converter should be simplified and the state-space equation should be obtained. Then, constructing the objective function for the crow search algorithm (CSA) and several parameters like inductance, on-resistance of the metal-oxide-semiconductor field-effect transistor (MOSFET), capacitance, and equivalent series resistance (ESR) of the capacitor are estimated based on CSA. Considering components degradation and variable operating conditions, several simulation experiments have been conducted to validate the presented approach. Finally, the prognosis for the capacitor of the boost converter has been conducted based on the least-square support vector machine (LSSVM) algorithm. The results show that this technique characterizes high computation efficiency and good estimation accuracy. Also, it will be a basis of further study for the circuit-level remaining useful life prediction.

11.2.2 Knowledge-Based Methods

On the other hand, knowledge-based methods are based on engineering experience and historical events, so the prognosis results obtained by these methods are more intuitive. Hence, it is easy to carry out efficiently when the process models can be easily obtained or the process knowledge has been readily accumulated. However, the formation of process knowledge is always time consuming, and knowledge-based models rely too much on the capability of the expert experience. Sometime it is even impossible to build such models due to the combinational explosion problem, and thus the prediction accuracy is greatly reduced and the application scope is largely limited.

11.2.3 Data-Driven Fault Prognosis

Apart from the aforementioned two types of prognosis methods, the data-based method mainly determines the health status of the system within a certain period of time by analyzing the previously observed data. Because they do not require the professional knowledge or mathematical models of industrial systems, the data-based methods are applicable when data are sufficiently abundant and have relatively low operating costs. With the development of modern information and digital technology, the collection of large amounts of system measurement data provides the basis for data-driven fault prognosis applications with a higher prediction accuracy, widely promoted in past decades. Even if enough observation data is not always available in some actual working environments, several specific data-driven methods can produce accurate prognosis results with few training samples or lack experimental samples [6]. The implementation procedures of a typical data-driven fault prognosis method are divided into four layers, which are shown in Figure 11.4,

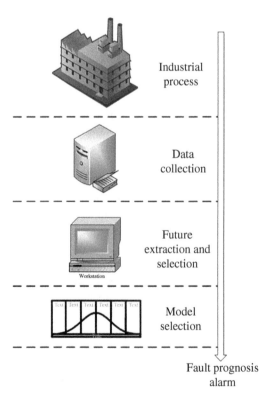

Industrial
process

Data
collection

Future
extraction and
selection

Workstation

Model
selection

Fault prognosis
alarm

Figure 11.4   Flow diagram of the model-based fault diagnosis.

including data collection and analysis layer, feature extraction and selection layer, model selection, training, and validation layer, and fault prognosis and health management layer. The data collection and analysis layer is the basis and data source of the data-based fault prognosis methodology. Generally, in this layer, the original measurement data is collected. The feature extraction and selection can be regarded as the pre-processing step for measurement data of the data-driven fault prognosis. Following the features extraction and the analysis results of process data characteristics, the intelligent prognosis model is then applied and the complexity is also evaluated.

As an example, substation automation is a key enabling technology for online monitoring, diagnosis, and prediction for the health condition of the substation assets. Circuit breakers (CBs) are one of the most vital components in a substation for the tripping action required during fault occurrence, line isolation, and other similar actions. It is critical to ensure that the CB is in a healthy state and can operate as expected. Enhanced automation and availability of various CB measurements make it possible to continuously monitor the health of all the components within a CB, including the trip coil assembly (TCA). Reference [7] presents the development of a new real-time diagnosis algorithm that runs at a substation and continuously monitors the health condition of a CB TCA and suggests maintenance actions, if necessary. The developed algorithm detects the abnormalities, finds their root causes, and predicts the possibility of potential health problems for the CB TCA. Additionally, the monitoring architecture also allows remote access of data for engineering access.

11.2.4 Fault Prognosis in Power Electronics

Electronic systems, such as power supplies, are complex multilayered devices consisting of different materials with inherent variability. Thermal gradient cycling occurs during system operation, which eventually results in thermomechanical fatigue-induced material failure. Such material failures can result in immediate electronic system shutdown with no advanced fault or warning signals. In the field of power electronics, fault prognosis is done in various types and levels. Both converter-level faults and element-level faults are investigated to ensure continuous operation of the system. Figure 11.5 shows the application of the temperature sensor on a heat sink in order to

Figure 11.5 Fault prognosis by a temperature sensor on the heat sink of a converter.

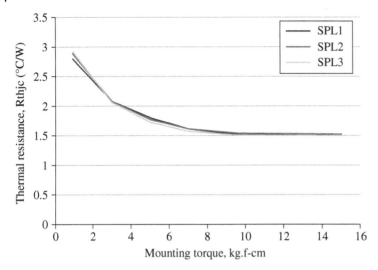

Figure 11.6 Effect of mechanical stability on the thermal resistance of a power module. *Source:* Used with permission from SCILLC dba onsemi.

ensure proper mechanical mounting of the transistors. As the mechanical mounting of the power devices has an important effect on their thermal resistance (shown in Figure 11.6), this temperature sensor prevents the thermal failure of the power devices due to mechanical vibration.

Reference [8] uses state-of-the-art material modeling to predict degradation of circuit-board elements as a means for "simulated fault detection." This effort has been focused on the specific aspect of solder fracture and fatigue since electronic industry statistics have attributed this failure issue as a driving factor in system reliability. This project demonstrates feasibility for using conventional sensing, combined with thermal modeling, to predict solder degradation due to thermal cycling as a means to prognosticate electronic power supply system reliability. Current status, as well as future plans, of electronic prognosis development is discussed.

11.3 Element-Level Internal Faults Prognosis

In modern reliability studies, it is mentioned that the expected lifetime of a power electronic converter is limited by its fragile components, such as semiconductor devices and capacitors [9]. These components are prone to wear-out failures depending on the converter mission profile and component thermal characteristics.

11.3.1 Passive Devices

Among the passive components, capacitors are mainly the weak point of the converters. Some of the remaining faults are related to the magnetic devices.

11.3.1.1 Capacitors

Capacitors are one type of reliability-critical components in power electronic systems. In the last two decades, many efforts in academic research have been devoted to the condition monitoring of capacitors to estimate their health status [10]. Industry applications are demanding more reliable power electronics products with preventive maintenance. Nevertheless, most of the developed capacitor condition monitoring technologies are rarely adopted by industry due to the complexity, increased cost, and other relevant issues. The capacitors are used in resonant converters such as the resonant tank, in the subbers, and in dc-links. Two first categories benefit the film capacitors that are very reliable. Based on the ripple current limitation of capacitors, a capacitor bank is usually placed at a dc-link based on a series–parallel configuration. Generally, three types of capacitors are used in dc-link applications, which are Aluminum Electrolytic Capacitors (Al-Caps), Metallized Polypropylene Film Capacitors (MPPF-Caps), and Multi-Layer Ceramic Capacitors (MLC-Caps). An overview of the prior-art research in this area is therefore needed to justify the required resources and the corresponding performance of each key method. It serves to provide a guideline for industry to evaluate the available solutions by technology benchmarking, as well as to advance the academic research by discussing the history development and the future opportunities. Therefore, reference [11] first classifies the capacitor condition monitoring methods into three categories and then the respective technology evolution in the last two decades is summarized. Finally, the state-of-the-art research and the future opportunities targeting for industry applications are given. Figure 11.7 shows the equivalent circuit of a real capacitor and the impedance diagram of the electrolyte capacitors to show the ESR as the minimum value of the impedance diagram. With the degradation of dc-link capacitors, a series of physical and chemical changes occur in the inside of capacitors, which will cause electrical parameters (e.g. ESR, C, etc.) and non-electrical parameters (e.g. weight, structure, internal temperature, internal pressure, etc.) to be changed. Therefore, the fault prognosis diagram for the capacitors is formed as shown in Figure 11.8. The calibration block necessary for compensating the effect of ambient temperature and the ESR value of the capacitors are shown in Figure 11.9. Table 11.1 shows the EOL criteria of the fault prognosis.

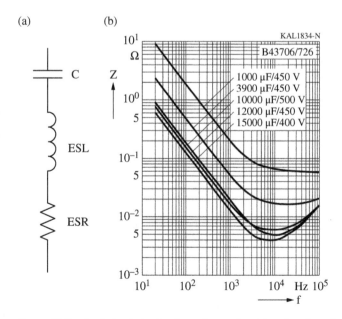

Figure 11.7 Equivalent circuit of a real capacitor (a) and an impedance diagram of the electrolyte capacitors (b). *Source:* TDK (with permission).

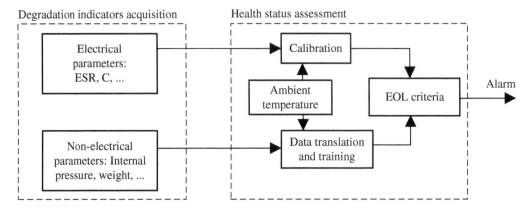

Figure 11.8 Flow diagram of fault prognosis in capacitors.

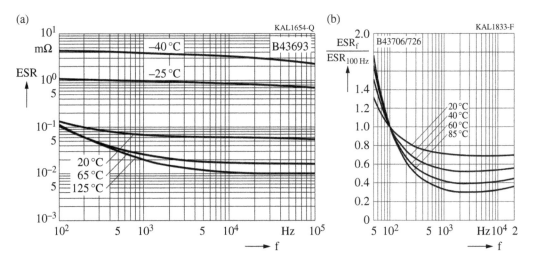

Figure 11.9 Variation of ESR in the capacitors: (a) ceramic capacitors, (b) electrolyte capacitors. *Source:* TDK (with permission).

Table 11.1 EOL criteria of various capacitors.

	Al-Caps	MPPF-Caps	MLC-Caps
EOL criteria	$C/C_0 < 80\%$	$C/C_0 < 95\%$	$C/C_0 < 90\%$
	$ESR/ESR_0 > 2$		

Electrolytic capacitors are ubiquitous in power electronic systems owing to their relatively large capacitance-to-volume ratio provided by virtue of their anodized electrode and electrolyte. As the capacitor ages, the electrolyte solution present in the capacitor is known to reduce and this leads to an increase in the ESR, as shown in Figure 11.10, and a reduction in the capacitance. The end-of-life for electrolytic capacitors is indicated by an increase in ESR by a factor of two. There are various methods for calculating the ESR. Figure 11.11 shows the method based on the power loss calculation.

A prognostic system, given in reference [12], proposes that it can measure the ESR of a capacitor while it is operating in a pulse-width modulation (PWM) circuit. By using lock-in amplifiers

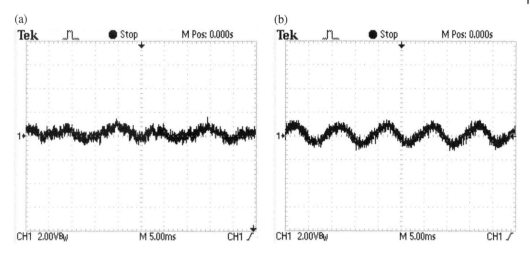

Figure 11.10 Effect of ESR increasing on the voltage ripple of the capacitors: (a) a fresh capacitor, (b) a capacitor after aging.

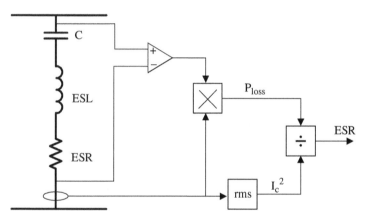

Figure 11.11 Power loss method for ESR calculation.

synchronized to the PWM fundamental frequency, it is possible to extract precise capacitor ripple voltage and ripple current. These signals are then combined to provide an estimate for the ESR. In reference [13], a novel scheme for the estimation of the ESR of the dc-link electrolytic capacitor in three-phase ac/dc PWM converters is proposed for condition monitoring. First, a controlled ac current component is injected into the input. Then, it induces ac voltage ripples on the dc output. By manipulating these ac voltage and current components with digital filters, the value of the ESR can be calculated, where the recursive least squares algorithm is used for reliable estimation results. In addition, the value of the ESR is corrected by considering the temperature effect, for which a simple temperature-sensing circuit has been designed. The simulation and experimental results show that the estimation error of the ESR is within a reasonable range, thereby enabling the determination of the appropriate time for the replacement of the capacitor. To set a predictive maintenance, an adaptive filter modeling-based method is presented in reference [14] using the least mean squares (LMS) algorithm. A signature of changes in capacitance and ESR will reflect in the capacitor ripple voltage because of aging, and these changes are monitored using adaptive filter modeling to predict the future status of the capacitor using only the input current of the system. A condition monitoring method based on the Artificial Neural Network (ANN) algorithm is therefore proposed in reference [15]. Implementation of the ANN to the dc-link capacitor condition monitoring in a back-to-back converter is presented. The error analysis of the capacitance estimation is also given. The presented method enables a pure software-based approach with high parameter estimation accuracy.

Reference [16] proposes a simple yet effective variable electrical network (VEN) condition monitoring method for dc-link capacitors in three-phase PWM ac–dc–ac power converters. The capacitance (C) and ESR of the dc-link capacitors are estimated through the designed VEN unit during the discharging process. This VEN method does not need the current sensor or injecting signals into the control loop. The above property is favorable for the hardware and controller design. In addition, the designed monitoring circuit can be an external unit or a built-in unit, which offers a flexible solution for electrical devices. When the designed VEN is used as an external unit, neither the hardware design nor the programs in the controller need to change, which is practical and economic for electrical devices in use.

11.3.1.2 Magnetic Devices

Fault prognosis of the large transformers is a well-known task in the power system. Power transformers are considered to be among the most costly pieces of equipment used in electrical systems. A major research effort has therefore focused on detecting failures in their insulating systems prior to unexpected partial discharge (PD), as shown in Figure 11.12. The failure due to PD can be very fast or acts in the long term, as shown in Figure 11.13.

Although several industrial methods exist for the online and offline monitoring of power transformers, all of them are expensive and complex, and require the use of specific electronic instrumentation. Reference [17] presents a transformer winding fault diagnosis and prognosis method based on a self-powered radio frequency identification (RFID) sensor tag. The proposed RFID sensor tag, which consists of an RFID tag, power management circuit, main control unit (MCU), and accelerometer, can acquire the vibration signals of transformer winding from the tank by the accelerometer, and then wirelessly transmits the signals to the RFID reader. An inductive energy harvester utilizing the surrounding magnetic field is optimized as the power supply for the proposed sensor tag, including the MCU and the accelerometer. A customized ac-dc converter together with a low-dropout voltage regulator is designed to provide stable dc voltage for the proposed sensor tag.

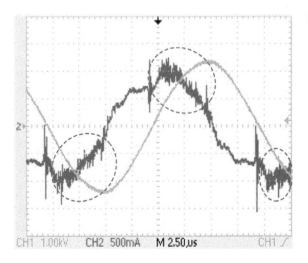

Figure 11.12 PD current in an insulator.

CH1 1.00kV   CH2 500mA   M 2.50μs   CH1

The RFID reader compiles the data from all the RFID sensor tags and then transmits them to the remote monitoring software, which is developed to display the diagnosis results and alert messages. The experimental results show that the proposed energy harvester can provide 197 µW power in a 50 Hz magnetic field, and the ac–dc converter is capable of providing 2.5 V dc voltage to power the circuitry. The measured maximum power consumption of the proposed sensor tag is 147 µW. Furthermore, the achieved reliable communication distance is 13 m in the test scenario. The experimental results show that the proposed method is effective in term of the diagnosis and prognosis of transformer winding. Reference [18] presents an online analysis of transformer leakage flux as an efficient alternative procedure for assessing machine integrity and detecting the presence of insulating failures during their earliest stages. A 12 kVA 400 V/400 V power transformer was specifically manufactured for the study. A finite-element model of the machine was designed to obtain the transient distribution of leakage flux lines in the machine's transversal section under normal operating conditions and when shorted turns are intentionally produced. Very cheap and simple sensors, based on air-core coils, were built in order to measure the leakage flux of the transformer and non-destructive tests were also applied to the machine in order to analyze pre- and post-failure voltages induced in the coils. Results point to the ability to detect very early stages of failure, as well as locating the position of the shorted turn in the transformer windings. Calculating the power loss is an indicator of the PD fault, as shown in Figure 11.14.

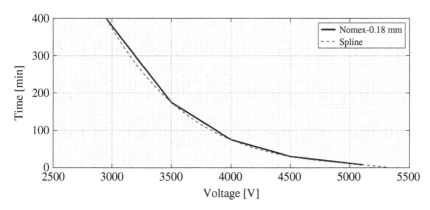

Figure 11.13 Effect of applied voltage on the lifetime of a Nomex insulator.

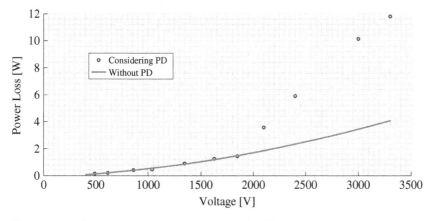

Figure 11.14 Power loss of an insulator considering PD.

11.3.2 Active Devices

The proposed monitoring systems are based on the online measurement of two damage indicators: the on-state voltage of the semiconductor and the voltage drop in the bond wires. The on-state voltage of a semiconductor can be employed for temperature estimation, in order to anticipate failures in the solder joints that increase the thermal resistance of the cooling path. Moreover, by measuring the voltage drop in the bond wires, the degradation of the bond wires can be detected [19].

11.3.2.1 IGBT

Reference [20] presents an overview of the major failure mechanisms of insulated-gate bipolar transistor (IGBT) modules and their handling methods in power converter systems improving reliability. The major failure mechanisms of IGBT modules are presented first and then the methods for predicting lifetime and estimating the junction temperature of IGBT modules are discussed. Subsequently, different methods for detecting open- and short-circuit faults are presented. Finally, fault-tolerant strategies for improving the reliability of power electronic systems under field operation are explained and compared in terms of performance and cost.

Wire Bond Problems

Bonding wire lift-off is a common failure in an IGBT module due to the thermomechanical fatigue under cyclical temperature and power swings, as shown in Figure 11.15. Consequently, this will lead to the chip-loss failure gradually in a high-power multi-chip IGBT module during long-term operation.

To prevent a catastrophic failure from happening, a self-testing technique for the IGBT module is proposed in reference [21] to detect the chip-loss failure at an initial stage. During the test, a shoot-through current is imposed on the device under test (DUT) by gate control and its peak value is selected as the failure criterion for a chip-loss failure prognosis. Since the IGBT switching behavior changes after failure, the shoot-through current will deviate from its initial value, which can be detected for a failure prognosis. For safety, the self-testing is conducted intermittently on-site during system downtime and the shoot-through current is limited to less than the IGBT rated current at a lowered dc voltage. To clarify the method, its test scheme, working principle, and implementation are firstly discussed in the reference. Then, experimental work was carried out on a test rig with 3.3 kV/800 A IGBT modules (emulating a metro traction drive system) to verify the method. Preliminary test results demonstrate that the proposed self-testing method is a simple and effective technique for detecting the gradually developed IGBT chip-loss failure.

In reference [22], an inverter characterization technique is used to improve the inverter fault diagnosis and prognosis techniques with the demonstrative bond wire lift-off fault. Knowledge of

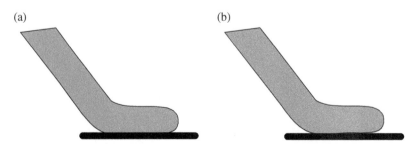

(a)   (b)

Figure 11.15   Wire bond lift off: (a) normal state, (b) cracked wire.

the nonlinear voltage-current characteristic of each inverter device is used to achieve these improvements. Three contributions are made in reference [22] to the field of inverter condition monitoring by using the characterization technique: use of the inverter device characteristics to improve a classifier-based diagnostic technique, direct use of the device characteristics for diagnosis, and improving the calculation of the remaining useful life with the device characteristics. The diagnostic method improvements are verified experimentally by adding resistances to create an artificial fault. Thermal and electrical simulations are used to demonstrate the improvements made to the prognostic method.

Reference [23] proposes an algorithm for the condition monitoring and prognosis of wire-bond fault in an IGBT extensively used power switch in inverter systems. The proposed algorithm employs both on-state collector emitter voltage in estimation of the wire bond lift-off, as shown in Figure 11.16, and collector current in the estimation of the IGBT junction temperature. These two signals complement each other as thermal sensitive electrical parameters and results in a better

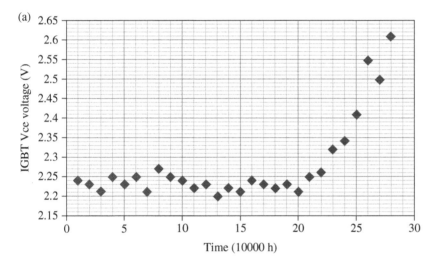

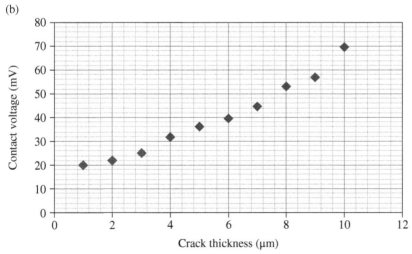

Figure 11.16 Effect of wire bond lift-off on the collector-emitter voltage of an IGBT: (a) variation of the collector-emitter voltage in the long term, (b) effect of crack thickness on the contact wire bond voltage.

estimation of the junction temperature. This algorithm has integrated a stochastic method and Coffin Manson model to introduce a criticality matrix, which is used to determine the necessity for IGBT replacement.

Reference [24] proposes a condition monitoring method of IGBT modules. A structure of a conventional IGBT module and a related parameter for the condition monitoring are explained. Then, a proposed real-time on-state collector-emitter voltage measurement circuit and condition monitoring strategies under different operating conditions are described. Finally, experimental results confirm the feasibility and effectiveness of the proposed method.

Reference [25] proposes a Monte Carlo-based analysis method to predict the lifetime consumption of bond wires of IGBT modules in a photovoltaic (PV) inverter. The variations in IGBT parameters (e.g. on-state collector-emitter voltage), lifetime models, and environmental and operational stresses are taken into account in the lifetime prediction. The distribution of the annual lifetime consumption is estimated based on a long-term annual stress profile of solar irradiance and ambient temperature. The proposed method enables a more realistic lifetime prediction with a specified confidence level compared to the state-of-the-art approaches. A study case of IGBT modules in a 10-kW three-phase PV inverter is given to demonstrate the procedure of the method. The obtained results of the lifetime distribution can be used to justify the selection of IGBTs for the PV inverter applications and the corresponding risk of unreliability.

Reference [26] presents a cost-effective prognostic method for the bond wires in the IGBT. Consider that the crack propagation in the wire bond leads to the bond wire lift-off; the corresponding state equation is established from the fracture mechanics theory, with the consideration of the uneven distribution of the temperature swings. Hence, the proposed model can work under different loading conditions. With the fact that the ON-state voltage ($V_{ce,on}$) of the IGBT shifts with the crack propagation, the historical $V_{ce,on}$ is used to predict the remaining useful lifetime (RUL), through which numerous power cycling tests are avoided and the low economical cost for doing the prognosis is fulfilled. The difficulty of the collector-emitter voltage measurement is its high amplitude swing. Therefore, the novel circuits, such as that in Figure 11.17, are used. In this article, the functional relationship between the increase of $V_{ce,on}$ and the crack length of each bond wire is obtained through finite-element simulations, while the effects of the temperature variation and metallization degradation to the $V_{ce,on}$ are compensated. Thus, the output equation can be obtained. Then, the unknown parameters of the aforementioned equations and the current crack length can be estimated by the particle-based marginalized resample-move algorithm. Finally, the RUL can be predicted effectively by evolving the particles obtained in the algorithm. The proposed method has been validated by the power cycling test.

Solder Fatigue

The temperature of a power semiconductor device is important for both its optimal operation and reliability. If the temperature is known during the operation of a converter, it can be used to monitor the health of power modules: a measurement of aging, scheduling of maintenance, or even implementation of active thermal control to reduce losses and increase lifetime can be performed given an accurate knowledge of the temperature. The thermal model of solid-state devices (Figure 11.18) is conventionally used to calculate the junction temperature. For fault prognosis proposals, Equation (11.1) is solved inversely for calculating the thermal resistance:

$$T_e - T_A = P_{loss} \cdot R_{th} \qquad (11.1)$$

Figure 11.17 Proposed circuit for detecting the long-term variation of the collector-emitter voltage.

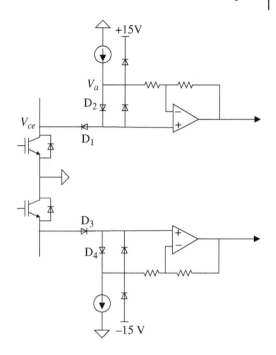

Thermal resistance is used to evaluate the solder quality of the device. However, it needs the junction temperature, but the conventional methods based on the mounting of a temperature sensor inside the power module have considerable errors. Figure 11.19 shows the position of the temperature sensor in a power module. It is seen that the sensor is far from the junction. Other direct methods based on fiber optics (Figure 11.20) cannot be used in industries and are mainly a research method.

Temperature measurements via thermosensitive electrical parameters (TSEP) are one way to carry out immediate temperature readings on fully packaged devices. However, successful implementation of these techniques during the actual operation of a device has not yet been achieved. Reference [27] provides an overview of the literature where the usage of TSEPs has been hypothesized or realized in realistic power electronic converter setups. Barriers and limitations preventing wider-scale implementation of these methods are discussed. Their potential use in the aforementioned goals in condition monitoring and active thermal control is also described.

An electrical method for junction temperature measurement of MOS-gated power semiconductor devices is presented in reference [28]. The measurement method involves detecting the peak voltage over the external gate resistor of an IGBT or MOSFET during turn-on. This voltage is directly proportional to the

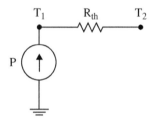

Figure 11.18 Thermal equivalent circuit in steady state used for thermal resistance calculation.

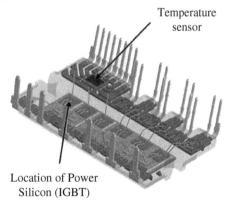

Figure 11.19 Position of the temperature sensor inside a power module. *Source*: Used with permission from SCILLC dba onsemi.

Figure 11.20 Direct measurement of the IGBT junction temperature by fiber optic.

peak gate current, and fluctuates with temperature due to the temperature-dependent resistance of the internal gate resistance. Primary advantages of the method include an immunity to load current variation and a good linear relationship with temperature. A measurement circuit can be integrated into a gate driver with no disruption to the operation and allows autonomous measurements controlled directly via the gate signal.

Reference [29] presents the use of a Kelvin-emitter resistor as a junction temperature sensor in IGBTs. The Kelvin-emitter resistor is placed directly on the IGBT die surface. The resistance is then evaluated using a sensing current injected between the power-emitter and Kelvin-emitter terminals. The Kelvin-emitter resistor is part of the overall gate resistance of the IGBT. Due to the resistor design, the temperature measured is a local temperature that correlates with the top of the resistor.

In reference [30], infrared measurements are used to assess the measurement accuracy of the peak gate current (I_{GPeak}) method for IGBT junction temperature measurements. Single IGBT chips with the gate pad in both the center and the edge are investigated, along with paralleled chips, as well as chips suffering partial bondwire lift-off. Results are also compared with a traditional electrical temperature measurement method: the voltage drop under low current ($V_{CE(low)}$). In all cases, the I_{GPeak} method is found to provide a temperature slightly overestimating the temperature of the gate pad. Consequently, both the gate pad position and chip temperature distribution influence whether the measurement is representative of the mean junction temperature. These results remain consistent after chips are degraded through bondwire lift-off. In a paralleled IGBT configuration with non-negligible temperature disequilibrium between the chips, the I_{GPeak} method delivers a measurement based on the average temperature of the gate pads.

Thermal loading of power devices is closely related to the reliability performance of the whole converter system. The electrical loading and device rating are both important factors that determine the loss and thermal behaviors of power semiconductor devices. In the existing loss and thermal models, only the electrical loadings are focused and treated as design variables, while the device rating is normally predefined by experience with limited design flexibility. Consequently, a more complete loss and thermal model is proposed in reference [31], which takes into account not only the electrical loading but also the device rating as input variables. The quantified correlation between the power loss, thermal impedance, and silicon area of IGBT is mathematically established. By this new modeling approach, all factors that have impacts to the loss and thermal profiles of the power devices can be mapped accurately, enabling more design freedom to optimize the efficiency and thermal loading of the power converter. The proposed model can be further improved by experimental tests, and it is well agreed by both circuit and finite element method (FEM) simulation results.

In reference [32], the static and dynamic thermal behavior of the IGBT module system mounted on a water-cooled heat sink is analyzed. Although the three-dimensional finite element method (3-D FEM) delivers very accurate results, its usage is limited by an imposed computation time in arbitrary load cycles. Therefore, an RC component model (RCCM) is investigated to extract thermal resistances and time constants for a thermal network. The uniqueness of the RCCM is an introduction of the time constants based on the Elmore delay, which represents the propagation delay of the heat flux through the physical geometry of each layer. The dynamic behavior predicted by the thermal network is equivalent to numerical solutions of the 3-D FEM. The RCCM quickly offers an insight into the physical layers of the components and provides useful information in a few minutes for the arbitrary or periodic power waveforms. This approach enables a system designer to couple the thermal prediction with a circuit simulator in order to analyze the electro-thermal behavior of IGBT module systems simultaneously.

Reference [33] adds to the second aspect by examining the effect of cyclic junction temperature variations ΔT_j of low amplitude in different stages of the power module aging process. It is found that such relatively minor stress cycles, which happen frequently during normal operation, may not be able to directly initiate a crack but can contribute to the development of damage due to stress concentration. This agrees with the observation that the aging process tends to accelerate towards the end of life. This study investigates the dependence of the aging effect on the amplitude of ΔT_j, the mean junction temperature T, and the present health condition of the module, and proposes a lifetime model focusing on die-attach solder fatigue. It is assumed that the future aging process is independent of the operational history that has led to the current state of health. The model is intended for operational management of converter systems that are subjected to frequent low ΔT_j stress cycles and are supposed to be in service reliably for a long time with a slow aging process. Experimental results validate the model.

Reference [34] presents a series of experimental results on the aging effects of cyclic junction temperature variations (ΔT_j) of low amplitudes in power modules, to help in the capture of module reliability characteristics and the derivation of lifetime models in the future. Power cycling tests, for non-aged and aged modules, are designed to illustrate the failure mechanisms. IGBT modules in actual converters are usually operated in a ΔT_j range up to 40 °C; therefore, tests are carried out to observe the effects of such narrow ΔT_j stress cycles on the module lifetime. It is found that the relatively minor stress cycles may not be able to directly initiate a crack but can contribute to the development of damage in the die attach solder layer due to stress concentration. Finite element analysis modeling is utilized to verify the stress concentration effect. The experiment results show that the effects of the narrow ΔT_j stress cycles are affected by the aging status of the module and the stress level itself.

Reference [35] proposes a novel on-board condition monitoring of the aging of solder layers in IGBTs for electric vehicle applications. The diagnostic technique makes use of the chip itself as a temperature sensor while current sensors are already in place for control purposes. An auxiliary power supply unit that can be created from the 12 V battery and an in situ data-logger circuit is developed for condition monitoring. The novel aspect of the proposed technique relates to monitoring IGBTs when the electric vehicle is operating during stop-and-go traffic conditions or at routine services. The accelerated aging tests are performed on the test vehicles and the condition monitoring system is validated using simulation and thermoelectrical experimentation. The thermal performance of the thermal resistance/impedance and junction temperature of the IGBTs demonstrates the effectiveness of the proposed technique for IGBT health monitoring.

Reference [36] discusses the estimation of possible device destruction inside power converters in order to predict failures by means of simulation. The study of IGBT thermal destruction under short-circuits is investigated. An easy experimental method is presented to estimate the temperature decay in the device from the saturation current response at low gate-to-source voltage during the cooling phase. A comparison with other classical experimental methods is given. Three one-dimensional thermal models are also studied: the first is a thermal equivalent circuit represented by series of resistance-capacitance cells; the second treats the discretized heat-diffusion equation; and the third is an analytical model developed by building an internal approximation of the heat-diffusion problem. It is shown that the critical temperature of the device just before destruction is larger than the intrinsic temperature, which is the temperature at which the semiconductor becomes intrinsic. The estimated critical temperature is above 1050 K, so it is much higher than the intrinsic temperature. The latter value is underestimated when multidimensional phenomena are not taken into account. The study is completed by results showing the threshold voltage and the saturation current degradation when the IGBT is submitted to a stress (a repetitive short circuit).

11.3.2.2 MOSFET
Reference [37] presents a novel real-time power-device temperature estimation method that monitors the power MOSFET's junction temperature shift arising from thermal aging effects and incorporates the updated electrothermal models of power modules into digital controllers. Currently, the real-time estimator is emerging as an important tool for active control of the device junction temperature as well as online health monitoring for power electronic systems, but its thermal model fails to address the device's ongoing degradation. Because of a mismatch of coefficients of thermal expansion between layers of power devices, repetitive thermal cycling will cause cracks, voids, and even delamination within the device components, particularly in the solder and thermal grease layers. Consequently, the thermal resistance of power devices will increase, making it possible to use thermal resistance (and junction temperature) as key indicators for condition monitoring and control purposes. In this reference, the predicted device temperature via threshold voltage measurements is compared with the real-time estimated ones, and the difference is attributed to the aging of the device. The thermal models in digital controllers are frequently updated to correct the shift caused by thermal aging effects. Experimental results on three power MOSFETs confirm that the proposed methodologies are effective to incorporate the thermal aging effects in the power-device temperature estimator with good accuracy. The developed adaptive technologies can be applied to other power devices such as IGBTs and SiC MOSFETs, and have significant economic implications.

11.4 External Faults Prognosis

The load faults cause shutdown of the converters, which leads to the non-resilient operation of the system, especially in multi-load systems. Therefore, load fault prognosis is important to prevent the effect of load failure. As an example, the motor fault prognosis is important in mission-critical motor drives. In modern drives, stator winding is usually controlled by a high-frequency PWM controller. This leads to a capacitive parasitic current in stray capacitors between winding and the motor core. In addition, bearings are the only mechanical connection between the rotor and stator in an electric motor. Therefore, bearings are affected by a high-frequency current and the resulting losses cause failure in the bearings. Bearing temperature and current help to detect the problem before failure. Figure 11.21 shows the position of a bearing in an electric motor, while Figure 11.22 shows the high-frequency current path. This current passes through a parasitic capacitor between the stator winding and rotor.

Rotor windings in squirrel cage induction motors are manufactured from aluminum alloy, copper, or copper alloy. Larger motors generally have rotors and end-rings fabricated out of these whereas motors with ratings less than a few hundred horsepower generally have die-cast aluminum alloy rotor cages. Broken rotor bars rarely cause immediate failures, especially in large multi-pole (slow-speed) motors. However, if there are enough broken rotor bars, the motor may not start as it may not be able to develop sufficient accelerating torque. Regardless, the presence of broken rotor bars precipitates deterioration in other components that can result in time-consuming and expensive fixes.

Replacement of the rotor core in larger motors is costly; therefore, by detecting broken rotor bars early, such secondary deterioration can be avoided. The rotor can be repaired at a fraction of the cost of rotor replacement, not to mention averting production revenue losses due to unplanned downtime.

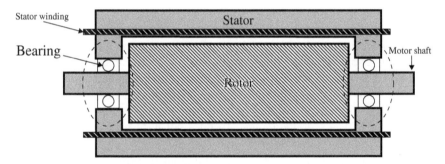

Figure 11.21 Schematic diagram of the bearing position in an electric motor.

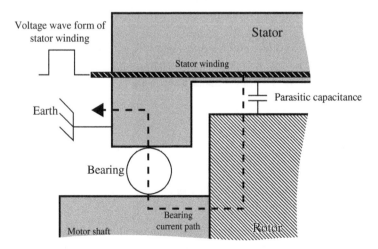

Figure 11.22 High-frequency harmonic current path through the bearing.

Some of the more common secondary effects of broken rotor bars are:

- Broken bars can cause sparking, a serious concern in hazardous areas.
- If one or more rotor bars are broken, the healthy bars are forced to carry additional current leading to rotor core damage from persistent elevated temperatures in the vicinity of the broken bars and current passing through the core from broken to healthy bars.
- Broken bars cause torque and speed oscillations in the rotor, provoking premature wear of bearings and other driven components.
- Large air pockets in die-cast aluminum alloy rotor windings can cause non-uniform bar expansion, leading to rotor bending and imbalance that causes high vibration levels from premature bearing wear.
- As the rotor rotates at a high radial speed, broken rotor bars can lift out of the slot due to centrifugal force and strike against the stator winding, causing catastrophic motor failure.
- Rotor asymmetry (the rotor rotating off-center), both static and dynamic, could cause the rotor to rub against the stator winding, leading to rotor core damage and even a catastrophic fault.

Motor current signature analysis technology has existed for many years to help diagnose problems in induction motors related to broken rotor bars, air gap eccentricity, drive-train wear analysis, and shaft misalignment. The technology relies on the fact that each of these problems produces recognizable frequency patterns in the motor load current that can be predicted by using empirical formulae and measurements. These problems give rise to magnetic asymmetry in the rotor air gap that produces current components at specific frequencies in the load current.

A trace of the motor supply current is obtained by using a clamp-on current probe either from one of the main phase leads to the motor or from the secondary side of a motor current transformer (CT). A fast Fourier transform is performed on the time-domain data to obtain a frequency spectrum. Depending on the device used, this can be done either by the datalogger itself or by computer software.

Once the frequency spectrum is obtained and stored, empirical formulae are used to look for frequency signatures in the spectrum within various frequency ranges, depending on the problem to be diagnosed. For example, broken rotor bar frequencies (also called sidebands or pole-passing frequencies) usually can be found within ± 5 Hz of the motor supply frequency; for air gap eccentricity a wider range is required for the search, from a few hundred Hz up to a few kHz. If the predicted frequency patterns are present in the spectrum, a positive diagnosis is returned. In all cases, accurate estimate of the operating slip of the motor is a prerequisite to reliable diagnosis as the predictor equations require operating slip as one of the input parameters. In an induction motor, slip is dependent on the load and increases with increased load. In most cases, the only knowledge a tester would have regarding slip is that at full load; the motor nameplate data contains the rated speed at a rated horsepower and the slip can therefore be easily derived when the motor is running at the full rated load. However, as motors rarely operate at exactly full load, determining the operating slip becomes a challenge.

11.5 Summary and Conclusions

In this chapter, the methods of fault prognosis in power electronic systems are described. The results of this study are listed as the following.

1) System fault prognostic techniques have been the considerable subjects of condition-based maintenance systems in recent times due to the potential advantages that could be gained from

reducing downtime, decreasing maintenance costs, and increasing machine availability. In the maintenance of engineering systems, condition monitoring, fault diagnosis, and fault prognosis constitute some of the principal tasks. Fault prognosis prevents the unscheduled shutdowns in the system with difficult availability.

2) As per the literature, current fault prognosis methods can be generally classified as three types: model-based methods, knowledge-based methods and data-based methods. Model-based methods are approaches that build an accurate mathematical description of systems using physics or first-principle. Knowledge-based methods are based on engineering experience and historical events, so the prognosis results obtained by these methods are more intuitive. Apart from the aforementioned two types of prognosis methods, the data-based method mainly determines the health status of the system within a certain period of time by analyzing the previously observed data.

3) In the field of power electronics, fault prognosis is done in various types and levels. Both converter-level faults and element-level faults are investigated to ensure the continuous operation of the system. The expected lifetime of a power electronic converter is limited by its fragile components, such as semiconductor devices and capacitors. These components are prone to wear-out failures depending on the converter mission profile and component thermal characteristics. Capacitors and active elements are the most important elements that must be checked by the fault prognosis methods.

4) Capacitor fault prognosis is based on the long-term variation of its parameters, such as ESR and capacitance. In power modules, wire bond life-off is checked by the collector-emitter voltage and the solder fatigue is checked by the thermal resistance of the module.

5) Load fault prognosis prevents the non-resilient operation of the system. especially in multi-load systems.

References

1 Schwan, M., Schilling, K., Zickler, U., and Schnettler, A. (2006). Component reliability prognosis in asset management methods. In: *2006 International Conference on Probabilistic Methods Applied to Power Systems*, Stockholm, Sweden, 1–6.

2 Teng, W., Han, C., Hu, Y. et al. (2020). A robust model-based approach for bearing remaining useful life Prognosis in wind turbines. *IEEE Access* 8: 47133–47143.

3 Montanari, G.C., Hebner, R., Morshuis, P., and Seri, P. (2019). An approach to insulation condition monitoring and life assessment in emerging electrical environments. *IEEE Transactions on Power Delivery* 34 (4): 1357–1364.

4 Gao, Z., Cecati, C., and Ding, S.X. (2015). A survey of fault diagnosis and fault-tolerant techniques – Part I: Fault diagnosis with model-based and signal-based approaches. *IEEE Transactions on Industrial Electronics* 62 (6): 3757–3767.

5 Sun, Q., Wang, Y., Jiang, Y., and Shao, L. (2017). Condition monitoring and prognosis of power converters based on CSA-LSSVM. In: *2017 International Conference on Sensing, Diagnostics, Prognostics, and Control (SDPC)*, Shanghai, China, 524–529.

6 Gao, Z., Cecati, C., and Ding, S.X. (2015). A survey of fault diagnosis and fault-tolerant techniques – Part II: Fault diagnosis with knowledge-based and hybrid/active approaches. *IEEE Transactions on Industrial Electronics* 62 (6): 3768–3774.

7 Biswas, S.S., Srivastava, A.K., and Whitehead, D. (2015). A real-time data-driven algorithm for health diagnosis and prognosis of a circuit breaker trip assembly. *IEEE Transactions on Industrial Electronics* 62 (6): 3822–3831.

8 Nasser, L. and Curtin, M. (2006). Electronics reliability prognosis through material modeling and simulation. In: *2006 IEEE Aerospace Conference*, Big Sky, MT, USA, 1–7.

9 Spagnuolo, G., Xiao, W., and Cecati, C. (2015). Monitoring, diagnosis, prognosis, and techniques for increasing the lifetime/reliability of photovoltaic systems. *IEEE Transactions on Industrial Electronics* 62 (11): 7226–7227.

10 Wang, H. and Blaabjerg, F. (2014). Reliability of capacitors for DC-link applications in power electronic converters – An overview. *IEEE Transactions on Industry Applications* 50 (5): 3569–3578.

11 Soliman, H., Wang, H., and Blaabjerg, F. (2016). A review of the condition monitoring of capacitors in power electronic converters. *IEEE Transactions on Industry Applications* 52 (6): 4976–4989.

12 Kankanamalage, R.R., Foster, M.P., and Davidson, J.N. (2019). Online electrolytic capacitor prognosis system for PWM drives. In: *2019 21st European Conference on Power Electronics and Applications (EPE '19 ECCE Europe)*, Genova, Italy, P.1–P.8.

13 Pu, X., Nguyen, T.H., Lee, D. et al. (2013). Fault diagnosis of DC-link capacitors in three-phase AC/DC PWM converters by online estimation of equivalent series resistance. *IEEE Transactions on Industrial Electronics* 60 (9): 4118–4127.

14 Imam, A.M., Habetler, T.G., Harley, R.G., and Divan, D.M. (2005). Condition monitoring of electrolytic capacitor in power electronic circuits using adaptive filter modeling. In: *2005 IEEE 36th Power Electronics Specialists Conference, Dresden, Germany*, 601–607.

15 Soliman, H., Wang, H., Gadalla, B., and Blaabjerg, F. (2015). Condition monitoring for DC-link capacitors based on artificial neural network algorithm. In: *2015 IEEE 5th International Conference on Power Engineering, Energy and Electrical Drives (POWERENG)*, Riga, Latvia, 587–591.

16 Wu, Y. and Du, X. (2019). A VEN condition monitoring method of DC-link capacitors for power converters. *IEEE Transactions on Industrial Electronics* 66 (2): 1296–1306.

17 Wang, T., He, Y., Luo, Q. et al. (2017). Self-powered RFID sensor tag for fault diagnosis and prognosis of transformer winding. *IEEE Sensors Journal* 17 (19): 6418–6430.

18 Cabanas, M.F., Melero, M.G., Pedrayes, F. et al. (2007). A new online method based on leakage flux analysis for the early detection and location of insulating failures in power transformers: Application to remote condition monitoring. *IEEE Transactions on Power Delivery* 22 (3): 1591–1602.

19 Gonzalez-Hernando, F., San-Sebastian, J., Garcia-Bediaga, A. et al. (2019). Wear-out condition monitoring of IGBT and MOSFET power modules in inverter operation. *IEEE Transactions on Industry Applications* 55 (6): 6184–6192.

20 Choi, U., Blaabjerg, F., and Lee, K. (2015). Study and handling methods of power IGBT module failures in power electronic converter systems. *IEEE Transactions on Power Electronics* 30 (5): 2517–2533.

21 Liu, Y., Dawei, X., and Fu, Y. (2016). Prognosis of chip-loss failure in high-power IGBT module by self-testing. In: *IECON 2016 – 42nd Annual Conference of the IEEE Industrial Electronics Society*, Florence, Italy, 6836–6840.

22 Babel, A., Muetze, A., Seebacher, R. et al. (2014). Condition monitoring and failure prognosis of IGBT inverters based on on-line characterization. In: *2014 IEEE Energy Conversion Congress and Exposition (ECCE)*, Pittsburgh, PA, USA, 3059–3066.

23 Haque, M.S., Baek, J., Herbert, J., and Choi, S. (2016). Prognosis of wire bond lift-off fault of an IGBT based on multisensory approach. In: *2016 IEEE Applied Power Electronics Conference and Exposition (APEC)*, Long Beach, CA, USA, 3004–3011.

24 Choi, U., Blaabjerg, F., Jørgensen, S. et al. (2017). Reliability improvement of power converters by means of condition monitoring of IGBT modules. *IEEE Transactions on Power Electronics* 32 (10): 7990–7997.

25 Reigosa, P.D., Wang, H., Yang, Y., and Blaabjerg, F. (2016). Prediction of bond wire fatigue of IGBTs in a PV inverter under a long-term operation. *IEEE Transactions on Power Electronics* 31 (10): 7171–7182.

26 Hu, K., Liu, Z., Du, H. et al. (2020). Cost-effective prognostics of IGBT bond wires with consideration of temperature swing. *IEEE Transactions on Power Electronics* 35 (7): 6773–6784.

27 Baker, N., Liserre, M., Dupont, L., and Avenas, Y. (2013). Junction temperature measurements via thermo-sensitive electrical parameters and their application to condition monitoring and active thermal control of power converters. In: *IECON 2013 – 39th Annual Conference of the IEEE Industrial Electronics Society*, Vienna, Austria, 942–948.

28 Baker, N., Munk-Nielsen, S., Iannuzzo, F., and Liserre, M. (2016). IGBT junction temperature measurement via peak gate current. *IEEE Transactions on Power Electronics* 31 (5): 3784–3793.

29 Baker, N., Iannuzzo, F., Beczkowski, S., and Kristensen, P.K. (2019). Proof-of-concept for a Kelvin-emitter on-chip temperature sensor for power semiconductors. In: *2019 21st European Conference on Power Electronics and Applications (EPE '19 ECCE Europe)*, Genova, Italy, P.1–P.8.

30 Baker, N., Dupont, L., Munk-Nielsen, S. et al. (2017). IR camera validation of IGBT junction temperature measurement via peak gate current. *IEEE Transactions on Power Electronics* 32 (4): 3099–3111.

31 Ma, K., Bahman, A.S., Beczkowski, S., and Blaabjerg, F. (2015). Complete loss and thermal model of power semiconductors including device rating information. *IEEE Transactions on Power Electronics* 30 (5): 2556–2569.

32 Chan-Su Yun, P., Malberti, M.C., and Fichtner, W. (2001). Thermal component model for electrothermal analysis of IGBT module systems. *IEEE Transactions on Advanced Packaging* 24 (3): 401–406.

33 Lai, W., Chen, M., Ran, L. et al. (2016). Low ΔT_j stress cycle effect in IGBT power module die-attach lifetime modeling. *IEEE Transactions on Power Electronics* 31 (9): 6575–6585.

34 Lai, W., Chen, M., Ran, L. et al. (2017). Experimental investigation on the effects of narrow junction temperature cycles on die-attach solder layer in an IGBT module. *IEEE Transactions on Power Electronics* 32 (2): 1431–1441.

35 Ji, B., Pickert, V., Cao, W.P., and Xing, L. (2013). Onboard condition monitoring of solder fatigue in IGBT power modules. In: *2013 9th IEEE International Symposium on Diagnostics for Electric Machines, Power Electronics and Drives (SDEMPED)*, Valencia, Spain, 9–15.

36 Ammous, A., Allard, B., and Morel, H. (1998). Transient temperature measurements and modeling of IGBT's under short circuit. *IEEE Transactions on Power Electronics* 13 (1): 12–25.

37 Chen, H., Ji, B., Pickert, V., and Cao, W. (2014). Real-time temperature estimation for power MOSFETs considering thermal aging effects. *IEEE Transactions on Device and Materials Reliability* 14 (1): 220–228.

12

Fault Diagnosis

12.1   Tools and Considerations of Fault Diagnosis

In the previous chapter, the fault prognosis methods of the power electronic converters were reviewed. In those methods, the preliminary signs of the faults are detected by using novel methods. However, a fault may occur in the converter. In the current chapter, we take one further step and assume that a fault occurs in the converter but there is a short time interval between the fault occurrence and catastrophic damage to the converter. Protection methods are the techniques using methods for saving the converter in this condition. The prerequisite requirement of the protection system is fault diagnosis. Fault diagnosis is determining:

1) Which fault occurred?
2) Determining the root cause of the out-of-control status.

The first goal is used to isolate the fault as shown in Figure 12.1. Early diagnosis of process faults, while the plant is still operating in a controllable region, can help to avoid event progression and reduce the amount of productivity losses during abnormal events. The second goal is used for the post-fault analysis. Process fault diagnosis involves interpreting the current status of the plant given sensor readings and process knowledge [1]. Fault diagnosis in industrial processes is a challenging task that demands effective and timely decision-making procedures under extreme conditions of noisy measurements, highly interrelated data, large numbers of inputs and complex interaction between the symptoms and faults [2].

The fault diagnosis can be done in a simpler way with the existing sensors of the converters. Figure 12.2 shows a step-down dc to dc converter. In this converter the voltage regulation algorithm works based on the output voltage feedback of the converter. This sensor can also be used as an indicator of the converter health. For example, failure of the converter capacitor leads to increasing the output voltage ripple, as shown in Figure 12.3. This fault can be used by the existing voltage sensor.

Fault detection (FD) and diagnosis of inverters are extremely necessary for improving power electronic system reliability. This is motivated by solving the uncertainty problem in the fault diagnosis of converters, which is caused by various factors, such as noise of sensors. Reference [3] proposes a Bayesian network-based data-driven fault diagnosis methodology of three-phase inverters. Two output line-to-line voltages for different fault modes are measured, the signal features are extracted using fast Fourier transform, the dimensions of samples are reduced using a principal component analysis, and the faults are detected and diagnosed using Bayesian networks.

Resilient Power Electronic Systems, First Edition. Shahriyar Kaboli, Saeed Peyghami, and Frede Blaabjerg.
© 2022 John Wiley & Sons Ltd. Published 2022 by John Wiley & Sons Ltd.
Companion website: www.wiley.com/go/kaboli/resilientpower

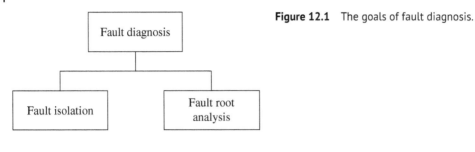

Figure 12.1 The goals of fault diagnosis.

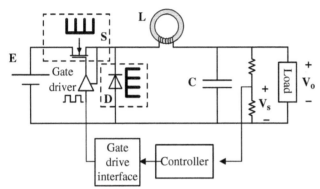

Figure 12.2 A buck converter with the output voltage sensor.

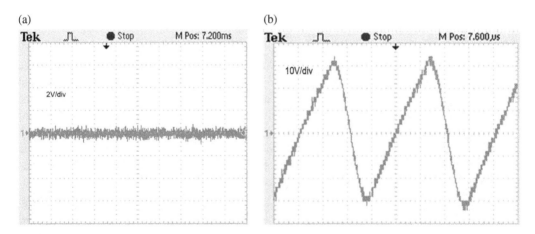

Figure 12.3 Effect of failure in the capacitor of a power supply on the ripple of the output voltage: (a) before failure, (b) after failure.

The main part of the studies of fault diagnosis of the converters focuses on the open or short fault of the switches. A novel fast-diagnostic method for open-switch faults in inverters without sensors is proposed to improve the reliability of the power electronic system in reference [4]. The presented method works by analysis of the switching function model of the inverter under both healthy and faulty conditions. Due to the different voltages drop of each power switch under a healthy condition from open-switch faults in some cases, open-circuit faults can be detected by sensing the collector-emitter voltages of the lower power switches in each leg. The diagnostic scheme employs the hardware circuit to obtain indirectly the voltages of lower power switches.

Also, this method minimizes the detection time and is available for the inverter faults in the systems with open-loop or closed-loop current control strategies.

As the converter becomes complex and is constructed with a huge number of switches, its fault diagnosis becomes difficult. For example, the modular multilevel converter (MMC) is attractive for medium- or high-power applications because of the advantages of its high modularity, availability, and high-power quality. However, reliability is one of the most important issues for MMCs, which are made of a large number of power electronic submodules (SMs). Reference [5] proposes an effective FD and localization method for MMCs. An MMC fault can be detected by comparing the measured state variables and the estimated state variables with a Kalman filter. The fault localization is based on the failure characteristics of the SM in the MMC. The proposed method can be implemented with less computational intensity and complexity, even in the case where multiple SM faults occur in a short time interval.

The fault diagnosis is done at the element-level and system-level. The wire-bond-related failure, one of the most commonly observed packaging failures, is investigated by analytical and experimental methods using the on-state voltage drop as a failure indicator. In reference [6], a sophisticated test bench is developed to generate and apply the required current/power pulses to the device under test. The proposed method is capable of detecting small changes in the failure indicators of the IGBTs and freewheeling diodes and its effectiveness is validated experimentally. The novelty of the work lies in the accurate online testing capacity for diagnostics and prognostics of the power module, with a focus on the wire bonding faults, by injecting external currents into the power unit during the idle time. Test results show that the IGBT may sustain a loss of half the bond wires before the impending fault becomes catastrophic. The measurement circuitry can be embedded in the IGBT drive circuits and the measurements can be performed in situ when the electric vehicle stops in stop-and-go, red light traffic conditions, or during routine servicing. In reference [7], the change rate of the dc reactor voltage with predefined protection voltage thresholds is proposed to provide fast and accurate dc FD in a meshed multiterminal HVDC system. This is equivalent to the measurement of the second derivative of the dc current but has better robustness in terms of electromagnetic-interference noise immunization. In addition to fast dc FD, the proposed scheme can also accurately discriminate the faulty branch from the healthy ones in a meshed dc network by considering the voltage polarities and amplitudes of the two dc reactors connected to the same converter dc terminal. Fast FD leads to lower fault current stresses on dc circuit breakers and converter equipment. The proposed method requires no telecommunication, is independent from power-flow direction, and is robust to fault-resistance variations.

For slow varying parameters, the fault diagnosis system has enough time to check the correct data. Figure 12.4 shows the temperature rise of three transformers in a power supply. It is seen that even the temperature in a faulty transformer reaches its limit in several minutes. However, a fault diagnosis system should usually be fast in order to detect the faults before their catastrophic effects and this fast response should be as smart as the correct command is generated. This topic is explained in the next section as a case study.

12.2 Fault Isolation with Resilience Considerations

Short-circuit fault (SCF) detection is mandatory in a high-voltage dc power supply (HVPS) to prevent fatal damage. The majority of converters employ a single sensor to detect the SCF. This attribute increases the interference vulnerability of the FD system in the presence of noise. Therefore, missed detections and false alarms are possible to occur. Missed detections and false alarms are

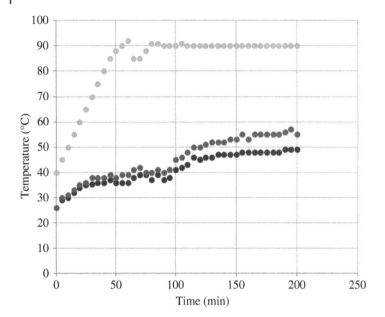

Figure 12.4 Temperature rise of three different transformers in a power supply.

harmful catastrophes in most applications. A commonly used method to suppress noise impacts is to use a low-bandwidth low-pass filter. However, the use of the low-bandwidth low-pass filter reduces the speed of FD due to filter delay. This section presents a fast FD algorithm based on multisensor data fusion. The presented FD system consists of three FD subsystems with high-bandwidth low-pass filters, where each one is individually implemented by sensing three distinct variables and reaches a local decision about the SCF occurrence. Finally, all local decisions are gathered in a decision fusion center and a global decision regarding the converter situation will be made based on the weighted decision fusion.

12.2.1 Definition of the Problem

The high-voltage dc power supplies are essential components in various industries and medical applications such as testing equipment, water purification, and vacuum tubes [8–11]. Among these applications, the vacuum tubes are inherently susceptible to the SCF due to the possibility of an arc occurrence in the vacuum tube. When the SCF occurs in the vacuum tube, the output current increases drastically. Hence, in the HVPS, the correct operation of the protection and the FD system is essential to prevent serious damage both in the converter and the vacuum tube [12]. The structure of the HVPS with the protection system and the FD system is depicted in Figure 12.5. The conventional FD system consists of a load current feedback that is measured by the output current sensor and the protection system is made up of a circuit breaker in the input and a crowbar. When the SCF occurs in the HVPS, the output current increases sharply. Thus, the FD system diagnoses the SCF. Subsequently, the protection system is activated and the crowbar is triggered to divert the stored energy in the output capacitor from the faulty load and the circuit breaker interrupts the input current flow. Many proposals have addressed the FD in different topologies of power electronics converters [5, 13–17]. In references [13] and [14], the SCF detection is based on monitoring the gate-emitter voltage of the IGBT. Due to the sharp increase in the output current in

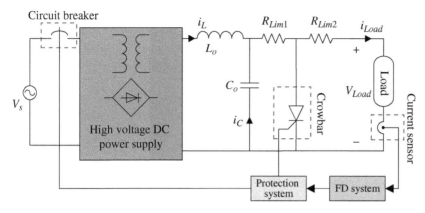

Figure 12.5 Structure of the HVPS with the protection and the FD system.

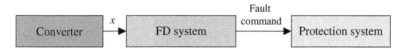

Figure 12.6 Conventional FD system.

the SCF condition, the current changing rate is an appropriate parameter for employing in the FD system. Hence in reference [15], the voltage at a parasitic inductor is used for SCF detection. Several papers have addressed the observer-based [5] and the model-based [16] FD methods. Another diagnosis method based on the duty cycle and the inductor current slope is presented in reference [17]. Most FD methods use a single FD system. The structure of a conventional FD system is depicted in Figure 12.6, where a single variable of the converter, x, is used to process the FD. Generally, the FD system may declare two different incorrect states about the converter situation; first, miss detections during the faulty conditions and, second, false alarms during normal operation of the HVPS [18]. Miss detections may stem from sensors and FD system components failures [19, 20]. In the case of a faulty condition, miss detections can be excessively hazardous due to the possibility of harmful damage in both the converter and the load. False alarms can be the result of noise impacts in the FD system [21]. Disturbing the correct function of the converter without a rational reason is one of the major consequences of false alarms. Moreover, the frequent occurrence of false alarms reduces the availability of the converter. In noise-free environments, all methods can detect the SCF correctly and do not report any false alarm. However, in the presence of noise, the FD system is exposed to interference due to the use of a single sensor. As a result, false alarms are repeatedly reported and the availability of the converter decreases considerably. In this section, in order to overcome false alarm problems, a proposed algorithm based on multisensor data fusion is presented to reduce the noise impacts. Multisensor data fusion theory can be applied to increase accuracy and reliability of data. For example, a smart car is equipped with multiple sensors in collision avoidance applications [22]. Furthermore, multisensor data fusion can be applied to suppress noise impacts. For example, two different kinds of sensors, including a depth camera and an inertial body sensor are applied to improve the human action recognition in the presence of noise [23]. The proposed FD method in this section, which consists of multiple FD subsystems, is shown in Figure 12.7. By increasing the number of FD subsystems, the probability that all FD subsystems are simultaneously interfered with by the noise will decrease [24, 25]. As shown in Figure 12.7, each FD subsystem diagnoses the SCF based on different converter variables, x_i, and makes its local decision, D_i. The subscript i denotes the number of the FD subsystem, where $i = \{1, 2, 3, \ldots, n\}$ and n is the total number of FD subsystems. Ultimately, all the local decisions of subsystems will be gathered in the decision fusion center and the global decision, D_g, about the converter's state will be made based on the weighted decision fusion scheme.

The structure of the conventional FD system is illustrated in Figure 12.8. This FD system consists of a current sensor that measures the load current, a buffer for isolating the sample signal, a low-pass filter for suppressing the noise impacts, a comparator for detecting the current increment, and a latch. In a normal condition, the output of the comparator is "-1" and the protection system is inactive because the current is less than *Threshold*. When the SCF occurs, the load current increases severely and exceeds *Threshold*. Thus, the output of the comparator becomes "+1" and the SCF is detected and the protection system is activated. The conventional FD system operates accurately in noise-free environments. By contrast, the FD system may be interfered with by the noise in noisy environments. As shown in Figure 12.8, in order to prevent false alarm problems, the low-pass filter is used in the FD system to suppress the noise impacts. As the bandwidth of the low-pass filter decreases, the probability of the noise impacts and false alarms reduces, while the speed of the FD system decreases due to the low-pass filter delay. Quick FD is extremely vital in high-voltage applications. Hence a fast novel FD scheme is mandatory in the presence of noise.

Multisensor data fusion algorithms have recently been investigated to be employed in a variety of important applications, such as surveillances, medical applications, and autonomous vehicles [26]. In data fusion theory, multiple sensor data about a specific phenomenon are combined [27]. Improving the data precision and reliability and increasing the noise immunity are

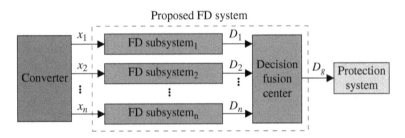

Figure 12.7 Proposed FD system.

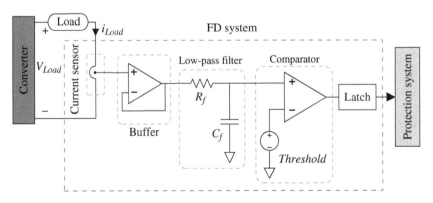

Figure 12.8 Conventional FD system for the HVPS.

the outstanding purposes of multisensor data fusion [28]. Data fusion can be classified into three levels. In the raw data-level fusion, the raw data from sensors are directly combined. In order to combine the raw data, all the sensors should measure an identical physical variable. In the feature-level fusion, the data from sensors are initially processed and features are extracted for each sensor data. Afterwards, the derived features are combined. The feature-level fusion is used when the data from the sensors do not measure an identical physical variable [24]. In decision-level fusion, the data from each sensor are completely processed to make a local decision about a phenomenon. Afterwards, local decisions are combined to make a global decision.

As shown in Figure 12.7, because each FD subsystem makes a decision regarding the HVPS situation, the information of FD subsystems should be combined at the decision level. Accordingly, the weighted decision fusion is reviewed elaborately to be used as a decision fusion algorithm. The structure of the weighted decision fusion scheme is illustrated in Figure 12.9. This algorithm combines n local decisions, D_i, and D_i is denoted by

$$D_i = \begin{cases} -1, \rightarrow \text{if the fault doesn't occur} \\ +1, \rightarrow \text{if the fault occurs} \end{cases} \tag{12.1}$$

In reference [21], the optimum decision fusion rule for making the global decision, D_g, is described as

$$D_g = \begin{cases} +1, \rightarrow \text{if } W_0 + \sum_{i=n}^{i=n} W_i D_i > 0 \\ -1, \qquad \text{otherwise} \end{cases} \tag{12.2}$$

where W_i and W_0 are weights, defined as

$$W_i = \log \left(\frac{\left(1 - P_i^F\right)\left(1 - P_i^M\right)}{P_i^F P_i^M} \right) \tag{12.3}$$

where P_i^M and P_i^F are the probabilities of miss detections and false alarms for the ith FD subsystem, respectively, and

$$W_0 = \log \left(\frac{P_1}{P_0} \right) + \frac{1}{2} \sum_{i=1}^{i=n} \log \left(\frac{P_i^M \left(1 - P_i^M\right)}{P_i^F \left(1 - P_i^F\right)} \right) \tag{12.4}$$

where P_1 and P_0 ($P_0 + P_1 = 1$) are the probabilities of a fault occurrence and a fault non-occurrence, respectively.

Figure 12.9 Weighted decision fusion algorithm.

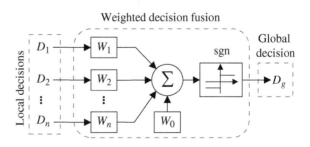

12.2.2 Increasing the Resilience of the Converter

The proposed FD system consists of a number of FD subsystems and the decision fusion center, as depicted in Figure 12.7. Generally, increasing the number of FD subsystems leads to a reduction in the probabilities of false alarms and miss detections [24, 25]. However, it is impractical to increase the number of FD subsystems arbitrarily. Thus, we add just one sensor to the existing sensors that are normally used in the HVPS. The output current sensor and the output voltage sensor are two existing sensors that are normally used for control purposes. The third sensor (which we added) measures the output capacitor current. Accordingly, three ($n = 3$) separate FD subsystems are employed to diagnose the SCF. The first FD subsystem diagnoses the SCF based on the output current, i_{Load}, that is measured by the output current sensor. This sensor is normally used for a control purpose in the HVPS. As shown in Figure 12.10, the first FD subsystem includes a low-pass filter to suppress the noise impacts and a comparator. The output of the comparator is the first local decision, D_1, of this FD subsystem. Eventually, D_1 is sent to the decision fusion center. The second FD subsystem detects the SCF based on the load voltage, V_{Load}, which is measured by the output voltage sensor. V_{Load} is normally used in the converter control loop to regulate it. In addition to the control purpose, the V_{Load} sample can be applied to detect the SCF. The second FD subsystem consists of a low-pass filter to reduce the noise impacts and a comparator. V_{Load} decreases drastically when the arc fault occurs in vacuum tubes [29] and, as a result, the drastic decrease is an appropriate signature for FD. Finally, D_2 is transported to the decision fusion center. The third FD subsystem diagnoses the SCF based on the output capacitor current, i_c, which is measured by the extra sensor that was added. The currents i_{L_F}, i_{C_F}, and i_{Load_F} are the output inductor current, the output capacitor current, and the SCF current in the faulty condition, respectively. The current i_{L_F} grows slightly in the first microseconds after the SCF occurrence. In comparison, i_{C_F} rises extremely fast immediately after the SCF occurrence [30, 31]. Accordingly, it can be concluded that the change in the inductor current is negligible in the first microseconds after the SCF occurrence and i_{C_F} mainly supplies the huge SCF current immediately after the SCF, while Δi_{Load_F} and Δi_{C_F} are the transient changes of the load current and the transient change of the capacitor current immediately after the SCF, respectively. As a result, the third FD subsystem diagnoses the SCF when i_c increases severely. Eventually, D_3 is sent to the decision fusion center to be combined with D_1 and D_2 to make a global decision, D_g.

The decision fusion center receives D_1, D_2, and D_3 regarding the converter situation to make D_g. The weighted decision fusion scheme is used to combine D_1, D_2, and D_3. The structure of the weighted decision fusion scheme is depicted in Figure 12.9. The advantages of the weighted decision fusion scheme are simple implementation and fast reactions. However, P_i^M, P_i^F, P_1, and P_0 should be empirically defined in order to determine the weights, W_0, W_1, W_2, and W_3 [32].

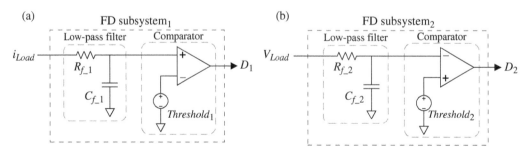

Figure 12.10 First (a) and second (b) FD subsystems.

The probability of the arc occurrence in vacuum tubes is between 0.001 and 0.05 [33], which is the worst case considered in this study. Misdetections can be the result of sensors and FD system component failures [19, 20]. Thus, P_i^M is equal to the failure probability of the ith FD subsystem, P_i^{failure}, which is derived as follows [34]:

$$P_i^M = P_i^{\text{failure}} = 1 - e^{-\lambda_i t} \tag{12.5}$$

where t is the total time that an FD subsystem has operated from time '0' and λ_i is the failure rate of the ith FD subsystem. Since the failure of each component of an FD subsystem leads to complete failure of that FD subsystem, λ_i is equal to the sum of the failure rate of each component of that FD subsystem as follows:

$$\lambda_i = \sum_{k_i=1}^{k_i=h_i} \lambda_{ki} \tag{12.6}$$

where λ_{ki} is the failure rate of the kth component of the ith FD subsystem and $k_i = \{1, 2, 3, \ldots, h_i\}$, where h_i is the total number of components of the ith FD subsystem. The component failure rates are calculated based on MIL-HDBK-217 [34].

P_i^F is derived as follows:

$$P_i^F = \frac{T_i^F}{T} \tag{12.7}$$

where T is the interval that the HVPS uses in the presence of noise. T_i^F is the total interval where the ith FD subsystem declares false alarms and is calculated as

$$T_i^F = \sum_{j=1}^{j=m} t_j \tag{12.8}$$

where $j = \{1, 2, 3, \ldots, m\}$ and m is the total number of false alarms in the T interval and t_j is the interval of each false alarm. Note that the HVPS supplies a dummy load instead of the vacuum tube to guarantee that all FDs are false alarms in the T interval. After defining P_i^M, P_i^F, P_1, and P_0, it is possible to tune W_i and W_0, respectively (Figures 12.11–12.14).

12.2.3 Results of Resilient Operation

In this section, the numerical simulation in PSIM software is executed to verify the performance of the proposed FD system in the presence of noise. The simulation setups for both the conventional FD and the proposed FD with similar situation are depicted in Figure 12.15. When N_1, N_2, and N_3 are random noises, they are added to i_{Load}, V_{Load}, and i_c, respectively. A fast FD is essential in

Figure 12.11 Equivalent circuit of the HVPS in the SCF condition.

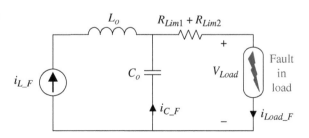

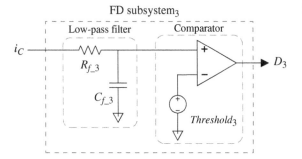

FD subsystem₃ is replaced by a figure.

Figure 12.12 Third FD subsystem.

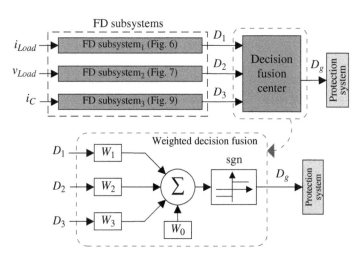

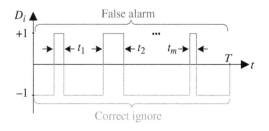

Figure 12.13 Decision fusion center.

Figure 12.14 Time diagram of false alarms announcement.

high-voltage applications. Hence the bandwidth of the fault diagnosis of all FD subsystems and the conventional FD system are considered to be similar when $f_c = 35.5\,\text{kHz}$ reaches a desirable FD time. The bandwidth of an RC low-pass filter is defined by

$$f_c = \frac{1}{2\pi \cdot RC} \tag{12.9}$$

where R and C are the resistor and the capacitor of the RC low-pass filter. The simulation conditions and the FD parameters are shown in Table 12.1 and the converter specifications are given in Table 12.2. The simulation procedure is executed in two following steps.

As mentioned in Section 12.2.1 P_i^M, P_i^F, P_1, and P_0 should be defined to tune W_i and W_0. The probability of an arc occurrence in the vacuum tube is between 0.001 and 0.05 [33], where the

(a)

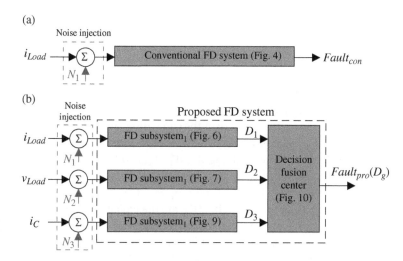

(b)

Figure 12.15 Block diagram of FD system: (a) Conventional FD system, (b) proposed FD system.

Table 12.1 Simulation conditions and FD parameters.

Symbol	Parameter	Value
SNR_1	Signal-to-noise ratio of 1st signal (i_{Load})	6.9
SNR_2	Signal-to-noise ratio of 2nd signal (V_{Load})	4.7
SNR_3	Signal-to-noise ratio of 3rd signal (i_c)	5.4
$1/(2\pi.R_{f-1}. \cdot C_{f-1})$	Bandwidth of 1st FD subsystem	35.5 kHz
$1/(2\pi.R_{f-2}.C_{f-2})$	Bandwidth of 2nd FD subsystem	35.5 kHz
$1/(2\pi.R_{f-3}. \cdot C_{f-3})$	Bandwidth of 3rd FD subsystem	35.5 kHz
$1/(2\pi.R_f. \cdot C_f)$	Bandwidth of the conventional FD system	35.5 kHz

Table 12.2 Converter specifications.

Symbol	Parameter	Value
P_o	Output power	400 W
V_{Load}	Output voltage	4 kV DC
V_s	Input voltage	220 V AC (50 Hz)
i_{Load}	Output current	0.1 A
$i_{Load-Fault}$	Output current in faulty condition	1 A
C_o	Output filter capacitor	23 μF
$R_{Lim1} + R_{Lim2}$	Current limiter resistor$_1$ and resistor$_2$	4 kΩ

worst case is considered in the simulation $P_1 = 0.05$. The probability of the arc fault non-occurrence is defined by

$$P_0 = 1 - P_1 = 1 - 0.05 = 0.95 \qquad (12.10)$$

P_i^M for all FD subsystems are assumed to be similar, $P_1^M = P_2^M = P_3^M = 0.005$, while P_i^F is defined by processing D_1, D_2, and D_3. In the first step, D_1, D_2, and D_3 are recorded to calculate P_i^F during the simulation in the presence of noise. Note that in the first step of the simulation, any deliberate SCF is not applied to the HVPS to guarantee that all FDs of subsystems are false alarms.

In the second step, W_i and W_0 are calculated. After defining W_0, W_1, W_2, and W_3, the weights of the decision fusion center can be tuned. Afterwards, in order to confirm the considerable progress of the proposed FD system, both the conventional FD and the proposed FD are tested through simulation in exactly the same noise level. Fault$_{con}$ and Fault$_{pro}(D_g)$ are the global decisions of the conventional FD and the proposed FD regarding the converter situation, respectively. Fault$_{con}$ and Fault$_{pro}(D_g)$ are recorded during the simulation interval, T_{sim}, to calculate the probability of false alarms of the conventional FD system, $P^{FA}_{Conventional}$, and the probability of false alarms of the proposed FD system, $P^{FA}_{Proposed}$. The simulation results are reported in Figure 12.16, which verify the considerable progress of the proposed FD scheme. According to the simulation results, the proposed FD reduces the probability of false alarms by 93%.

Note that it is possible to decrease the noise impacts in the conventional FD system by decreasing the bandwidth of the low-pass filter of the conventional FD system. In the simulation setup, we

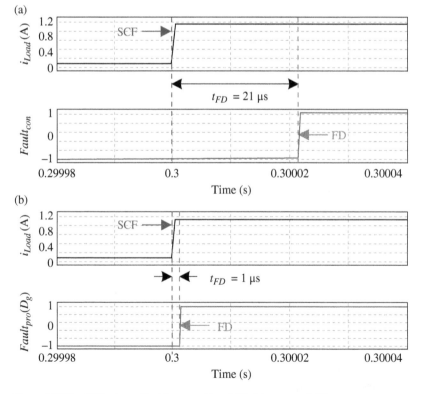

Figure 12.16 FD test results: (a) conventional FD, (b) proposed FD.

decreased only the bandwidth of the low-pass filter of the conventional FD system (without changing all other parameters including the noise level and the low-pass filters of FD subsystems of the proposed FD system) to reduce the probability of false alarms of the conventional FD system by 93% (the same improvement was achieved by using the proposed FD system). In order to reach such decrement (93%), the low-pass filter with a 1.7 kHz bandwidth is needed for the conventional FD. In this new condition, $P^{FA}_{Conventional}$ and $P^{FA}_{Proposed}$ are 2.37×10^{-4}. However, the paramount problem of the low-bandwidth low-pass filter is a considerable delay. The conventional FD system detects the SCF in $t_{FD} = 21\,\mu s$ due to the low-pass filter delay while the proposed FD system detects the SCF in $t_{FD} = 1\,\mu s$, where t_{FD} is the FD time.

Experimental tests were carried out to evaluate the performance of the proposed FD scheme in the presence of noise. In references [35] and [36], an analog random noise is generated based on an avalanche breakdown phenomenon in a zener diode. In the first step of this test, a deliberate SCF is applied to the load to evaluate the performance of each FD subsystem in response to the SCF. Figure 12.17a shows that the first FD subsystem diagnoses the SCF based on the rising i_{Load}. Figure 12.17b shows that the second FD subsystem diagnoses the SCF based on the falling V_{Load}. Figure 12.17c shows that the third FD subsystem diagnoses the SCF based on the rising i_c. It is obvious that the FD time of all FD subsystems is $t_{FD} = 1\,\mu s$.

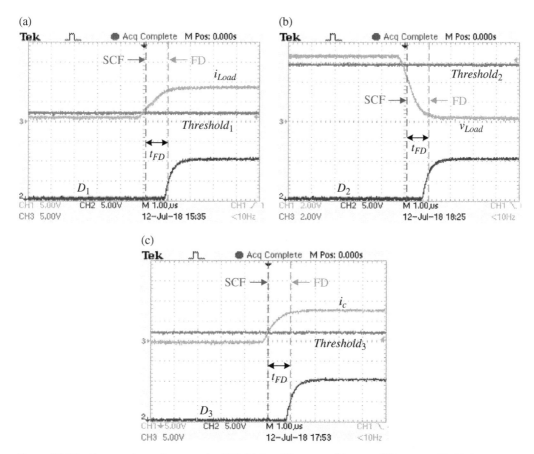

Figure 12.17 Measured waveforms under SCF: (a) first FD system, (b) second FD system, (c) third FD system.

After adjusting the weighted decision fusion center, the second step of the noise immunity test can be executed in the presence of deliberate noise, N_1, N_2, and N_3. In this step, Fault$_{con}$ and Fault$_{pro}(D_g)$ are recorded and processed to calculate $P^{FA}_{Conventional}$ and $P^{FA}_{Proposed}$ to show the considerable improvement of the proposed FD system.

The probability of false alarms is decreased by 83% using the proposed FD system with respect to the conventional FD system. D_1, D_2, D_3, and Fault$_{pro}(D_g)$ of the proposed FD system in the presence of deliberate noise are shown in Figure 12.18 during 10 ms. For example, all FD subsystems declare false alarms repeatedly due to the presence of deliberate noise, but thanks to the proposed FD system, no false alarm is reported, as shown in Figure 12.18a. As shown in Figure 12.18b, the FD system declares a false alarm, because two FD subsystems declare a false alarm simultaneously. However, the probability of simultaneous occurrence of false alarms in at least two FD subsystems

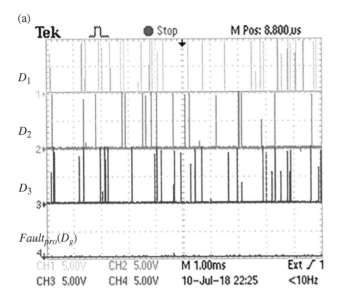

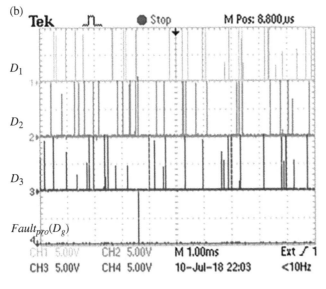

Figure 12.18 D_1, D_2, D_3, and D_g in the presence of noise: (a) with no false alarm, (b) with one false alarm.

is considerably less than the probability of the occurrence of false alarms in the conventional FD scheme that consists of a single FD system.

12.3 Post-Fault Analysis

One of the valuable parts of the fault diagnosis is the post-fault analysis. Investigation of the damaged parts helps the designers to develop new fault diagnosis schemes, tune the existing fault diagnosis parameters, and determine the protection thresholds. In a post-fault analysis, microscopy and probing are the effective methods. During microstructure analysis of metals and alloys, a microscopic examination is conducted to study the microstructural features of the material under magnification. The properties of a material determine how well it will perform under a given application, and these properties are dependent on the structure of the material.

12.3.1 Case Study: Crack in Ceramic

Ceramics are widely used in the package of power electronic modules. Fracture of ceramics is one of the most important failure factors in these modules, as shown in Figure 12.19. In this section, a microscopic assessment is presented of the fracture process of ceramics in the power module. The presented investigation shows a twinning phenomenon in the fractured ceramic. The reason for initiating and propagating the twins is investigated. Increasing the ceramic temperature decreases the required stress level for generating the twins.

Because of the widespread use of ceramics in new industries, its mechanical and electrical properties have been studied over the previous decades. Crack growth is the final step of ceramic fracture under a mechanical stress [37]. Some deformations are seen in the ceramic structure before fracture. The deformation of the ceramic structure under a mechanical stress may be in forms of diffusional processes, dislocation glide, or deformation twinning. All of these deformation processes may have occurred at the same time. The applied strain rate consists of contributions from dislocation motion and twinning, such as the following [38, 39]:

$$\dot{\varepsilon}_{\text{applied}} = \dot{\varepsilon}_{\text{dislocation}} + \dot{\varepsilon}_{\text{twin}} \tag{12.11}$$

where $\dot{\varepsilon}_{\text{applied}}$ is the applied strain rate, $\dot{\varepsilon}_{\text{dislocation}}$ is the dislocation motion, and $\dot{\varepsilon}_{\text{twin}}$ is the twining rate.

Figure 12.19 A broken ceramic of the power module.

Twinning is a phenomenon in materials to handle a stress without fracture. Deformation twinning is the homogeneous shear of a part of a crystal structure such that the resulting structure is unchanged. Therefore, twin growth can be a firm sign of the fracture limit. Both rhombohedral and basal twinning are observed in ceramics [39]. The broken sample is inspected by a field emission scanning electron microscope (FESEM). Figure 12.20 shows a part of the inspected region. The twinning phenomenon is seen in various grains of the ceramic. The average length of the twins is about 20 mm. The twins are in the same direction in a single grain.

(a) (b)

Figure 12.20 Twinning in the broken ceramic: (a) three regions with twinning phenomenon, (b) region A in a higher magnification.

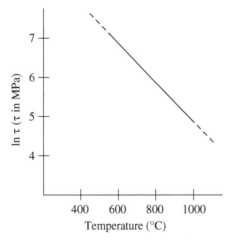

Figure 12.21 The required stress for twin propagation versus increasing the temperature.

The tendency of twinning in ceramics is very low and deformation twins and partial dislocations can only be observed at very high mechanical stress [40]. When the impact velocity is such that the applied stress is lower than the Hugoniot elastic limit (HEL), deformation twins are not activated [40]. To justify the twin growth in the broken sample, a secondary factor is needed to activate the twinning process. Plastic deformation of ceramic crystals has occurred at intermediate temperatures under a considerable pressure to activate basal slip and both basal and rhombohedral deformation twinning [41]. The required stress, t, for twin boundary propagation is strongly dependent on the temperature, as shown in Figure 12.21. It can be seen that the required stress for the twin propagation deceases with increasing temperature. In this condition, a small mechanical stress leads to propagate the twins and cracks. Cracking along twinning interfaces has been observed in a number of brittle materials such as alumina ceramic [41]. Therefore, if there are signs of overtemperature in the studied sample, the proposed hypothesis may be possible: overtemperature reduces the required stress for twin propagation and the crack propagates along the twin boundaries.

To verify this scenario, the studied power module is inspected for proof of overtemperature. The signs of overtemperature conditions can be seen in the surface of the module. Figure 12.22 shows the inner surface of the studied power module. The twinning effect is observed on the surface. This observation indicates that the temperature of copper of the waveguide wall surface increases to the annealing temperature to cause the twinning effect.

12.3.2 Melted Part

In this section, the failure process of the metal part of a power module is investigated for the transient interval of the power supply. Figure 12.23 shows a precise view of the inspected region of the damaged sample. A multi-layer formation is seen in the melted region. This multilayer formation shows that the studied part melted and then was solidified several times.

The studied part is made of annealed 4J34 Kovar alloy. Figure 12.24 shows the microstructures of the alloy. A few annealing twins are seen, which are indicated by white arrows. The damaged part was inspected with the optical microscope OLYMPUS SZX12. Figure 12.25 shows the studied cross-sections of the sample. The sample was cut along the cross-section, which was polished by

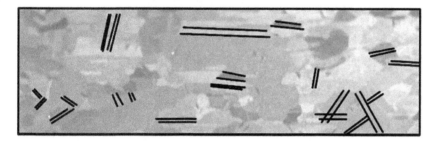

Figure 12.22 Twin growth on the surface of a power module.

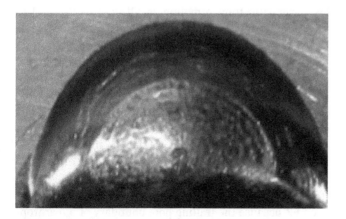

Figure 12.23 Multilayer formation of the melted region.

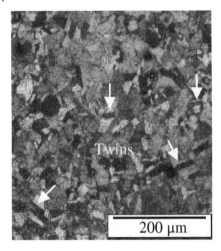

Figure 12.24 Original microstructure of the annealed Kovar 4J34.

sandpaper up to 2000 grit to remove the oxide film and was cleaned with acetone. The section was finally etched to give better recognition of their grain structure. Figure 12.26 shows the grain structure of the section. The structure can be divided into three parts: fusion zone (FZ), heat-affected zone (HAZ), and base material (BM). At the edge of the FZ, the liquid metal is fully mixed.

The presence of a clear boundary line between HAZ and FZ, as shown in Figure 12.26, is a strong sign of high-speed melting. If the sample melted in a slow way, the interface between HAZ and FZ disappeared [41], as was reported in the post-annealed samples [41]. Figure 12.27 shows the section C. There are some semi-melted grains in this figure. This phenomenon occurs when the melting speed is high [42].

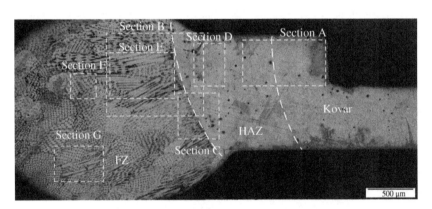

Figure 12.25 Three zones, BM, HAZ, and FZ, in the cross-section.

Figure 12.28 shows the section F in cross-section 1(R) with a clear martensite. The solidification process of the FZ is done by martensitic transformation in all cross-sections. The solid-phase transformation is a diffusional process, which needs both enough time and temperature [41]. In the case of rapid melting and consequent cooling, there is not enough time for diffusive solid-phase transformation. In this condition, the solid phase is directly generated from the γ phase by martensitic transformation [41]. This is another sign of rapid melting in the studied case.

Figure 12.29 shows the section D. Some coarse twins are seen in HAZ. The original grains beside the fusion line form HAZ. As the γ to α phase transformation occurs, coarse twins are generated in HAZ [41]. On the other hand, the direct γ to α phase transformation is a sign of rapid melting and consequently cooling. Therefore, coarse twins in HAZ are a sign of rapid melting.

Figure 12.30 shows the section E. There are some columnar grains at the edge of FZ. The columnar grains grow perpendicularly to the FZ boundary. The straight shape of these columnar grains is a sign of a teardrop-shaped FZ because the trailing pool boundary of a teardrop-shaped FZ is essentially straight. On the other hand, a teardrop-shaped FZ is a sign of high-speed

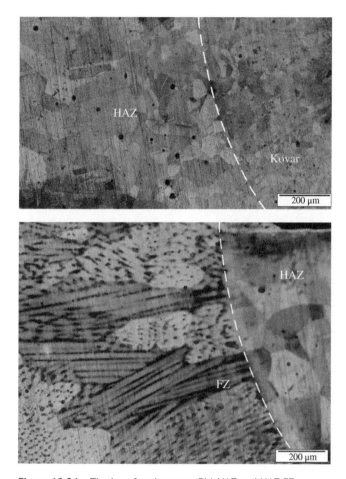

Figure 12.26 The interface between BM-HAZ and HAZ-FZ.

Figure 12.27 Semi-melted grains in section C.

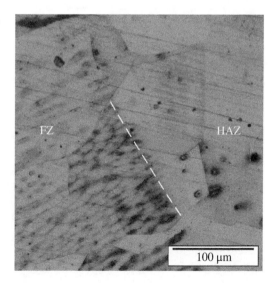

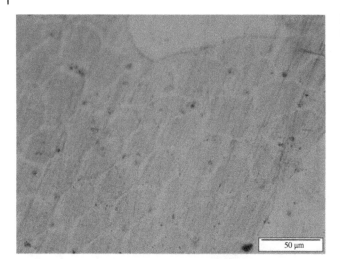

Figure 12.28 Martensitic transformation in section F.

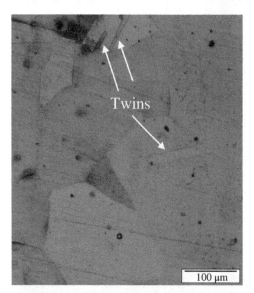

Figure 12.29 Coarse twining in HAZ in section D.

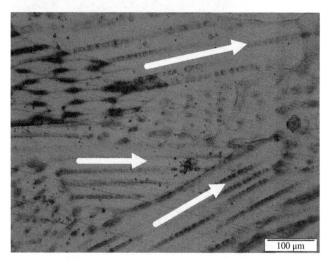

Figure 12.30 Straight columnar formation in section E.

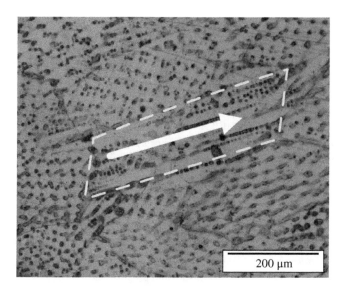

Figure 12.31 Columnar formation inside FZ in section G.

melting [42]. Therefore, straight columnar grains at the edge of the FZ is a sign of high-speed melting.

Figure 12.31 shows the section G. This is a very interesting photo and shows some columnar grains inside FZ. The columnar grains are formed at the edge of FZ. Therefore, the presence of these columnar grains far from the FZ boundary is a strong sign of multimelting of the sample.

12.4 Summary and Conclusions

In this chapter, fault diagnosis in the power electronic converters was described. The main results of the presented study are listed as the following:

1) Fault diagnosis is done for two goals: fault isolation and fault root analysis. Early diagnosis of process faults, while the plant is still operating in a controllable region, can help to avoid events progression and reduce the amount of productivity losses during abnormal events.
2) A fault diagnosis system should usually be fast in order to detect the faults before their catastrophic effects. However, this fast response should be as smart as the right command is generated.
3) Fault diagnosis in industrial processes is a challenging task that demands effective and timely decision-making procedures under extreme conditions of noisy measurements, highly interrelated data, a large number of inputs, and complex interaction between the symptoms and faults. Application of artificial intelligent methods helps to provide resilience in the power electronic converters, especially in a noisy environment.
4) Investigation into damaged parts helps the designers to develop new fault diagnosis schemes, tune the existing fault diagnosis parameters, and determine the protection thresholds. In a post-fault analysis, microscopy and probing are the effective methods.

References

1 Song, Y. and Wang, B. (2013). Survey on reliability of power electronic systems. *IEEE Transactions on Power Electronics* 28 (1): 591–604.

2 Riera-Guasp, M., Antonino-Daviu, J.A., and Capolino, G. (2015). Advances in electrical machine, power electronic, and drive condition monitoring and fault detection: state of the art. *IEEE Transactions on Industrial Electronics* 62 (3): 1746–1759.

3 Cai, B., Zhao, Y., Liu, H., and Xie, M. (2017). A data-driven fault diagnosis methodology in three-phase inverters for PMSM drive systems. *IEEE Transactions on Power Electronics* 32 (7): 5590–5600.

4 An, Q., Sun, L., Zhao, K., and Sun, L. (2011). Switching function model-based fast-diagnostic method of open-switch faults in inverters without sensors. *IEEE Transactions on Power Electronics* 26 (1): 119–126.

5 Deng, F., Chen, Z., Khan, M.R., and Zhu, R. (2015). Fault detection and localization method for modular multilevel converters. *IEEE Transactions on Power Electronics* 30 (5): 2721–2732.

6 Ji, B., Pickert, V., Cao, W., and Zahawi, B. (2013). In situ diagnostics and prognostics of wire bonding faults in IGBT modules for electric vehicle drives. *IEEE Transactions on Power Electronics* 28 (12): 5568–5577.

7 Li, R., Xu, L., and Yao, L. (2017). DC fault detection and location in Meshed multiterminal HVDC systems based on DC reactor voltage change rate. *IEEE Transactions on Power Delivery* 32 (3): 1516–1526.

8 Hwang, F., Shen, Y., and Jayaram, S.H. (2006). Low-ripple compact high-voltage DC power supply. *IEEE Transactions on Industry Applications* 42 (5): 1139–1145.

9 Sun, J., Ding, X., Nakaoka, M., and Takano, H. (2000). Series resonant ZCS-PFM DC-DC converter with multistage rectified voltage multiplier and dual-mode PFM control scheme for medical-use high-voltage X-ray power generator. *IEE Proceedings – Electric Power Applications* 147 (6): 527–534.

10 Barbi, I. and Gules, R. (2003). Isolated DC–DC converters with high-output voltage for TWTA telecommunication satellite applications. *IEEE Transactions on Power Electronics* 18 (4): 975–984.

11 Jang, S.R., Ryoo, H.J., Ahn, S.H. et al. (2012). Development and optimization of high-voltage power supply system for industrial magnetron. *IEEE Transactions on Industrial Electronics* 59 (3): 1453–1461.

12 Joshi, T.G.S. and John, V. (2017). Performance comparison of ETT- and LTT-based pulse power crowbar switch. *IEEE Transactions on Plasma Science* 45 (11): 2994–3000.

13 Rodríguez, M.A., Claudio, A., Theilliol, D., and Vela, L.G. (2007). A new fault detection technique for IGBT based on gate voltage monitoring. In: *2007 IEEE Power Electronics Specialists Conference*, 1001–1005.

14 Rodríguez-Blanco, M.A., Claudio-Sanchez, A., Theilliol, D. et al. (2011). A failure-detection strategy for IGBT based on gate-voltage behavior applied to a motor drive system. *IEEE Transactions on Industrial Electronics* 58 (5): 1625–1633.

15 Wang, Z., Shi, X., Tolbert, L.M. et al. (2014). A di/dt feedback-based active gate driver for smart switching and fast overcurrent protection of IGBT modules. *IEEE Transactions on Power Electronics* 29 (7): 3720–3732.

16 Izadian, A. and Khayyer, P. (2010). Application of Kalman filters in model-based fault diagnosis of a DC-DC boost converter. In: *IECON 2010 - 36th Annual Conference on IEEE Industrial Electronics Society*, 369–372.

17 Shahbazi, M., Jamshidpour, E., Poure, P. et al. (2013). Open- and short-circuit switch fault diagnosis for nonisolated DC-DC converters using field programmable gate Array. *IEEE Transactions on Industrial Electronics* 60 (9): 4136–4146.

18 Fowler, K.R. and Land, H.B. (2004). System design that minimizes both missed detections and false alarms: a case study in arc fault detection. In: *Proceedings of the 21st IEEE Instrumentation and Measurement Technology Conference (IEEE Cat. No.04CH37510)*, vol. 3, 2213–2216.

19 Hussain, S., Mokhtar, M., and Howe, J.M. (2015). Sensor failure detection, identification, and accommodation using fully connected Cascade neural network. *IEEE Transactions on Industrial Electronics* 62 (3): 1683–1692.

20 Liu, Y., Stettenbenz, M., and Bazzi, A.M. (2019). Smooth fault-tolerant control of induction motor drives with sensor failures. *IEEE Transactions on Power Electronics* 34 (4): 3544–3552.

21 Vu, V.T., Pettersson, M.I., Machado, R. et al. (2017). False alarm reduction in wavelength-resolution SAR change detection using adaptive noise canceler. *IEEE Transactions on Geoscience and Remote Sensing* 55 (1): 591–599.

22 Labayrade, R., Royere, C., and Aubert, D. (2005). A collision mitigation system using laser scanner and stereovision fusion and its assessment. In: *IEEE Proceedings. Intelligent Vehicles Symposium, 2005*, 441–446.

23 Chen, C., Jafari, R., and Kehtarnavaz, N. (2015). Improving human action recognition using fusion of depth camera and inertial sensors. *IEEE Transactions on Human-Machine Systems* 45 (1): 51–61.

24 Dasarathy, B.V. (1997). Sensor fusion potential exploitation-innovative architectures and illustrative applications. *Proceedings of the IEEE* 85 (1): 24–38.

25 Yuan, Y. and Kam, M. (2004). Distributed decision fusion with a random-access channel for sensor network applications. *IEEE Transactions on Instrumentation and Measurement* 53 (4): 1339–1344.

26 Luo, R.C., Yih, C.-C., and Su, K.L. (2002). Multisensor fusion and integration: approaches, applications, and future research directions. *IEEE Sensors Journal* 2 (2): 107–119.

27 Hall, D.L. and Llinas, J. (1997). An introduction to multisensor data fusion. *Proceedings of the IEEE* 85 (1): 6–23.

28 Chair, Z. and Varshney, P.K. (1986). Optimal data fusion in multiple sensor detection systems. *IEEE Transactions on Aerospace and Electronic Systems* AES-22 (1): 98–101.

29 Pronko, S.G.E. and Harris, T.E. (2001). A new crowbar system for the protection of high power gridded tubes and microwave devices. *2nd IEEE International Vacuum Electronics Conference*, Noordwijk, April 2001.

30 Farhadi, M. and Mohammed, O.A. (2016). A new protection scheme for multi-bus DC power systems using an event classification approach. *IEEE Transactions on Industry Applications* 52 (4): 2834–2842.

31 Krone, T., Xu, C., and Mertens, A. (2017). Fast and easily implementable detection circuits for short circuits of power semiconductors. *IEEE Transactions on Industry Applications* 53 (3): 2871–2879.

32 Mirjalily, G., Luo, Z.-Q., Davidson, T.N., and Bosse, E. (2003). Blind adaptive decision fusion for distributed detection. *IEEE Transactions on Aerospace and Electronic Systems* 39 (1): 34–52.

33 Koch, L., Lesche, A., and Maring, W. (1996). Fighting arcing and field emission in medical X-ray tubes. In: *Proceedings of 17th International Symposium on Discharges and Electrical Insulation in Vacuum*, vol. 2, 1077–1081.

34 Department of Defense (US) (1995). MIL-HDBK-217F Notice 2: Reliability prediction of electronic equipment. In: *Military Handbook*. Washington, DC.

35 Abdipour, A., Moradi, G., and Saboktakin, S. (2008). Design and implementation of a noise generator. In: *2008 IEEE International RF and Microwave Conference*, 472–474.

36 Arslan, S. and Yıldırım, B.S. (2018). A broadband microwave noise generator using Zener diodes and a new technique for generating white noise. *IEEE Microwave and Wireless Components Letters* 28 (4): 329–331.

37 Ishihara, A., Kondo, S., Tochigi, E. et al. (2014). Observations of crack propagation along a Zr-doped alumina grain boundary. *Microscopy* 63 (1): 20–21.

38 Chen, M.W., McCauley, J.W., Dandekar, D.P., and Bourne, N.K. (2006). Dynamic plasticity and failure of high-purity alumina under shock loading. *Nature Materials* 5 (8): 614–618.

39 Zhao, F., Wang, L., Fan, D. et al. (2016). Macrodeformation twins in single-crystal aluminum. *Physical Review Letters* 116 (7): 075501-1–075501-5.

40 Castaing, J., He, A., Lagerlof, K.P.D., and Heuer, A.H. (2004). Deformation of sapphire (a-Al_2O_3) by basal slip and basal twining below 700°C. *Philosophical Magazine* 84 (11): 1113–1125.

41 Chen, G., Zhang, G., Shu, X. et al. (2018). Effect of annealing on the microstructure and properties of electron beam welded 4J34 Kovar alloy. *Journal of Materials Engineering and Performance* 27: 6758–6764.

42 Kou, S. (2003). *Welding Metallurgy*. Wiley.

Part IV

Methods of Resilience in Power Electronic Systems

13

Resilience Against Internal Faults

13.1 Stress Reduction as a Tool of Resilience

Following the materials presented in previous chapters about mechanisms of fault in power converters, any method that reduces those failure factors can be considered as a technique for reliability improvement. The failure mechanism of a converter is started immediately after starting. In the beginning, the converter operates normally but under stress of failure factors:

- Power losses in the converter cause the temperature to rise in various parts of the converter
- Applied voltage causes an electric field to be applied to insulators
- Mechanical forces lead to vibrations
- Environmental factors

These factors act from the beginning of a converter application. In the long term, they cause the converter to age and its failure [1]. The time interval for changing a stress factor to a failure factor is directly related to the value of the stress. A higher stress leads to a shorter time to failure and vice versa [2, 3]. Thus, the first method of achieving resilience is to reduce the stress factors on the converter. Figure 13.1 shows the stress-strength diagram of a system under two different conditions: low stress and high stress. It is seen that the low stress operation gives more time to the system for failure. This extra time can be used for resilience against the faults. Mission profiles such as environmental and operational conditions together with the system structure, including energy resources, grid, and converter topologies induce stress on different converters and thereby play a significant role in power electronic systems reliability. Temperature swing and maximum temperature are two of the critical stressors on the most failure-prone components of converters, i.e. capacitors and power semiconductors. Temperature-related stressors generate electrothermal stress on these components, ultimately triggering high potential failure mechanisms. Failure of any component may cause converter outage and system shutdown. Reference [4] explores the reliability performance of different converters operating in a power system and indicates the failure-prone converters from a wear-out perspective. It provides a system-level reliability insight for design, control, and operation of multiconverter systems by extending the mission-profile-based reliability estimation approach. The analysis is provided for a dc microgrid due to the increasing interest that dc systems have been gaining in recent years; however, it can be applied for reliability studies in any multiconverter system. The outcomes can be worthwhile for maintenance and risk management as well as security assessment in modern power systems.

Resilient Power Electronic Systems, First Edition. Shahriyar Kaboli, Saeed Peyghami, and Frede Blaabjerg.
© 2022 John Wiley & Sons Ltd. Published 2022 by John Wiley & Sons Ltd.
Companion website: www.wiley.com/go/kaboli/resilientpower

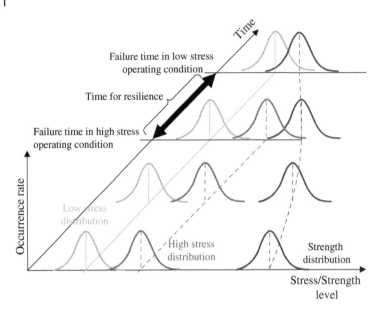

Figure 13.1 Stress-strength diagram of the systems under low and high stress operating conditions.

13.2 Methods of Stress Reduction

There are various stressors in power electronic systems. Any method that reduces one of these stressors helps the system reliability.

13.2.1 Junction Temperature

The overtemperature is one of the most important parameters affecting the failure rate of a component. Ambient and operating temperatures have a major impact on the failure rate prediction results of electronic equipment, especially equipment involving semiconductors and integrated circuits. For example, increasingly thermally stressful environments are seen in applications such as electric vehicles, where ambient temperatures under the hood exceed 150 °C, while some wind turbine applications can place large temperature cycling conditions on the system [5]. A thermal analysis should be a part of the design and reliability analysis process for electronic equipment.

In MIL-HDBK-217 Equation (13.1) is determined for the MOSFET failure rate:

$$\lambda_p = \lambda_b \times \pi_T \times \pi_A \times \pi_Q \times \pi_E \text{ Failure} / 10^6 \text{ hours} \tag{13.1}$$

where π_T is the temperature factor, which is deduced from

$$\pi_T = \exp\left\{-1925 \times \left(\frac{1}{T_j + 273} - \frac{1}{298}\right)\right\} \tag{13.2}$$

where T_j is the junction temperature that is related to MOSFET power losses. Therefore, for determination of π_T, one important step is evaluation of MOSFET power losses. Reduction of power losses results in the decrement of π_T and therefore the failure rate comes down. The effect of a temperature rise on the failure rate of the MOSFETs and diodes is shown in Figure 13.2.

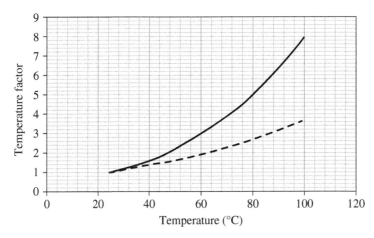

Figure 13.2 A comparison between temperature factor in MIL-217 between diode (dashed) and MOSFET (solid).

A novel real-time power-device temperature estimation method is presented in reference [6], which monitors the power MOSFET's junction temperature shift arising from thermal aging effects and incorporates the updated electrothermal models of power modules into digital controllers. Currently, the real-time estimator is emerging as an important tool for active control of the device junction temperature as well as online health monitoring for power electronic systems, but its thermal model fails to address the device's ongoing degradation. Because of a mismatch of coefficients of thermal expansion between layers of power devices, repetitive thermal cycling will cause cracks, voids, and even delamination within the device components, particularly in the solder and thermal grease layers. Consequently, the thermal resistance of power devices will increase, making it possible to use thermal resistance (and junction temperature) as key indicators for condition monitoring and control purposes. In this reference, the predicted device temperature via threshold voltage measurements is compared with the real-time estimated ones, and the difference is attributed to aging of the device.

13.2.2 Temperature Swing

The junction temperature varies in power cycling whose frequency is comparable with the thermal time constant of the module. Figure 13.3 shows the temperature swing of an IGBT while the case temperature is almost constant. If the power cycling has slow variations, both the junction and the case temperatures vary simultaneously, as shown in Figure 13.4. This phenomenon can be predicted by the thermal equivalent circuit of the insulated gate bipolar transistor (IGBT) in Figure 13.5. In this figure, the variations in junction temperature are attenuated and the case temperature remains constant. The thermal impedance of the switch is shown in Figure 13.6. In reference [7], the effect of a junction temperature swing duration on the lifetime of transfer molded power IGBT modules is studied and a relevant lifetime factor is modeled. This study is based on 39 accelerated power cycling test results under six different conditions by an advanced power cycling test setup, which allows tested modules to be operated under more realistic electrical conditions during the power cycling test. The analysis of the test results and the temperature swing duration-dependent lifetime factor under different definitions and confidence levels are presented. This study enables the temperature swing effect on a lifetime model of IGBT modules to be included for

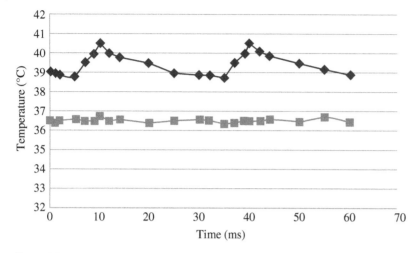

Figure 13.3 Temperature swing of the junction (up) of an IGBT while the case temperature (down) is constant.

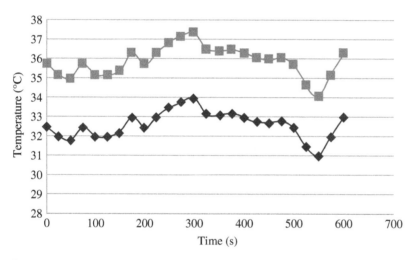

Figure 13.4 Temperature swing of the junction (up) of an IGBT and the case temperature (down) in the slow power cycling.

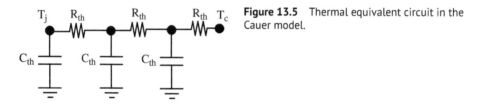

Figure 13.5 Thermal equivalent circuit in the Cauer model.

its lifetime estimation and may result in an improved lifetime prediction of IGBT modules under given mission profiles of converters. A post-failure analysis of the tested IGBT modules is also performed. In reference [8], the lifetime prediction of power device modules based on the linear damage accumulation is studied in conjunction with simple mission profiles of converters. Superimposed power cycling conditions, which are called simple mission profiles in this reference,

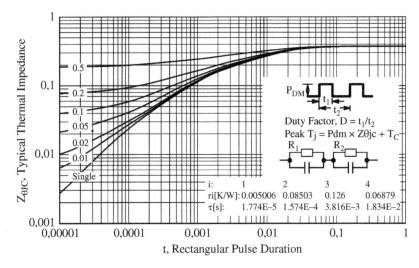

Figure 13.6 Thermal impedance of an IGBT. *Source:* Used with permission from SCILLC dba onsemi.

are based on a lifetime model in respect to junction temperature swing durations. This model has been built based on 39 power cycling test results of 600 V, 30A three-phase-molded IGBT modules. Six tests are performed under three superimposed power cycling conditions using an advanced power cycling test setup. The experimental results validate the lifetime prediction of the IGBT modules based on the linear damage accumulation by comparing the test results with the predicted lifetime from the lifetime model. Furthermore, the test results show the importance of the junction temperature swing duration effect for the lifetime prediction of IGBT modules under power converter applications. Reference [8] presents an experimental study on the aging of IGBT power modules. The aim is to identify the effects of power cycling on these devices with high baseplate temperatures (60–90 °C) and wide temperature swings (60–100 °C). These values for thermal stresses have been defined according to automotive applications. The test conditions are provided by two types of test benches that will be described here. The changes in electrical and thermal indicators are observed regularly by a monitoring system. At the end of the test (reaching damage criterion or failure), different analyses are performed (acoustic scanning and SEM imaging) and the damage is listed systematically. Nineteen samples of 600 V, 200 A IGBT modules were thus aged using five different power-cycling protocols. The final summary of results shows that aging mechanisms mainly concern wire bonds and emitter metallization, with a gradual impact depending on protocol severity.

Reference [9] presents an apparatus and methodology for an advanced accelerated power cycling test of IGBT modules. In this test, the accelerated power cycling test can be performed under more realistic electrical operating conditions with online wear-out monitoring of a tested power IGBT module. The various realistic electrical operating conditions close to real three-phase converter applications can be achieved by the simple control method. Furthermore, by the proposed concept of applying temperature stress, it is possible to apply various magnitudes of temperature swing in a short cycle period and to change the temperature cycle period easily. Thanks to a short temperature cycle period, test results can be obtained in a reasonable test time. A detailed explanation of apparatus such as configuration and control methods for the different functions of accelerated power cycling test setup is given. Then, an improved in situ junction temperature estimation method using on-state collector-emitter voltage V_{CE_ON} and load current is proposed. In addition, a procedure of advanced accelerated power cycling test and test results with 600 V, 30 A transfer molded IGBT modules are presented in order to verify the validity and effectiveness of the proposed apparatus and methodology. Finally, a physics-of-failure analysis of tested IGBT modules is provided.

To prevent the temperature swing, the thermal active control method is proposed. Reference [10] introduces an advanced gate driver used as the thermal swing control method for the reduction of ac load current-related ΔT_j in IGBTs. A switchable gate resistor network is applied to the advanced gate driver, so that the switching power losses can be changed according to the amplitude of the ac current. Accordingly, a closed-loop thermal control method including the functions of a root-mean-square calculation and phase analysis is proposed. Hence ΔT_j can be reduced by means of changing losses-related gate resistors on the basis of output fundamental frequency and amplitude of the ac load current. As a result, a longer device useful life duration can be achieved. Furthermore, the maximum junction temperature under high-temperature operation can be reduced by means of the proposed method. Simulations and experiments are provided to validate the effectiveness of the proposed active gate driver. Active thermal control techniques make it feasible to regulate the steady state and transient thermal-mechanical stress in power electronic modules for applications such as motor drives. Online junction temperature estimation and manipulation of the switching frequency and current limit to regulate the losses are used to prevent overtemperature and power cycling failures in IGBT power modules. The techniques developed in reference [11] are used to actively control the junction temperature of the power module. This control strategy improves power module reliability and increases utilization of the silicon thermal capacity by providing sustained operation at maximum attainable performance limits

13.2.3   Electric Field

A voltage sharing circuit that allows use of low-voltage semiconductor devices to obtain higher output voltages is made up of a series stack of the low-voltage devices with diode limiters provided to bias the low-voltage devices such that the maximum voltage across any one of these devices is limited to a predetermined value. If required, the modules can be connected in series for higher voltage ratings or in parallel for higher current ratings, although when connected in parallel the combined current ratings should be reduced by 20%. Figure 13.7 shows voltage sharing between a series connection of diode in a high-voltage diode stack.

Corona discharge usually forms at highly curved regions on electrodes, such as sharp corners, projecting points, edges of metal surfaces, or small diameter wires. The high curvature causes a

Figure 13.7   A high-voltage series-connected diode stack.

high potential gradient at these locations, so that the air breaks down and forms plasma there first. In order to suppress corona formation, terminals on high-voltage equipment are frequently designed with smooth large-diameter rounded shapes like balls or toruses, and corona rings are often added to insulators of high-voltage parts. In industrial applications, high-voltage power supply spikes with durations ranging from a few microseconds to hundreds of milliseconds are commonly encountered. The electronics within these systems must not only survive transient voltage spikes, but in many cases also operate reliably throughout the event. Application of voltage suppressors as shown in Figure 13.8 helps to keep the device voltage at a low level. Figure 13.9 shows the application of varistors as the voltage suppressors in a high-voltage series connected rectifier stack.

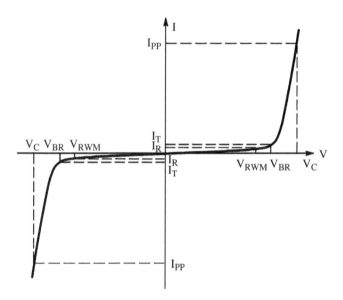

Figure 13.8 I–V curve of a varistor. *Source:* Little Fuse Co. (with permission).

Figure 13.9 Application of varistors in a high-voltage rectifier.

13.2.4 Environmental Factors

Environmental failure factors enter the converters via non-electrical paths. To protect the converters against these factors, the main solution is closing the input channel.

- Humidity
Humidity affects both the components and the systems. Aluminum oxide can dissolve in either acid or alkaline solutions, and leads to corrosion failures on a bonding-pad or internal line. In order to evaluate the reliability of bonding wires inside the dc/dc power, the reliability test structure of bonding wire in the dc/dc power has been designed in reference [12], which has been made by connecting several bonding wire in series, where all of these bonding wires are bonded onto the surface of the aluminum slice, and all the aluminum slice is mounted on the surface of a ceramic substrate. Two ampere electrical currents have been put into the test structure chain to simulate the bonding wire working condition under the operation environment, so the reliability of the dc/dc power under the humidity condition can be assessed. The resistor degradation process of the bonding wire has been conducted through the acceleration of temperature and humidity under the 85 °C/85% RH environment. After 2185 hours of the accelerating lifetime experiment, the reliability of the dc/dc power under the humidity condition has been evaluated. The result is that if the process is well controlled, the bonding wires are robust to the humidity environment and no failure takes place even if there is 5000 PPM of water vapor existing in the package. Variations in the behavior of power supplies caused by environmental conditions require accurate characterization of the electrical behavior dependence with environmental conditions. Reference [13] introduces models that facilitate the prediction of the influence of RH and other environmental factors on the sensitive circuitry in power electronic systems. An application example of a high power density, high voltage dc–dc converter is used to verify the results. Application of conformal coating to protect the power electronic systems against the humidity effect is recommended, as shown in Figure 13.10.
- Mechanical factors
A shock absorber is a mechanical device designed to smooth out or damp a shock impulse, and convert kinetic energy to another form of energy (usually thermal energy, which can be easily dissipated). In a vehicle, shock absorbers reduce the effect of traveling over rough ground, leading to improved ride quality and vehicle handling. While shock absorbers serve the purpose of limiting excessive suspension movement, their intended sole purpose is to damp spring

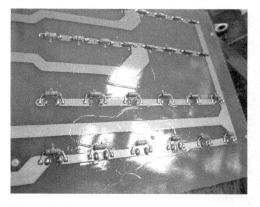 

Figure 13.10 Application of conformal coating to protect the power converters against the humidity effect.

oscillations. Shock absorbers use valving of oil and gasses to absorb excess energy from the springs. Spring rates are chosen by the manufacturer based on the weight of the vehicle, loaded and unloaded. Some people use shocks to modify spring rates but this is not the correct use. Along with hysteresis in the tire itself, they damp the energy stored in the motion of the unsprung weight up and down. Effective wheel bounce damping may require tuning shocks to an optimal resistance. Spring-based shock absorbers commonly use coil springs or leaf springs, though torsion bars are used in torsional shocks as well. Ideal springs alone, however, are not shock absorbers, as springs only store and do not dissipate or absorb energy. Vehicles typically employ both hydraulic shock absorbers and springs or torsion bars. In this combination, "shock absorber" refers specifically to the hydraulic piston that absorbs and dissipates vibration. Figure 13.11 shows the application of a room-temperature-vulcanizing (RTV) adhesive for mechanical stability of electronic parts.

- Dust

Dust is an annoying factor in the power systems. Photovoltaic (PV) systems operating in an outdoor environment are vulnerable to various factors, especially dust impact. Abnormal operations lead to massive power losses, and severe faults such as a short circuit may cause safety problems and fire hazards [14]. However, dust mainly causes a decrease in the voltage strength of the equipment. Reference [15] selects the rod-plate gap as the research object and builds the dust simulation experiment. The platform is used to conduct experimental research on ac discharge characteristics, and the influence of factors such as gap distance and dust volume fraction on the ac-discharge of the rod-plate gap is obtained. When the dust volume fraction is relatively small, the ac breakdown voltage of the tip-plate gap decreases with the increase in the dust volume fraction and the voltage increases with the increase of the dust volume fraction when the sand dust volume fraction is larger. Application of a dust filter is a solution for preventing the dust effect, as shown in Figure 13.12.

13.3 Derating

Derating is a method for allowing the faulty converter to continue its mission with an acceptable performance. Derating is a technique usually employed in power electronic devices, wherein the devices are operated at less than their rated maximum power dissipation, taking into account the case/body temperature, the ambient temperature, and the type of cooling mechanism used. Derating increases the margin of safety between part design limits and applied stresses, thereby providing extra protection for the part. By applying derating in an electric or electronic component, its degradation rate is reduced. The resilience and life expectancy are improved. Although

Figure 13.11 Application of RTV adhesive for mechanical stability of electronic parts.

Figure 13.12   A dust filter with EMI cover.

the derating method is mainly a method for faulty systems, it can also be used for extending life in normal cases. Component failure rates generally decrease as applied stress levels decrease. Thus, derating or operating components at levels below their ratings will increase reliability. Usually, failure happens when the applied load exceeds the strength. "Load" might refer to voltage, power, or an internal stress, such as junction temperature. "Strength" might refer to any resisting physical properties. Even for components made from the same materials and by the same processes, strength differences still exist due to the factors such as microscopic material defects or variations within a single manufacturing process. Therefore, the strength of a component is considered to be a random variable. The load applied to electronic parts, such as power, temperature, or humidity, is also a random variable. Thus, statistical distributions are usually used to describe the load and strength. Derating is a technique through which either stresses acting on a part are reduced or the strength of that part is increased by replacing it with a component with higher rated values. The derating method is well known in the power system. Transformers are normally designed and built for utilizing at a rated frequency and prefect sinusoidal loads. Supplying non-linear loads by a transformer leads to higher losses, early fatigue of insulation, and reduction in the useful life of the transformer. To prevent these problems, the rated capacity of the transformer supplying non-linear loads must be reduced. In reference [16], at first, a useful model of a transformer under harmonic conditions is presented. The model, in addition to ordinary parameters, includes a potential difference defined as the second derivative of the load current that represents eddy current losses in windings and the other stray losses represented as a resistor in series with the leakage inductance and dc resistance. Finally, the losses and capacity of a 50 KVA transformer under a harmonic load current is calculated. The stator winding in a large electric machine must be able to carry the rated current without exceeding specified thermal limits. A medium material like water is usually used to transfer the heat generated in the stator bar. Therefore, it is necessary to place some channels for cooling water inside the stator bar. On the other hand, an ac current carrying a conductor embedded in a narrow slot in a magnetic material drives the magnetic flux around itself and this alternating leakage flux induces alternating voltages along the length of the strips. If solid conductors are used, these voltages would cause circulating currents around the bar, resulting in unacceptable eddy current loss and heating. In order to minimize this effect, the conductor is divided into strips lightly insulated and arranged in

Figure 13.13 A loading curve for derating a generator.

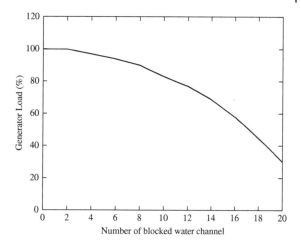

a number of stacks along the bar width. The strips are transposed along the length of the bar using the Roebel method. Some of these strands are hollow in order to pass the cooling water. These hollow strands are blocked because of many factors, such as the existence of external particles in the water and oxidation of the strand. In these cases, the water cannot pass through the channel and the temperature increases. Therefore, the steady-state load of the generator should be decreased to reduce the temperature. There are some operation programs to determine the load of the generator in a faulty case, as shown in Figure 13.13.

13.4 System Derating

One of the most important causes of failure in wind power systems is due to the failures of the power converter and to one of its most critical components, the power semiconductor devices. Reference [17] proposes a novel derating strategy for the wind turbine system based on the reliability performance of the converter and the total energy production throughout its entire lifetime. An advanced reliability design tool is first established and demonstrated, in which the wind power system together with the thermal cycling of the power semiconductor devices are modeled and characterized under a typical wind turbine system mission profile. Based on the reliability design tools, the expected lifetime of the converter for a given mission profile can be quantified under different output power levels, and an optimization algorithm can be applied to extract the starting point and the amount of converter power derating that is necessary in order to obtain the target lifetime requirement with a maximum energy production capability. A nonlinear optimization algorithm has been implemented and various case studies of lifetime requirements have been analyzed. Finally, an optimized derating strategy for the wind turbine system has been designed and its impact has been highlighted. In reference [18], the issues of derating the design of an integrated circuit are discussed. The disadvantage of the traditional derating design method is discussed. This reference puts forward the method using Saber to simulate the electric stress of the electronic product, which can improve the accuracy of the result and the speed of the transient stress calculation. The method that uses the derating analysis based on a Saber simulation is presented. By using this new method, the accuracy of the derating analysis results is greatly improved and the instantaneous stress value can be simulated. To illustrate the method, the implementation to a power supply module of an IC controller is taken as an example. Multiphase machines offer inherent tolerance to faults such as open converter legs (OCLs), which are especially frequent. Because of this, they are particularly attractive for applications where fault tolerance is important, such as offshore wind energy or aerospace, naval and military vehicles. It has been previously shown that, under an OCL, certain stator winding configurations (SWCs) different from star yield a smaller stator copper loss (SCL) and a larger maximum achievable torque (MAT) than star SWC for the same torque command and machine. This advantage comes just at the expense of a moderate increase in converter rating. However, only the case of a single OCL was studied in general. The SCL, MAT, and required converter rating for two OCLs are currently unknown for different combinations of phase number, faulted legs, and SWCs. Actually, under two OCLs (unlike for one OCL) it may be possible to actively modify the order of the faulted/healthy legs to enhance the performance in these terms, but this possibility has not been studied so far in spite of its potential. Reference [19] addresses the post-fault performance of multiphase drives with two OCLs, for various SWCs. The MAT (derating factor), SCL, and necessary converter rating are assessed in numerous possible scenarios. The most convenient alternatives are established. Most importantly, in view of the conclusions of this analysis, a novel method is proposed to improve substantially the MAT and SCL by actively altering the connections between the converter and machine terminals. Experimental results with 6-phase and 12-phase setups are provided. In reference [20], a derated MPPT (Maximum Power Peak Tracking) scheme is proposed for a grid integrated solar PV system, and, due to a reverse power flow toward the grid, the voltage on the PCC (Point of Common Coupling) increases. According to the norms of the grid, only up to a 10% increment in the PCC voltage is acceptable. In this situation, when the PCC voltage exceeds the limit, then a single option is to derate the MPPT operation. In this work, a derated MPPT scheme is introduced, which is a modified form of the most popular perturb and observe (P&O) MPPT algorithm. The Derated P&O (DPO) MPPT algorithm operates for the MPP (Maximum Power Peak) when the PCC voltage is under the limit. However, when the PCC voltage exceeds the limit, then the DPO operates on a specific point where the PCC voltage is just below the upper limit. The derating practice combined with the genetic optimization technique is used on a full bridge converter circuit for attaining targeted reliability at a minimum possible cost. In reference [21], all kinds of stress factors (voltage stress factor, current stress factor, power stress factor, and temperature stress factor) applied on components of a simulated module have been factored in. Reliability optimization problems involve a complex method of selection of components with multiple choices that produce the desired result. A Genetic Algorithm method has been applied on a Full Bridge Converter Circuit to demonstrate its usefulness and efficiency in achieving set reliability at a minimum cost. This proposed method is more promising and efficient than other methods of reliability optimization, such as redundancy allocation, as it does not increase the overall complexity and weight of the system.

13.5 Component Derating

Manufacturers will sometimes point out that there is a strong relationship between the junction temperature and the failure rate, frequently modeling this as an Arrhenius curve and predicting perhaps a 10 : 1 increase in the failure rate for a rise in the junction temperature from 130 to 160 °C, based on a 1 eV activation energy. There will be other evidence of derating; for example, high-current devices may be recommended as a "soft start" circuit, in order to prevent damage from an inrush current. Figure 13.14 shows the derating curve of some power devices.

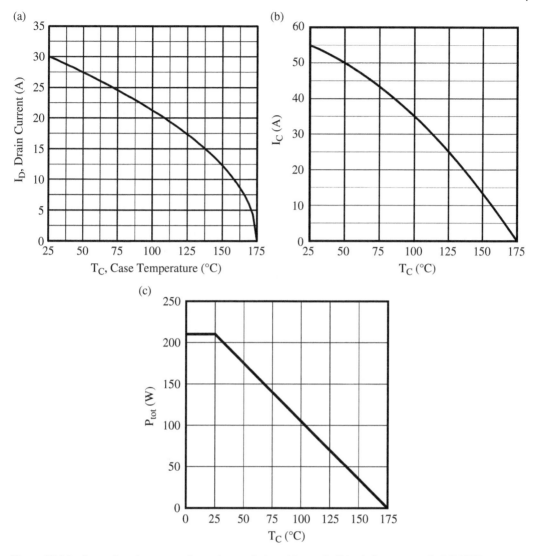

Figure 13.14 Some derating curves for various switches: (a) permissible drain current of a MOSFET versus temperature, (b) permissible drain current of an IGBT versus temperature, (c) recommended derating curve of a MOSFET. *Source:* Used with permission from SCILLC dba onsemi.

Film capacitors are widely assumed to have a superior reliability performance than aluminum electrolytic capacitors in a dc-link design of power electronic converters. However, the assumption needs to be critically judged, especially for applications under high humidity environments. Reference [22] proposes a humidity-dependent lifetime derating factor for a type of plastic-boxed metallized dc film capacitors. It overcomes the limitation that the humidity impact is not considered in the state-of-the-art dc film capacitor lifetime models. The lifetime derating factor is obtained based on a total of 8700 hours of accelerated testing of film capacitors under different humidity conditions, enabling a more justified lifetime prediction of film capacitors for dc-link applications under specific climatic environments. The analysis of the testing results and the detailed discussion on the derating factor with different lifetime definitions and confidence levels are presented.

13.6   Summary and Conclusions

In this chapter, requirements for operation of an electric power converter in low stress conditions were described. Based on the background of previous chapters, stress reduction approaches lead to a long useful life of the systems. The main topics of this chapter are summarized as follows:

1) Any method that reduces the stress of factors affecting reliability can be used for better system design. Some of the methods for reliability improvement act at the hardware level. The aim of these methods is usually reduction of a hot-spot temperature or reduction of an electric field applied to devices. Other methods act as control strategies for low-stress operation of the converter.
2) Application of insulators with high electric strength and an increase of space between points with a high potential difference help to reduce the risk of electric breakdown. Using series and parallel connections of devices helps to reduce voltage and thermal stress, respectively. This method is a commonly used tool for designing a reliable system.
3) Application of mechanical dampers prevents damages due to mechanical vibration and shock. A shock absorber is a mechanical device designed to smooth out or damp a shock impulse, and convert kinetic energy to another form of energy. Isolator coatings are useful tools for the reduction of undesired influences of environmental parameters on devices.
4) Derating is a tool for utilization of a faulty system with degraded performance. A derating factor is affected by environmental and operating points of the system. A derating algorithm can be used for increasing some system functional indices while decreasing some others. The degree of derating is dependent on an acceptable reduction in system performance.

References

1 Chiodo, E. and Mazzanti, G. (2006). Bayesian reliability estimation based on a Weibull stress-strength model for aged power system components subjected to voltage surges. *IEEE Transactions on Dielectrics and Electrical Insulation* 13 (1): 146–159.
2 Zhang, J., Ma, X., and Zhao, Y. (2017). A stress-strength time-varying correlation interference model for structural reliability analysis using copulas. *IEEE Transactions on Reliability* 66 (2): 351–365.
3 Roy, D. and Dasgupta, T. (2001). A discretizing approach for evaluating reliability of complex systems under stress-strength model. *IEEE Transactions on Reliability* 50 (2): 145–150.
4 Peyghami, S., Wang, H., Davari, P., and Blaabjerg, F. (2019). Mission-profile-based system-level reliability analysis in DC microgrids. *IEEE Transactions on Industry Applications* 55 (5): 5055–5067.
5 Baker, N., Liserre, M., Dupont, L., and Avenas, Y. (2014). Improved reliability of power modules: A review of online junction temperature measurement methods. *IEEE Industrial Electronics Magazine* 8 (3): 17–27.
6 Chen, H., Ji, B., Pickert, V., and Cao, W. (2014). Real-time temperature estimation for power MOSFETs considering thermal aging effects. *IEEE Transactions on Device and Materials Reliability* 14 (1): 220–228.
7 Choi, U., Blaabjerg, F., and Jørgensen, S. (2017). Study on effect of junction temperature swing duration on lifetime of transfer molded power IGBT modules. *IEEE Transactions on Power Electronics* 32 (8): 6434–6443.

8 Choi, U., Ma, K., and Blaabjerg, F. (2018). Validation of lifetime prediction of IGBT modules based on linear damage accumulation by means of superimposed power cycling tests. *IEEE Transactions on Industrial Electronics* 65 (4): 3520–3529.

9 Choi, U., Jørgensen, S., and Blaabjerg, F. (2016). Advanced accelerated power cycling test for reliability investigation of power device modules. *IEEE Transactions on Power Electronics* 31 (12): 8371–8386.

10 Luo, H., Iannuzzo, F., Ma, K. et al. (2016). Active gate driving method for reliability improvement of IGBTs via junction temperature swing reduction. In: *2016 IEEE 7th International Symposium on Power Electronics for Distributed Generation Systems (PEDG)*, Vancouver, BC, Canada, 1–7.

11 Murdock, D.A., Torres, J.E.R., Connors, J.J., and Lorenz, R.D. (2006). Active thermal control of power electronic modules. *IEEE Transactions on Industry Applications* 42 (2): 552–558.

12 Zhang, X. and He, X. (2012). The reliability evaluation of the bonding wire in the DC/DC power under the environment of humidity. In: *2012 13th International Conference on Electronic Packaging Technology & High Density Packaging*, Guilin, China, 1357–1359.

13 Ciprian, R. and Lehman, B. (2009). Modeling effects of relative humidity, moisture, and extreme environmental conditions on power electronic performance. In: *2009 IEEE Energy Conversion Congress and Exposition*, San Jose, CA, USA, 1052–1059.

14 Huang, J., Wai, R., and Yang, G. (2020). Design of hybrid artificial bee colony algorithm and semi-supervised extreme learning machine for PV fault diagnoses by considering dust impact. *IEEE Transactions on Power Electronics* 35 (7): 7086–7099.

15 Wang, J., He, Z., Tu, Z. et al. (2019). Study on ac breakdown characteristics of rod – plate gap in sand – dust environment. In: *2019 11th Asia-Pacific International Conference on Lightning (APL)*, Hong Kong, China, 1–5.

16 Sadati, S.B., Tahani, A., Jafari, M., and Dargahi, M. (2008). Derating of transformers under non-sinusoidal loads. In: *2008 11th International Conference on Optimization of Electrical and Electronic Equipment*, Brasov, Romania, –263, 268.

17 Vernica, I., Ma, K., and Blaabjerg, F. (2018). Optimal derating strategy of power electronics converter for maximum wind energy production with lifetime information of power devices. *IEEE Journal of Emerging and Selected Topics in Power Electronics* 6 (1): 267–276.

18 Yan, H. and Wang, T. (2016). Derating design for power supply module based on saber simulation. In: *2016 Prognostics and System Health Management Conference (PHM-Chengdu)*, Chengdu, China, 1–5.

19 Yepes, A.G. and Doval-Gandoy, J. (2021). Study and active enhancement by converter reconfiguration of the performance in terms of stator copper loss, derating factor and converter rating of multiphase drives under two open legs with different stator winding connections. *IEEE Access* 9: 63356–63376.

20 Satapathy, S.S. and Kumar, N. (2019). Derated MPPT scheme for grid Integrated solar PV energy conversion system. In: *2019 International Conference on Computing, Power and Communication Technologies (GUCON)*, New Delhi, India, 975–980.

21 Tripathi, H. and Pradhan, N. (2016). Reliability optimization of electronics module by derating using genetic algorithm. In: *2016 International Conference on Control, Computing, Communication and Materials (ICCCCM)*, Allahbad, India, 1–6.

22 Wang, H., Diaz Reigosa, P., and Blaabjerg, F. (2015). A humidity-dependent lifetime derating factor for DC film capacitors. In: *2015 IEEE Energy Conversion Congress and Exposition (ECCE)*, Montreal, QC, Canada, 3064–3068.

14

Resilience Against External Faults

14.1 Resilient Protection System

Protection systems act when a fault occurs in the converter. Their performance is very important; isolation of the converter is not always the best choice because this strategy has a bad effect on their availability. Availability means the probability that a system is operational at a given time, i.e. the amount of time a device is actually operating as the percentage of total time it should be operating. Availability features allow the system to stay operational even when faults occur. A highly available system would disable the malfunctioning portion and continue operating at a reduced capacity. In contrast, a less capable system might crash and become totally non-operational. Any failure factor needs to have a time interval before damaging the converter. If the protection system is very fast, it protects the converter but conflict with noise is possible. Therefore, protection systems should be fast enough to protect the converter but not so fast as to cause an incorrect operation.

14.1.1 Case Study: Resilient Series Protection Switch

This section presents a high-voltage fault current limiter (HVFCL) for high-voltage dc power supplies (HVDCPSs), which limits the current of the power supply automatically in the short-circuit fault (SCF). The proposed HVFCL is based on series-connected insulated gate bipolar transistors (IGBTs). The main achievements of the chapter are balanced voltage sharing and a very low value of the short-circuit current near to the load nominal current for the series-connected IGBTs during the SCF. These achievements result in a longer maximum permissible short-circuit time (MPSCT) and are obtained by a control strategy that puts the series-connected IGBTs in a specific operating point in the active region during the SCF. HVDCPSs are widely used in industries. Due to the high-voltage output of these power supplies, they are probably subjected to the SCF more than other low-voltage power supplies [1, 2]. The SCF may occur due to voltage breakdown inside the load. Vacuum tubes are one of the most important loads of HVDCPSs and unavoidably exhibit short-circuit behavior [3]. The output SCF could cause catastrophic failure for both the HVDCPS components and the load [4]. Therefore, an effective protection mechanism is essential to limit the fault energy that is deposited into the load during faulty conditions [5]. The block diagram of an HVDCPS and its protection system is shown in Figure 14.1a. When the fault occurs in the load, the fault detection system detects the short circuit and sends the protection command to the HVDCPS and the circuit breaker (CB). Subsequently, the power supply is turned off and the circuit breaker

Resilient Power Electronic Systems, First Edition. Shahriyar Kaboli, Saeed Peyghami, and Frede Blaabjerg.
© 2022 John Wiley & Sons Ltd. Published 2022 by John Wiley & Sons Ltd.
Companion website: www.wiley.com/go/kaboli/resilientpower

isolates the power supply from the grid [6]. Applying an opening switch at the power supply output is usually recommended to prevent releasing the residual stored energy of the power supply into the load, as shown in Figure 14.1b.

The timing diagram of the protection process during an SCF is shown in Figure 14.2. Three time intervals can be distinguished during an SCF at the output of an HVDCPS. The first time interval is a rest time when the amplitude of the HVDCPS output current increases sharply to a high value. In this time interval, the protection system of the HVDCPS is not activated. The main reason for this delay is the suppression of the effect of noise. In a noisy environment, a fast protection system is probably able to send false alarms. Hence, it decreases the availability of the power supply. Therefore, the protection system is activated with some delays in the second time interval to ensure that the detected fault is real. There is another delay interval that is for the performance of the fault terminator unit (HVDCPS switching elements are turned off). Finally, the fault terminates in the third part of the time interval. The amount of short-circuit current value in the first and second intervals is a destructive factor for both the power supply and the load. The effects of the SCF on the lifetime and the reliability of the semiconductor devices have been deeply investigated in reference [7]. The main root cause of the failure is related to the high level of the current circulating in the power supply during the SCF [8]. On the other hand, there are several high voltage applications/loads where a limited short-circuit current is an absolute requirement. These applications are related to the supply of vacuum tubes [9]. When the dc breakdown occurs inside the vacuum tube, an arc on the negative electrode of the tube is generated. This arc is normally known as a vacuum arc [10] and can be terminated rapidly if the SCF current is kept at a limited value [11]. Therefore, limiting the short-circuit current value at the delay intervals is very important. By so doing, the vacuum arc can be interrupted without disconnecting the HVDCPS from the load and bringing it back for normal operation. This mandatory process for the vacuum arc conditions debilitates the availability of the HVDCPS considerably [12].

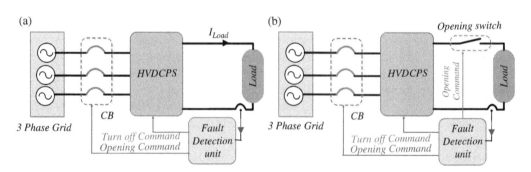

Figure 14.1 Diagram of the high-voltage dc power supply and its protection system: (a) without and (b) with output opening switch.

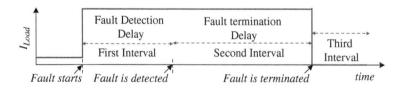

Figure 14.2 Timing diagram of protection against a SCF in a HVDCPS.

For DC applications, different strategies for limiting the fault current have been reported in the state of the art in the field. Regarding the extant schemes, fault current limiting can be performed based on control strategies or with hardware-based methods. In the first category, there are several control strategies used to limit the fault current in the short-circuit interval [13]. Using this approach, the power supply will be current-controlled during the SCF. Regarding the power supply dynamic, the fault current limiting is carried out with a considerable delay. In the second category, a fault current limiter (FCL) device is served. During the SCF, the impedance of FCL increases to limit the short-circuit current value. As an example, a saturable core in series with the target system is used as an FCL [14]. The main shortcoming of this strategy is a mandatory external biasing coil with excessive power loss [15]. To solve this issue, superconducting coils have been proposed [16]. However, their maintenance is a difficult task [17]. All the mentioned solutions need an activation command for the fault current limiting. For the field of high-voltage power supplies (HVPSs), the activation command is subjected to a high level of noise [18]. This will result in a probable malfunction of the FCL. Hence, the availability is sacrificed for the fault current limiting. In addition, the extant FCLs will react with some delays. They also return to the normal condition with some delays [12]. The shortcomings mentioned earlier hinder the practitioners from applying the extant FCL schemes to the applications where a high level of availability and a fast fault current limiting are the absolute necessities. To remove the activation command, reference [19] proposes an FCL structure based on the series-connected IGBTs for pulsed power applications. The IGBTs have a limited short-circuit current due to their current-voltage characteristics in the active region. Hence, the work presented in reference [19] takes advantage of the inherent SCF current limiting of the IGBTs. The series-connected IGBTs have a negligible voltage in the normal condition and bear the high voltage power supply output in the SCF. The important issue in such an FCL is the severely unbalanced voltage for the series-connected IGBTs [19, 20]. The origin of the unbalanced voltage is related to different I–V characteristics of the IGBTs due to intrinsic differences. In reference [19], only a safe operating condition for the series-connected IGBTs is provided for a very small time (less than 20 μs). This fact makes the mentioned approach suitable just for pulsed power applications. Moreover, there is not any accurate control on the short-circuit current value of the series-connected IGBTs in reference [19]. Hence, this approach lacks competence for HVDCPS where the SCF may last for a time interval of more than 20 μs. Hence, using the proposed strategy, the need for the exclusive FCL activator is removed. Moreover, the powers absorbed from the power supply in the normal and short-circuit intervals are approximately equal, due to the approximate similarity between the currents in these intervals. Consequently, the HVDCPS does not experience any disturbance in the short-circuit interval. Hence, as another outcome, the load can be normally fed very quickly after the short-circuit removal, which can be interpreted as a high level of availability for the HVDCPS. The block diagram of the HVFCL based on the series-connected IGBTs is depicted in Figure 14.3. In the nominal condition, the series-connected IGBTs are in the complete on-state. Hence, their voltages are negligible. Consequently, the load voltage (V_{DC} in Figure 14.3) and the power supply output voltage (V_{bus} in Figure 14.3) are approximately the same. When an SCF occurs, the series-connected IGBTs limit the short-circuit current based on their active region characteristics, using a proper control scheme.

In an SCF condition, the voltage and current of the IGBTs increase considerably, which result in a high level of power. Hence, in the passage of time, if the short-circuit energy violates the specific value known as the critical energy, the IGBTs fail. There are several works that have paid extensive attention to find the IGBT's short-circuit tolerable energy [4]. Investigations in reference [21] show that if the energy in the short circuit is below the critical energy, the IGBT can survive the short circuit more than 10000 times. In addition, the critical energy is more than the value reported in

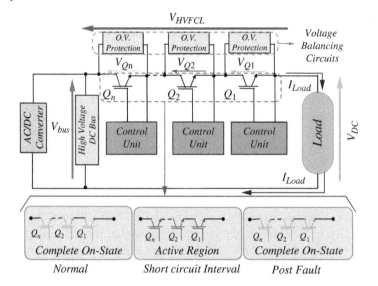

Figure 14.3 The schematic of the HVFCL based on series-connected IGBTs.

the component datasheet [21]. The IGBTs short-circuit energy of each IGBT ($E_{sc,IGBTi}$) can be written as

$$E_{sc,IGBTi} = \int_{0}^{t_{SC}} v_{Qi} i_{SCi} \mathrm{d}t \approx V_{Qi} I_{SCi} t_{SC} = P_{SCi} t_{SC} \tag{14.1}$$

where I_{SCi} is the short-circuit current of each IGBT in the series structure, V_{Qi} is the voltage of each IGBT in the series structure, t_{sc} is the short-circuit time interval, and P_{sc} is the short-circuit average power. The approximation in equation (14.1) is rational since the voltages and currents of the IGBTs are almost constant during the SCF. According to equation (14.1), the product of voltages and currents with a specified value for the short-circuit energy determines the MPSCT. The unequal voltage sharing for the series-connected IGBTs in the SCF interval will result in unbalanced short-circuit energy and a dangerous overvoltage condition [19]. Two methods can be evidenced for providing safe voltage conditions for the series-connected IGBTs. The first proposes active voltage clamping [22]. According to Figure 14.4, all IGBTs have the same and maximum possible short-circuit current. Additionally, their voltages are clamped to a voltage level (V_{clamp}), which is more than the balanced voltage (V_{bus}/n). Accordingly, their short-circuit power (P_{SC}) is very high and considering a limited short circuit energy, the MPSCT is about 10 µs.

The second method proposes an overvoltage protection scheme using the clamp mode resistor, capacitor, and diode (CMRCD) snubbers [19]. The brief of the operation principle of the CMRCD snubbers is provided in Figure 14.5. In this approach, the safe voltage for the series-connected IGBTs is guaranteed but a balanced voltage is not provided for them.

Providing a balanced voltage for the series-connected IGBTs in the short-circuit interval and limiting the short-circuit current to a value less than the saturated current simultaneously increases the short-circuit tolerable time interval. In such a desirable condition, the short-circuit current can be set to values much less than the saturated current. Moreover, the balanced voltage equalizes the instantaneous power loss of the IGBTs during the short-circuit interval. Hence, the short-circuit time can be expanded to a value in which the maximum short-circuit energy of the device is

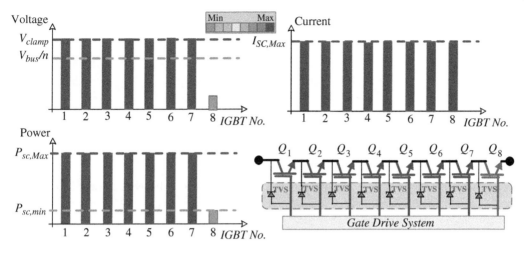

Figure 14.4 The brief operation principle of the active clamp method.

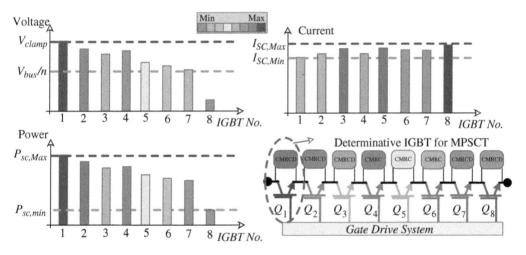

Figure 14.5 The brief operation principle of the CMRCD snubbers.

reported in the datasheet permits. Accordingly, Figure 14.6 shows the expanded permissible short-circuit time attained by the balanced voltage for the series-connected IGBTs. The higher permissible short-circuit time has two main achievements:

- The first is for the temporary SCF, where extending this time increases the resilience of the HVDCPS.
- The second is very practically desirable for the sustained SCF conditions since the opening command of the series-connected IGBTs can be heavily filtered to suppress the effect of noise.

In order to increase the permissible short-circuit time, this section presents a control strategy providing a balanced voltage for the series-connected IGBTs. This strategy can control the short-circuit current of the HVDCPS to a very low level approximately equal to the load nominal current. Figure 14.7 presents the desired result of the proposed strategy. Figure 14.8 represents the MPSCT

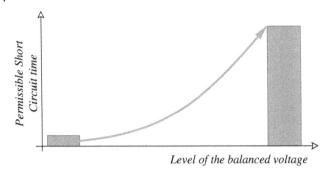

Figure 14.6 Increment of the MPSCT interval by balancing the voltage of the series-connected IGBTs.

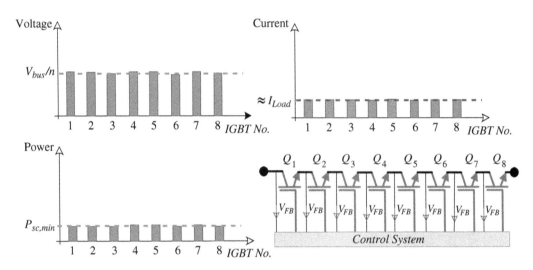

Figure 14.7 The improved results achieved by the proposed structure.

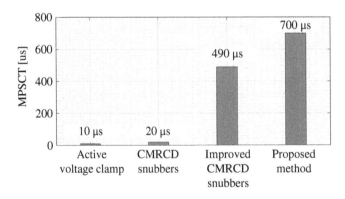

Figure 14.8 MPSCT achieved by different approaches.

achieved by CMRCD snubbers, active voltage clamp, the considered improved CMRCD approaches, and that of the proposed method. Based on the results of Figure 14.8, the MPSCT for the series-connected obtained by the proposed method is more than the extant approaches.

14.1.2 Case Study: Resilient Parallel Protection Switch

High-voltage power supplies (HVPSs) have wide industrial applications, including medical processes, water treatment, and high-power vacuum tubes. In the last application, HVPSs drive high-power microwave tubes. Considering high voltage and high power, these types of microwave tubes face some issues in their reliable operation. One of the most potential issues is the vacuum arc fault in microwave tubes that can lead to irrecoverable damages to the tubes. Regarding the importance and price of high-power microwave tubes, the existence of reliable and fast protection mechanism in the output stage of the HVPS is mandatory. According to information provided by microwave tubes manufacturers, it is recommended that after an arc vacuum fault occurrence, the fault current should be interrupted within $10\,\mu s$ and the injected arc energy (AE) into the tube should not generally exceed a limit of 10 J. The protection mechanisms for protecting the microwave tubes against vacuum arc faults are divided into two categories, namely the opening switch and closing switch, and shunt the crowbar. The series-connected IGBTs are utilized as a series switch to protect the microwave tubes against arc fault and to decrease the restoration time of the HVPS after the fault occurrence. However, some unwanted transients occur during the opening process of the series switch that may be harmful for the microwave tubes. A shunt crowbar shortens the microwave tube. Thus, no transient occurs for the tube during the operation of the protection system. The HVPS block diagram with a shunt crowbar is shown in Figure 14.9.

The crowbar can be made with both solid-state switches and gas discharge switches. Nevertheless, the thyristor-based crowbar, as a solid-sate switch, is more preferable than the others because of its more reliable operation and its long life. However, the thyristor-based crowbars used for protecting microwave tubes suffer from their limited critical rate of rise of an on-state current, which necessitates using a series inductor in the path of the crowbar. This series inductor and the FCL resistors determine the time constant of the crowbar current and, in turn, the fault current interruption time (FCIT) after the fault occurrence. In this section, the IGBT-based crowbar is employed to decrease the amounts of the FCIT and AE, as shown in Figure 14.10. This crowbar is capable of handling a very high slope of the turn-on current. In addition, one IGBT enjoys smaller turn-on delay time in comparison with a thyristor having the same electrical ratings. Therefore, taking the advantage of the IGBT switches, the IGBT-based crowbar not only will be turned on faster than the thyristor-based crowbar but also can divert the whole fault current through its path at once. By

Figure 14.9 The block diagram of the high-voltage dc power supply.

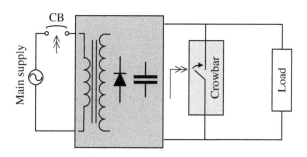

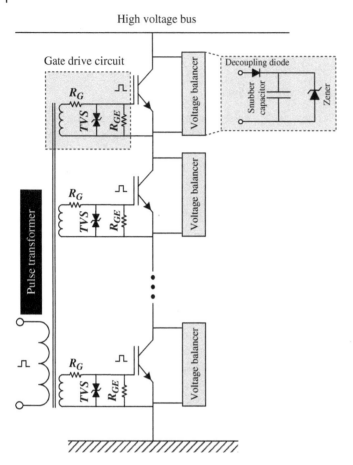

Figure 14.10 Structure of the IGBT-based crowbar including series connected IGBTs, gate drive circuits and voltage balancing circuits.

doing so, the IGBT-based crowbar can reduce the amounts of the FCIT and AE and, therefore, it can help to satisfy the protection requirements of the microwave tubes more appropriately.

The results of simulations for both thyristor- and IGBT-based crowbars are compared by the charts in Figure 14.11. Figure 14.12 depicts the waveforms of output voltage and current, crowbar current, and the trigger signal. As can be seen, after the detection delay (1 μs), the trigger signal is high to fire the crowbar. The crowbar then turns on after its turn-on delay (2 μs). Afterwards, the crowbar and load current increases and decreases exponentially. The settling time of the crowbar current is in the region of 7 μs. Finally, when the current of the crowbar reaches its peak value, a residual current, which is equal to 0.5 A, will flow through the load. All in all, the fault current will be interrupted with a thyristor-based crowbar 9 μs after receiving the fault signal. The SCF happens, plotted in Figure 14.13, the output voltage falls to zero, and the detection unit produces the fault signal after 1 μs. The IGBTs are turned on after their turn-on time, approximately 0.1 μs. In contrast to the thyristor-based crowbar, the crowbar current reaches its peak value after 1 μs. This type of crowbar was able to outdo the thyristor-based ones in the values of the FCIT and injected fault energy into the tube during the vacuum arc fault. These were because the IGBT-based crowbar does not need a series inductor to limit the rate of rise of its turn-on current.

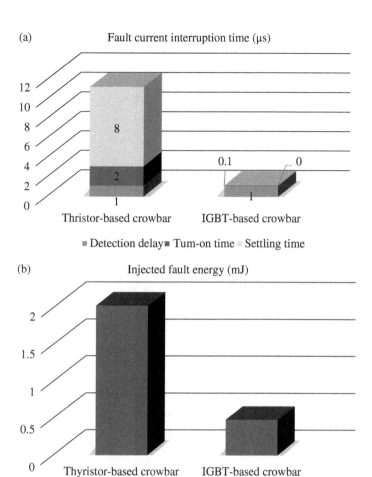

(a) Fault current interruption time (μs)

■ Detection delay ■ Turn-on time ■ Settling time

(b) Injected fault energy (mJ)

Figure 14.11 The comparison simulation results of FCIT and AE of both crowbar structure: (a) FCIT, (b) AE.

Figure 14.12 The waveforms of the variation of variables of HVPS with thyristor-based crowbar.

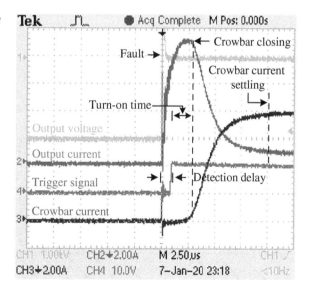

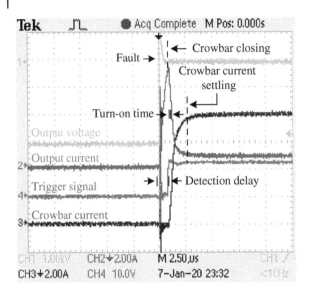

Figure 14.13 The waveforms of the variation of variables of HVPS with IGBT-based crowbar.

14.2 Electromagnetic Compatibility

One of the best examples is electromagnetic interference (EMI). Electromagnetic compatibility (EMC) is the branch of electrical sciences that studies the unintentional generation, propagation, and reception of electromagnetic energy with reference to the unwanted effects (EMI) that such energy may induce. The goal of EMC is correct operation, in the same electromagnetic environment, of different pieces of equipment that use electromagnetic phenomena, and the avoidance of any interference effects. Interference mitigation and hence EMC is achieved by addressing both emission and susceptibility issues, i.e. quieting the sources of interference and hardening the potential victims. The coupling path between the source and victim may also be separately addressed to increase its attenuation. Proper grounding and shielding are two important considerations in EMC of the power electronic converters.

14.2.1 Grounding

There are two primary reasons for grounding devices, cables, equipment, and systems. The first reason is to prevent shock and fire hazards in the event that an equipment frame or housing develops a high voltage due to lightning or an accidental breakdown of wiring or components. The second reason is to reduce EMI effects resulting from electromagnetic fields, common impedance, or other forms of interference coupling. Ground loops are a major cause of noise, hum, and

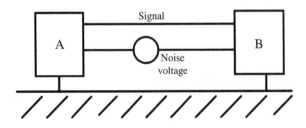

Figure 14.14 Noise voltage between different grounds.

interference in audio, video, and computer systems. They can also create an electric shock hazard, since ostensibly "grounded" parts of the equipment, which are often accessible to users, are not at ground potential. Loops in the ground path can cause currents in signal cable grounds by two main mechanisms. Figure 14.14 shows the effect of the difference between grounds of two connected systems.

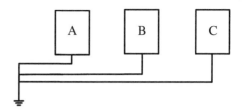

Figure 14.15 Single-point ground connection.

In general, the ground configurations will involve either a floating ground, a single point ground, a multipoint ground, or some hybrid combination of these. The single-point or star type of grounding scheme shown in the figure avoids problems of common-mode impedance coupling, discussed in the previous section. The only common path is in the earth ground (for earth-based structures), but this

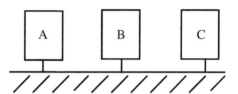

Figure 14.16 Multi-point ground connection.

usually consists of a substantial conductor of very-low impedance. Thus, as long as no or low ground currents flow in any low-impedance common paths, all subsystems or equipment are maintained at essentially the same reference potential. Figure 14.15 shows a system with a single-point ground. An important advantage of the single-point configuration is that it helps control conductively coupled interference.

Rather than have an uncontrolled situation, the other grounding alternative is multipoint grounding, as illustrated in Figure 14.16. For the example shown in Figure 14.17, each equipment or subsystem is bonded as directly as possible to a common low-impedance ground plane to form a homogeneous, low-impedance path. Thus, common-mode currents and other EMI problems will be minimized. The ground plane is then earthed for safety purposes.

Improper grounding leads to an increase in the voltage difference between the reference voltages of the parts of converters. In this regards, proper design of the system connections in Figure 14.18 shows a comparison between two grounding designs in two converters. In Figure 14.18a a two-layer PCB is selected. It is seen that the ground plane is not uniform. In Figure 14.18b a four-layer PCB is selected with a good ground plane.

14.2.2 Shielding

Electromagnetic shielding is the practice of reducing the electromagnetic field in a space by blocking the field with barriers made of conductive or magnetic materials. Shielding is typically applied to enclosures to isolate electrical devices from the "outside world" and to cables to isolate wires from the environment through which the cable runs. Electromagnetic

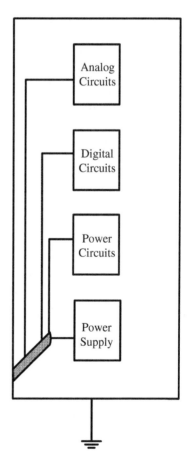

Figure 14.17 Grounding the system parts according to their noise behavior.

shielding that blocks radio frequency (RF) electromagnetic radiation is also known as RF shielding. The shielding can reduce the coupling of radio waves, electromagnetic fields, and electrostatic fields. A conductive enclosure used to block electrostatic fields is also known as a Faraday cage. The amount of reduction depends very much upon the material used, its thickness, the size of the shielded volume, the frequency of the fields of interest, and the size, shape, and orientation of apertures in a shield to an incident electromagnetic field. Figure 14.19 shows the effect of a shield on electric coupling via a parasitic capacitor, while Figure 14.20 shows a shielded circuit.

14.3 Application of Artificial Intelligent Methods

In this section, a case study is presented to show the advantages of artificial intelligent methods in the resilience topic. In this case study, the protection system of the high-voltage power supply presented in Chapter 12 is redesigned by a neural network-based diagnosis system. In this

(a) (b)

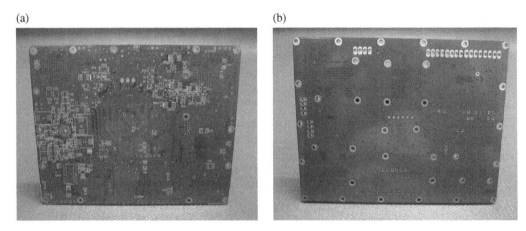

Figure 14.18 A comparison between two grounding designs: (a) poor design, (b) good design.

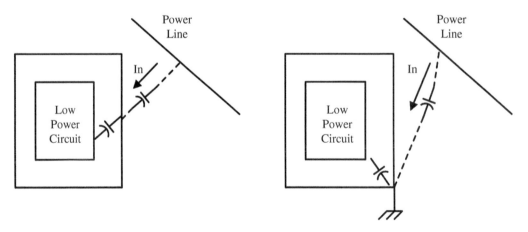

Figure 14.19 Operation principle of shielding.

Figure 14.20 A shielded circuit.

section, without imposing any additional sensors, a proposed FDS that utilizes the data of all existing sensors is presented to solve the problem of false alarms. In the proposed method, the data of both the output current and voltage sensors are combined to increase the reliability of the data, according to which the fault decision is made. A neural network, as a powerful algorithm for data fusion, is applied to combine the data of the sensors. The main reason that this method suppresses the noise impacts is the fact that when an SCF occurs, each variable has a unique signature of an SCF, thereby making it possible for the neural network-based FDS to distinguish an SCF from noise impacts. The proposed FDS consists of three parts: a sensor network, a neural network, and a decision-maker. The structure of the proposed FDS is shown in Figure 14.21. In the first part, several variables $(x_1 . . . x_n)$ are measured by the sensor network $(S_1 . . . S_n)$ where the total number of them is n. Then, their raw data $(y_1 . . . y_n)$ are transmitted to the neural network. Because the data of sensors are not processed in the first part, a negligible delay is imposed in this part of fault diagnosis. In the second part, the raw data of the sensor network are combined in the neural network to determine the probability that an SCF has just occurred, P_{NN}. As depicted in Figure 14.21, finally, the decision-maker takes a decision on an SCF occurrence, D_{NN}, based on P_{NN}, to activate the protection system. In the following subsections, how to select sensors, a brief review of the neural network theory, and the function of the decision-maker are thoroughly described.

In the FDS, the raw data of the sensors are gathered and combined to make a reasonable and reliable decision about an SCF. Because a neural network is an effective and powerful tool for data classification and pattern recognition, it is used to combine the data of sensors in the proposed FDS. As a result of its significant advantage, it can be trained to detect an SCF, the signature of which is unique, and to reject the noise impacts, the sign of which is hardly similar to an SCF. Generally, a neural network consists of three kinds of layers: an input layer, hidden layers, and an output layer. The structure of layers, each of which is made up of some multiple nodes, is shown in Figure 14.22a, where each node of a layer is connected with all nodes of the previous layer. Considering the nodes of the ith layer, the input data of this layer is a vector, H^i, which can be written as

$$H^i = \left[h_1^i \; h_1^i \cdots h_j^i \right] \tag{14.2}$$

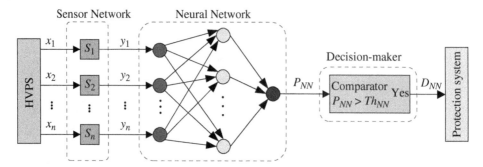

Figure 14.21 Neural network-based FDS.

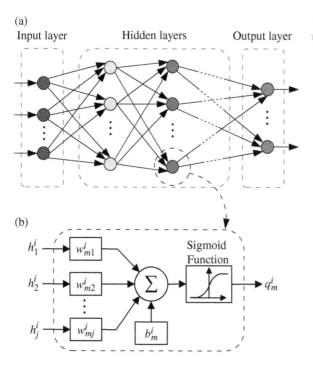

(a)
Input layer   Hidden layers   Output layer

(b)

Figure 14.22 Structure of a neural network: (a) different kinds of layers, (b) structure of a node of a neural network.

where j is the total number of nodes of the $(i-1)$th layer. In the ith layer, H^i is multiplied by the weight matrix, W^i, and is added with the bias vector, B^i. W^i can be formulated as

$$W^i = \begin{pmatrix} w^i_{11} & w^i_{21} & \cdots & w^i_{m1} \\ w^i_{12} & w^i_{22} & \cdots & w^i_{m2} \\ \vdots & \vdots & \ddots & \vdots \\ w^i_{1j} & w^i_{2j} & \cdots & w^i_{mj} \end{pmatrix} \tag{14.3}$$

where m is the number of nodes of the ith layer. Similarly, B^i is formulated as

$$B^i = \begin{bmatrix} b^i_1 \, b^i_1 \cdots b^i_m \end{bmatrix} \tag{14.4}$$

The use of biases is necessary since they make it possible for a neural network to be trained correctly and to operate effectively. In like fashion, the output of the *i*th layer, Q^i, that is the input data of the $(i+1)$th layer, can be written as

$$Q^i = H^{i+1} = \left[q_1^i\, q_2^i \cdots q_m^i \right] \tag{14.5}$$

With this in mind, as depicted in Figure 14.22b, one of the outputs of a layer, for example $q^i{}_m$, is calculated as follows:

$$q_m^i = g\left(\sum_j w_{mj}^i h_j^i + b_m^i \right) \tag{14.6}$$

where g is an activation function. Even though there are several different activation functions, the sigmoid function is one of the most popular ones. This is because it is a nonlinear function, and it is proper to classify multiple outputs. Furthermore, each class can be designated by different probabilities. The expression of the sigmoid function, σ, is as follows:

$$\sigma(x) = \frac{1}{1+e^{-x}} \tag{14.7}$$

Eventually, the output of a neural network is multiple classes of events with different probabilities.

In this application, a neural network that has three layers is used to combine the data of output current and voltage sensors. The number of nodes of the first layer, k_{input}, is defined by the number of sensors, or the number of variables. Therefore, since the proposed FDS utilizes two sensors, k_{input} becomes equal to two. Likewise, the number of nodes of the output layer, k_{output}, is determined by the number of fault types. Because the fault of the HVPS is only SCF, k_{output} becomes equal to one. Regarding the second layer, or the hidden layer, the number of nodes of this layer, k_{hidden}, can be defined based on the empirical formula that can be written as

$$k_{hidden} = \sqrt{k_{input} \times k_{output}} \tag{14.8}$$

Thus, k_{hidden} can be first defined as 1.42 ($\sqrt{2*1} = 1.42$), which is rounded up to two. Next, the neural network can be trained and tested to examine whether k_{hidden} is enough or not. If the result is not acceptable, k_{hidden} can be increased until an ideal result is achieved. The output of the neural network is P_{NN}. However, P_{NN} is needed to be translated to D_{NN}, to activate the protection system. To do so, the decision-maker compares P_{NN} with the threshold of the decision-maker, Th_{NN}, as illustrated in Figure 14.21. Whenever P_{NN} exceeds Th_{NN}, the decision-maker changes the D_{NN} from "0" to "1," and the protection system is activated. In this chapter, Th_{NN} is considered to be 0.8. The neural network is trained by the APPS/Neural Net Fitting tool of MATLAB software based on the Levenberg–Marquardt algorithm while the load is deliberately short-circuited. The training procedure is shown in Figure 14.23, where T_{SCF} is the signal of the deliberate SCF and is used as the teacher for the training algorithm.

Figure 14.24 displays the operation of the conventional and proposed FDS in this new condition. Despite the fact that the conventional FDS does not give any false alarms due to its low-bandwidth low-pass filter, it detects an SCF after $t_{con} = 34\,\mu s$, which is much more than $t_{CFD} = 10\,\mu s$, so there is a strong possibility that the vacuum tube suffers irreparable damage. Conversely, the proposed FDS both detects an SCF quickly ($t_{NN} = 4\,\mu s$) and does not give any false alarm. Because $t_{NN} = 4\,\mu s$

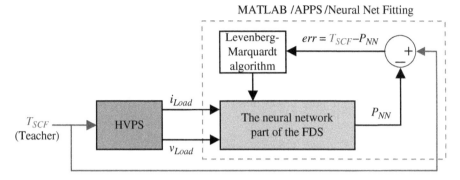

Figure 14.23 Neural network-based FDS training scheme.

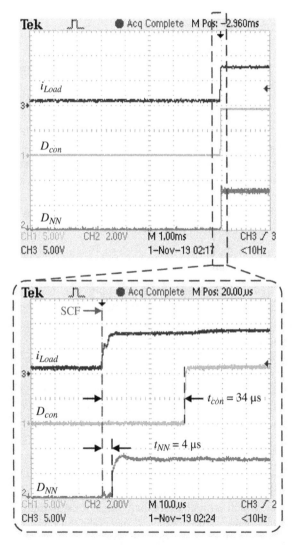

Figure 14.24 Experimental results of the conventional FDS and proposed FDS operation in the presence of noise from the aspect of fault detection delay.

is less than $t_{CFD} = 10\,\mu s$, the health of the vacuum tube is guaranteed. All in all, the experimental tests prove that not only is the neural network-based FDS able to diagnose an SCF quickly in the presence of noise, but it can also dismiss noise impacts and solve the problem of false alarms completely.

14.4 Fault Alarm Management

Alarm management is the application of human factors (or ergonomics as the field is referred to outside the US) along with instrumentation engineering and systems thinking to manage the design of an alarm system to increase its usability. Most often the major usability problem is that there are too many alarms annunciated in a plant upset, commonly referred to as an alarm flood (similar to an interrupt storm), since it is so similar to a flood caused by excessive rainfall input with a basically fixed drainage output capacity. However, there can also be other problems with an alarm system, such as poorly designed alarms, improperly set alarm points, ineffective annunciation, unclear alarm messages, etc.

The fundamental purpose of alarm annunciation is to alert the operator to deviations from normal operating conditions, i.e. abnormal operating situations. The ultimate objective is to prevent, or at least minimize, physical and economic loss through operator intervention in response to the condition that was alarmed. For most digital control system users, losses can result from situations that threaten environmental safety, personnel safety, equipment integrity, economy of operation, and product quality control, as well as plant throughput. A key factor in operator response effectiveness is the speed and accuracy with which the operator can identify the alarms that require immediate action.

By default, the assignment of alarm trip points and alarm priorities constitute basic alarm management. Each individual alarm is designed to provide an alert when that process indication deviates from normal. The main problem with basic alarm management is that these features are static. The resultant alarm annunciation does not respond to changes in the mode of operation or the operating conditions.

When a major piece of process equipment like a charge pump, compressor, or fired heater shuts down, many alarms become unnecessary. These alarms are no longer independent exceptions from normal operation. They indicate, in that situation, secondary, non-critical effects and no longer provide the operator with important information. Similarly, during start-up or shut-down of a process unit, many alarms are not meaningful. This is often the case because the static alarm conditions conflict with the required operating criteria for start-up and shut-down.

In all cases of major equipment failure, start-ups, and shut-downs, the operator must search alarm annunciation displays and analyze which alarms are significant. This wastes valuable time when the operator needs to make important operating decisions and take swift action. If the resultant flood of alarms becomes too great for the operator to comprehend, then the basic alarm management system has failed as a system that allows the operator to respond quickly and accurately to the alarms that require immediate action. In such cases, the operator has virtually no chance to minimize, let alone prevent, a significant loss.

In short, one needs to extend the objectives of alarm management beyond the basic level. It is not sufficient to utilize multiple priority levels because priority itself is often dynamic. Likewise, alarm disabling based on unit association or suppressing audible annunciation based on priority do not provide dynamic, selective alarm annunciation. The solution must be an alarm management system that can dynamically filter the process alarms based on the current plant operation and conditions so that only the currently significant alarms are annunciated.

The fundamental purpose of dynamic alarm annunciation is to alert the operator to relevant abnormal operating situations. They include situations that have a necessary or possible operator response to ensure:

- Personnel and Environmental Safety,
- Equipment Integrity,
- Product Quality Control.

The ultimate objectives are no different from the previous basic alarm annunciation management objectives. Dynamic alarm annunciation management focuses the operator's attention by eliminating extraneous alarms, providing better recognition of critical problems, and insuring a swifter, more accurate operator response.

Alarm management is usually necessary in a process manufacturing environment that is controlled by an operator using a control system, such as a distributed control system (DCS) or a programmable logic controller (PLC). Such a system may have hundreds of individual alarms that up until very recently have probably been designed with only limited consideration of other alarms in the system. Since humans can only do one thing at a time and can pay attention to a limited number of things at a time, there needs to be a way to ensure that alarms are presented at a rate that can be assimilated by a human operator, particularly when the plant is upset or in an unusual condition. Alarms also need to be capable of directing the operator's attention to the most important problem that he or she needs to act upon, using a priority to indicate the degree of importance or rank, for instance.

14.5   Summary and Conclusions

Protection systems save the converter from external faults. However, a safe but out-of-service converter is not the goal of resilience. In this chapter, availability of an electric power converter is described. The main topics of this chapter are summarized as follows:

1) Protection systems act when a fault occurs in the converter. Their performance is very important; isolation of the converter is not always the best choice because this strategy has a bad effect on availability of the converter. A highly available system would disable the malfunctioning portion and continue operating at a reduced capacity. In contrast, a less capable system might crash and become totally non-operational. Any failure factor needs a time interval for damaging the converter. If the protection system is very fast, it protects the converter but conflict with noise is possible. Therefore, protection systems should be fast enough to protect the converter but not so fast to cause incorrect operation.
2) Some failure factors may not damage the system but interfere with its proper operation. Noise is a common interfering factor in systems. Interference mitigation and hence EMC is achieved by addressing both emission and susceptibility issues, i.e. quieting the sources of interference and hardening the potential victims. The coupling path between the source and victim may also be separately addressed to increase its attenuation. Grounding has a key role for controlling the effect of noise and interference in circuits. The most important law in grounding is that all electric references must have the same potential in a power network. An EMI filter, or electromagnetic interference filter, is an electronic passive device that is used in order to suppress

conducted interference that is present on a signal or power line. EMI filters can be used to suppress interference that is generated by the device or by other equipment in order to make a device more immune to EMI signals present in the environment.

3) Artificial intelligent methods, as a powerful algorithm for data fusion, are applied to combine the data of the sensors in power electronic systems. The main reason that this method suppresses the noise impacts is the fact that when a fault occurs, each variable has a unique signature of a fault, thereby making it possible for the neural network-based FDS to distinguish a fault from noise impacts.

4) Some fault alarms are not effective for the immediate shutdown of systems.

References

1 Mohsenzade, S., Zarghany, M., Aghaei, M., and Kaboli, S. (2017). A high-voltage pulse generator with continuously variable pulsewidth based on a modified PFN. *IEEE Transactions on Plasma Science* 45 (5): 849–858.

2 Mohsenzade, S., Zarghani, M., and Kaboli, S. (2019). A high voltage series stacked IGBT switch with active energy recovery feature for pulsed power applications. *IEEE Transactions on Industrial Electronics* 1–1.

3 Gilmour, A.S. and Lockwood, D.L. (1975). The interruption of vacuum arcs at high DC voltages. *IEEE Transactions on Electron Devices* 22 (4): 173–180.

4 Wu, R., Blaabjerg, F., Wang, H. et al. (2013). Catastrophic failure and fault-tolerant design of IGBT power electronic converters An overview. In: *IECON 2013 – 39th Annual Conference of the IEEE Industrial Electronics Society*, 507–513.

5 Sanders, H.D., White, C., Dunham, C., and Warnow, D. (2014). Fast, solid-state crowbar switch to protect high power amplifier tubes. In: *2014 IEEE International Power Modulator and High Voltage Conference (IPMHVC)*, 504–507.

6 Joshi, T.G.S. and John, V. (2017). Performance comparison of ETT- and LTT-based pulse power crowbar switch. *IEEE Transactions on Plasma Science* 45 (11): 2994–3000.

7 Bahman, A.S., Iannuzzo, F., Uhrenfeldt, C. et al. (2017). Modeling of short-circuit-related thermal stress in aged IGBT modules. *IEEE Transactions on Industry Applications* 53 (5): 4788–4795.

8 Chen, Y., Li, W., Iannuzzo, F. et al. (2018). Investigation and classification of short-circuit failure modes based on three-dimensional safe operating area for high-power IGBT modules. *IEEE Transactions on Power Electronics* 33 (2): 1075–1086.

9 Jang, S.R., Seo, J.H., and Ryoo, H.J. (2016). Development of 50-kV 100-kW three-phase resonant converter for 95-GHz gyrotron. *IEEE Transactions on Industrial Electronics* 63 (11): 6674–6683.

10 Cunha, J.P.V.S., Begalli, M., and Bellar, M.D. (2012). High voltage power supply with high output current and low power consumption for photomultiplier tubes. *IEEE Transactions on Nuclear Science* 59 (2): 281–288.

11 Cobine, J.D. (1963). Recovery characteristics of vacuum arcs. *Transactions of the American Institute of Electrical Engineers, Part I: Communication and Electronics* 82 (2): 246–253.

12 Pouresmaeil, K. and Kaboli, S. (2019). A reopened crowbar protection for increasing the resiliency of the vacuum tube high-voltage DC power supply against the vacuum arc. *IEEE Transactions on Plasma Science* 47 (5): 2717–2725.

13 Zarei, S.F., Mokhtari, H., Ghasemi, M.A., and Blaabjerg, F. (2018). Reinforcing fault ride through capability of grid forming voltage source converters using an enhanced voltage control scheme. *IEEE Transactions on Power Delivery* 1–1.

14 Heidary, A., Radmanesh, H., Rouzbehi, K., and Pou, J. (2019). A DC-reactor-based solid-state fault current limiter for HVdc applications. *IEEE Transactions on Power Delivery* 34 (2): 720–728.

15 Abramovitz, A. and Smedley, K.M. (2012). Survey of solid-state fault current limiters. *IEEE Transactions on Power Electronics* 27 (6): 2770–2782.

16 Hoshino, T., Salim, K.M., Kawasaki, A. et al. (2003). Design of 6.6 kV, 100 a saturated DC reactor type superconducting fault current limiter. *IEEE Transactions on Applied Superconductivity* 13 (2): 2012–2015.

17 Nourmohamadi, H., Sabahi, M., and Abapour, M. (2018). A novel structure for bridge-type fault current limiter: capacitor-based nonsuperconducting FCL. *IEEE Transactions on Power Electronics* 33 (4): 3044–3051.

18 Ayoubi, R. and Kaboli, S. (2019). A robust short-circuit fault diagnosis for high voltage DC power supply based on multisensor data fusion. In: *2019 10th International Power Electronics, Drive Systems and Technologies Conference (PEDSTC)*, 659–664.

19 Mohsenzade, S., Zarghany, M., and Kaboli, S. (2018). A series stacked IGBT switch with robustness against short-circuit fault for pulsed power applications. *IEEE Transactions on Power Electronics* 33 (5): 3779–3790.

20 Fuhrmann, J., Eckel, H., and Klauke, S. (2016). Short-circuit behavior of series-connected high-voltage IGBTs. In: *2016 18th European Conference on Power Electronics and Applications (EPE'16 ECCE Europe)*, 1–10.

21 Lefebvre, S., Khatir, Z., and Saint-Eve, F. (2005). Experimental behavior of single-chip IGBT and COOLMOS devices under repetitive short-circuit conditions. *IEEE Transactions on Electron Devices* 52 (2): 276–283.

22 Bauer, F., Meysenc, L., and Piazzesi, A. (2005). Suitability and optimization of high-voltage IGBTs for series connection with active voltage clamping. *IEEE Transactions on Power Electronics* 20 (6): 1244–1253.

15

Inherently Resilient Power Electronic Systems

15.1 Immune Converter Against the Faults

The presented methods of reliability improvement as well as reliability calculation techniques help to have a safe converter without catastrophic failures. A review of the presented methods shows that the studied converters are faced with failure factors and the effect of these failure factors are sensed by the converter. It is true that the converter will be safe but this advantage has a cost. Extra parts, lower specifications than the nominal values and the time of recovery the costs for reliable operation. This cost is acceptable in many cases. However, in resilient operation, many of these costs are not acceptable. The derating method is usually a method exclusive of reliability improvement for a faulty converter. However, it usually means the de-rated converter continues to operate with new rated characteristics, which are less than the converter original nominal specifications. In many cases, this is not acceptable and there is a need to keep the original nominal rating of the converter. For example, consider a dc power distribution unit with several output voltage levels. In a power distribution unit there are several voltage regulators that provide some output voltage levels from a common dc input voltage source such as a battery. In this system, failure in one of the output voltage levels causes failure in the subsystem related to the failed output voltage level. It is true that just one of the output channels can fail but not all of the power distribution units fail. However, the system does not operate properly even though other output channels operate normally. Fault-tolerant converters are solutions of this drawback and we will discuss them in this chapter. Fault tolerance is the property that enables a system to continue to operate properly in the event of the failure of (or one or more faults within it) some of its components. Fault tolerance is particularly sought after in high-availability or life-critical systems. A fault-tolerant design enables a system to continue its intended operation, possibly at a reduced level, rather than failing completely, when some part of the system fails. Recovery from errors in fault-tolerant systems can be characterized as either roll-forward or roll-back. When the system detects that it has made an error, roll-forward recovery takes the system state at that time and corrects it, to be able to move forward. Roll-back recovery reverts the system state back to some earlier, correct version, for example using checkpointing, and moves forward from there. Some systems make use of both roll-forward and roll-back recovery for different errors or different parts of one error. Fault tolerance can be achieved by anticipating exceptional conditions and building the system to cope with them, and, in general, aiming for self-stabilization so that the system converges toward an error-free state. However, if the consequences of a system failure are catastrophic, or the cost of making it sufficiently reliable is very high, a better solution may be to use some form of duplication. In any case, if the consequence of a system failure is so catastrophic, the system must be able to use reversion to fall back to a safe

Resilient Power Electronic Systems, First Edition. Shahriyar Kaboli, Saeed Peyghami, and Frede Blaabjerg.
© 2022 John Wiley & Sons Ltd. Published 2022 by John Wiley & Sons Ltd.
Companion website: www.wiley.com/go/kaboli/resilientpower

mode. If each component, in turn, can continue to function when one of its subcomponents fails, this will allow the total system to continue to operate as well.

In special cases, the failure factor is applied to the converter but its effect is not sensed by the converter. As an example, the Z-source converter is described. Conventional pulse width modulation (PWM) inverters have been used widely for dc-to-ac or ac-to-dc power conversion. There are two types of PWM traditional inverters, voltage source and current source inverters (VSI and CSI), both of which have some limitations and theoretical barriers. The most important of these limitations are:

1) VSI and CSI act as a buck and boost inverter respectively for dc-to-ac power conversions.
2) Electromagnetic interference (EMI) noise problem. In VSI both switches in a leg cannot be switched on and in CSI cannot be switched off simultaneously. Therefore, because of EMI noise, a dead time should be considered in switching rules.

To overcome the limitations of conventional inverters, a new type of inverter (Z-source inverter or ZSI) was introduced by F.Z. Peng in 2003 [1] (Figure 15.1). In order to use shoot-through vector to control the dc boost of this new type of inverter, PWM methods were modified. Applying a ZSI to adjustable-speed drive (ASD) systems can eliminate some problems of conventional ASD systems, such as influence of voltage sag on the ASD system, inrush current transmission, and harmonic injection to the power network and low reliability of the system because of sensitivity to shoot-through. Because of voltage suppression during an output power increment of fuel cell stacks, the nonlinear characteristic of the photovoltaic (PV) model and ability of ZSI to boost and adjust the dc voltage to a desired level when applying this new type of inverter to fuel cell and PV systems has been widely investigated. As shown in reference [1], ZSI can boost the dc input voltage and therefore one of the control objects is controlling the dc boosted voltage.

Schematic configuration of a ZSI is shown in Figure 15.1. The impedance network, which is made of an X shape LC network, can boost the dc input voltage (V_o) in respect to the interval of the shoot-through zero state (T_0) during a switching cycle (T). In conventional VSI there are eight permissible switching states: six active and two zero states, while during the zero states there is no difference for the load if the upper three, the lower three, or all the six switches are gated (all the states short the output terminal of the inverter and produce zero voltage to the load). As discussed in reference [1] in ZSI, during the zero states all the switches are gated on (the shoot-through state) and this state is used to achieve boosting of the dc input voltage. Therefore, in ZSI, there are six active states and two zero states, which are the same as the conventional inverter and an additional

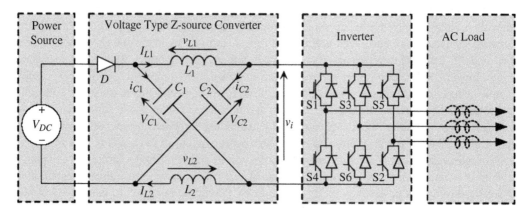

Figure 15.1 Schematic configuration of a Z-source inverter.

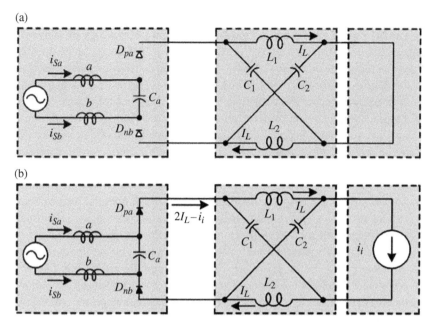

Figure 15.2 Equivalent circuits of ZSI: (a) shoot-through mode, (b) non-shoot-through mode.

shoot-through state (it is forbidden in conventional inverters), which is utilized advantageously to boost the dc-bus voltage. Two basic operation modes of ZSI are illustrated in Figure 15.2.

As verified in detail in reference [1], the basic relationships for ZSI are:

$$V_{C1} = V_{C2} = V_C = \frac{1 - \dfrac{T_0}{T}}{1 - 2\dfrac{T_0}{T}} V_o \tag{15.1}$$

$$\begin{cases} V_{in} = \dfrac{1}{1 - 2\dfrac{T_0}{T}} V_o = BV_o & \text{While non-shoot-through} \\ V_{in} = 0 & \text{While shoot-through} \end{cases} \tag{15.2}$$

where V_{C1} and V_{C2} are capacitor voltages of the impedance network, which are the same due to circuit symmetry. B is the boost factor of ZSI and V_o and V_{in} denote output and input of the impedance network dc voltage, respectively.

15.2 Elimination of Weak Elements

The active capacitor concept based on power electronic circuits has been proposed recently to exceed the physical limit of the passive capacitor. It retains the physical convenience of use as a passive capacitor and has the potential to increase either the power density or the lifetime, depending on the applications. However, the cost of the existing design by using ceramic or film capacitors to achieve extreme performance increases a lot, which must be taken into account in the design from the industry aspect. Reference [2] proposes a cost-constrained design of an active capacitor used for dc-link applications. It is implemented based on high-current electrolytic capacitors instead of film capacitors or ceramic capacitors. A model-based optimization design procedure is discussed in terms of performance factors of interest. A case study of a 5.5-kW single-phase inverter demonstrates a 38% volume reduction of the dc link with the proposed active capacitor under specific constraints of cost, volume, power loss, and lifetime. Reference [3] presents an active ripple cancelation (ARC) technique for reducing the switching harmonics of the output voltage of a dc–dc converter. An ARC module injects an ac current to accurately cancel the inductor current ripple while the switching converter provides the dc output current. As a result, the output spur of the switching converter can be reduced effectively. However, due to the ARC module, the proposed power converter includes two complex poles. To ensure the system stability and achieve accurate ripple cancelation, a proposed control method and an adaptive phase-lock circuit are utilized. Furthermore, the ripple cancelation capability is independent from the duty cycle, load current, and other converter parameters. The measurement results demonstrate that the switching spur reduction with the ARC technique can achieve 28 dB while the noise floor is kept the same.

Figure 15.3 shows the schematic diagram of a buck converter. In a buck converter, the output capacitor is the most probable failure factor. Therefore, redundant modules are used to improve the reliability of buck converters. In a buck converter, the inductor current ripple is the key parameter for determining the output capacitance. By reducing the output current ripple amplitude, we can reduce or even eliminate the output capacitor.

There are different ways to reduce inductor current ripple, such as increasing the inductor value and increasing the switching frequency. Among these methods, increasing the switching frequency is an attractive solution because it leads to reducing the converter dimensions. However, increasing the switching frequency leads to increasing the power loss. Interleaved topologies are suitable solutions for increasing the frequency of converter waveforms while the switching frequency is not increased. Due to the limitations of the frequency increment obtaining the desired level of the ripple, an output-series input-parallel interleaved converter can serve as an alternative. By appropriate time-shifting in the switching frequency, the output ripple frequency can be increased. In this chapter, interleaved topology of a buck converter is used to reduce the inductor current ripple. Therefore, the value of the output capacitor can be considerably decreased. It leads to a more reliable converter for PDU applications. In addition, the response time of the proposed converter is less than the one found in a classic buck converter due to the small capacitor of the proposed method. Figure 15.4 shows the topology of an N-phase buck converter. The summation of inductor currents, $i_{L1}, \ldots, i_{LN}$ is named the *total inductor current*. In this topology, the fundamental frequency of the total inductor current ripple will be $N.f_s$, where f_s is the switching frequency. The maximum amplitude of the total inductor current ripple will be N times smaller than that of the one-phase converter. By interleaving, the converter can even have less power loss. Figure 15.5 shows the amplitude of the output current ripple when the duty cycle varies between 0 and 1.

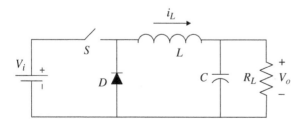

Figure 15.3 Schematic diagram of a buck converter.

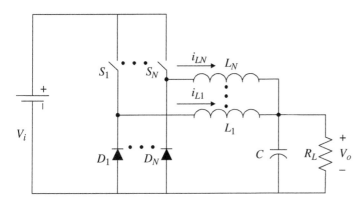

Figure 15.4 N-phase interleaved buck converter circuit.

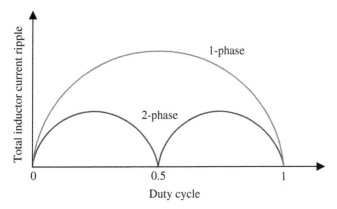

Figure 15.5 Variation of total inductor current ripple versus duty cycle in interleaved buck converters.

According to Figure 15.5, if the interleaved buck converter has N phases, the total inductor current ripple will be absolutely zero if the converter duty cycle equals to k/N, where $k = 0, \ldots, N$. Thus, if the converter supplies a dc load, the output capacitor can be removed. In this chapter, we have tried to design a buck converter that operates with these special duty cycles to eliminate the output current ripple completely. For example, suppose an application that uses a buck converter with a voltage gain equal to $1/3$. If we design a three-phase interleaved buck converter, the duty cycle must be equal to $1/3$ and the total inductor current ripple will theoretically be zero. Thus, we can reduce or even eliminate output capacitance. In addition, we can even reduce inductance of each phase to have a fast response converter. To derive the total inductor current ripple expression in a continuous conduction mode, the following equation expresses the total inductor current, $i_t(t)$:

$$i_t(t) = i_{L1}(t) + \ldots + i_{LN}(t) \tag{15.3}$$

In an interleaved N-phase buck converter, there is a phase shift equal to $2\pi/N$ between adjacent phases. Therefore, Equation (15.1) can be rewritten as

$$i_t(t) = i_{L1}(t) + \ldots + i_{L1}\left(t - \left\{(N-1).T_s / N\right\}\right) \tag{15.4}$$

It can be deduced that the period of $i_t(t)$ is T_s/N. The inductor current of phase 1, $i_{L1}(t)$, is triangular, as shown in Figure 15.6. Therefore, it can be deduced that $i_t(t)$ will also be triangular.

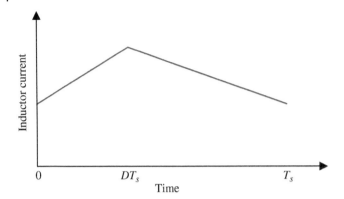

Figure 15.6 The inductor current of phase 1.

In the inductor current waveform of phase 1:

$$M_+ = di_{L1}(t)/dt, 0 < t < D.T_s \tag{15.5}$$

$$V_o = D.V_i \tag{15.6}$$

$$M_+ = (V_i - V_o)/L = V_o(1-D)/(DL) \tag{15.7}$$

$$M_- = di_{L1}(t)/dt, D.T_s < t < T_s \tag{15.8}$$

$$M_- = -V_o/L \tag{15.9}$$

where V_i is the input voltage, V_o is the output voltage, T_s is the switching period, L is the inductance of each phase, and D is the duty cycle.

Now consider the time interval between $m.T_s/N$ and $(m+1).T_s/N$ where $m = $ Floor $(N \cdot D)$. In this time interval, the following equations can be obtained:

$$di_t/dt = m.M_+ + (N - m - 1).M_- + M_{+=}S_1 \text{ if } m.T_s/N < t < D.T_s \tag{15.10}$$

$$di_t/dt = m.M_+ + (N - m - 1).M_- + M_- = S_2 \text{ if } D.T_s < t < (m+1).T_s/N \tag{15.11}$$

where $S_2 < S_1$. Finally,

$$OCR = S_1.T_s(D \quad m/N) \tag{15.12}$$

where OCR is the total inductor current ripple (peak-to-peak):

$$OCR = V_o N T_s [D - (m/N)][(m+1)/N - D]/L \tag{15.13}$$

According to Equation (15.13), if the buck converter has N phases and D equals k/N ($k = 0, \ldots, N-1$), OCR will be absolutely zero. It can be deduced from Equation (15.13) that if inductances reduce, OCR will remain zero if $D = k/N$. Thus, the response of the system can be faster. If a buck converter with voltage gain $= N_1/N_2$ is needed, to remove OCR, it is enough to design a buck converter with N_2 phases and the duty cycle, D, of each phase should be equal to N_1/N_2.

Figure 15.7 shows the schematic of the interleaved buck converter, Figure 15.8 shows the output voltage of the converter, and Figure 15.9 shows that the output voltage ripple is 100 mV (peak-to-peak). The current waveforms of the converter inductors and the phase shift between them are shown in Figure 15.10.

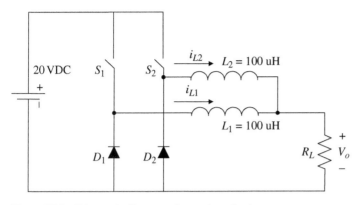

Figure 15.7 Schematic diagram of two-phase buck converter.

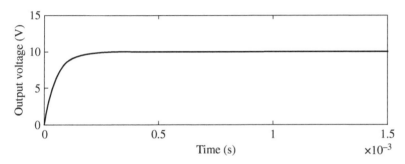

Figure 15.8 Output voltage of two-phase buck converter.

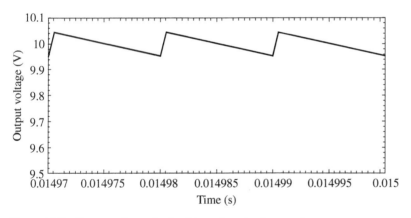

Figure 15.9 Output voltage ripple of two-phase buck converter.

To show the feature of the proposed converter, a single-phase buck converter is compared with the proposed method. Figure 15.11 shows the schematic diagram of the simulated single-phase converter. Both single-phase and two-phase converters have the same specifications. In the single-phase converter, the output capacitance has been tuned to reach the output voltage ripple and is equal to the output ripple of the two-phase converter (100 mV). Figure 15.12 shows the output voltage of a single-phase buck converter and Figure 15.13 shows the output voltage ripple of the

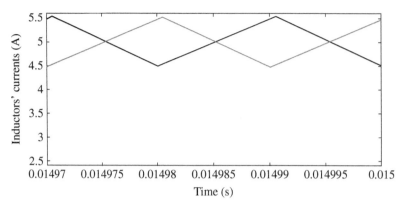

Figure 15.10 Current waveforms of inductors in two-phase buck converter.

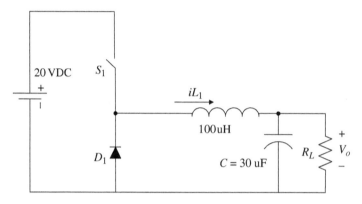

Figure 15.11 Schematic diagram of a single-phase buck converter.

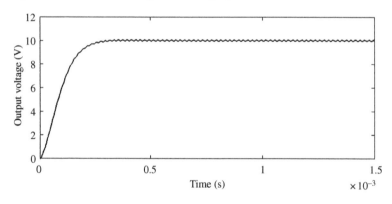

Figure 15.12 Output voltage of a single-phase buck converter.

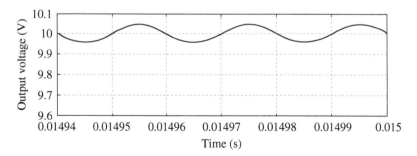

Figure 15.13 Output voltage ripple of a single-phase buck converter.

single-phase converter. The value of the output capacitor is 30 μF for a 100 mV ripple voltage. This 30 μF capacitance can considerably decrease the reliability of the converter. Figure 15.14 shows the inductor current in the single-phase converter.

15.3 Reconfigurable Converters

Reconfiguration is the method of resilience after the faulty conditions. In this method, the converter or system structure changes to keep the output the same as before the fault. Figure 15.15 shows a converter with reconfiguring capability used for wide ranges of input voltage. In this converter, increasing or decreasing the source voltage has no effect on the operation of the load because the converter can operate in both step-down and step-up modes.

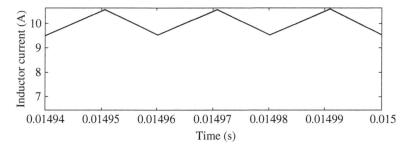

Figure 15.14 Inductor current of a single-phase buck converter.

Figure 15.15 A reconfigurable converter with up: original scheme, middle: step-down operation, down: step-up operation.

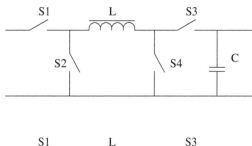

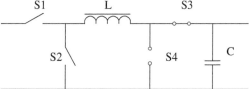

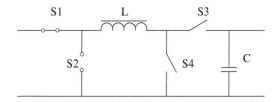

Reconfiguring needs extra switches/elements. Therefore, the reconfiguring schemes are seen more in complex converters than in simple converters. Reference [4] focuses on faults in a Modular Multilevel Converter (MMC) for use in high-voltage direct current (HVDC) systems by analyzing the vulnerable spots and failure mechanisms from device to system and illustrating the control and protection methods under a failure condition. At the beginning, several typical topologies of MMC–HVDC systems are presented. Then fault types such as capacitor voltage unbalance, unbalance between the upper and lower arm voltage are analyzed, and the corresponding fault detection and diagnosis approaches are explained. In addition, more attention is dedicated to control strategies, when running in MMC faults or grid faults. Fault-current handling capability of the MMCs under dc-cable short-circuit conditions is a major concern for the MMC applications on the HVDC transmission systems, where the MMCs based on half-bridge submodules (SMs) cannot block the fault currents to protect the converter devices. In reference [5], a comprehensive review is presented of the fault-ride-through capability of the HVDC transmission systems based on the MMCs adopting different SM schemes, where the MMCs can block the fault currents and compensate the reactive currents to the electric grid during the dc faults. An analysis of the dc short-circuit faults in the MMC is introduced and then the operation principle of different SM circuits building the MMC for blocking the fault currents is highlighted. The fault-tolerant operation of these MMC schemes as static synchronous compensators to enhance the ac grid stability during the dc faults is also investigated. Reference [6] introduces a galvanically isolated boost half-bridge dc–dc converter intended for modern power electronic applications where an ultra-wide input voltage regulation range is needed. A reconfigurable output rectifier stage performs a transition between the voltage doubler and the full-bridge diode rectifiers and, by this means, extends the regulation range significantly. The converter features a low number of components and resonant soft switching semiconductors, which result in a high-power conversion efficiency over a wide input voltage and load range. Reference [7] proposes a new series of resonant dc–dc converters with four configurable operation states depending on the input-voltage and output-voltage levels. It is well suited for the dc–dc stage of grid-connected photovoltaic systems with a wide-input voltage range and different grid voltage levels. The proposed converter consists of a dual-bridge structure on the primary side and a configurable half- or full-bridge rectifier on the secondary side. The root-mean-square (RMS) currents are kept low over a fourfold voltage-gain range; the primary-side MOSFETs and secondary-side diodes can achieve zero-voltage switching on and zero-current switching off, respectively. Therefore, the converter can maintain high efficiencies over a wide voltage gain range. A fixed-frequency pulse width-modulated control scheme is applied to the proposed converter, which makes the gain characteristics independent of the magnetizing inductance and thereby simplifies the design optimization of the resonant tank. A three-phase dual-active-bridge (3p-DAB) converter is an attractive topology for bidirectional power conversion in high-power applications. However, conduction and switching losses are two main loss mechanisms that severely affect its efficiency, and adoption of any single modulation scheme or topology cannot minimize these losses over a wide operating range. For this purpose, a reconfigurable topology of the 3p-DAB converter is proposed in reference [8] that utilizes a reconfigurable and tunable resonant network to offer multiple degrees of freedom in minimizing conduction and switching losses over a wide range of operating conditions. The converter is designed such that for 40–100% of the rated output power, it operates as a tunable 3p-DAB resonant immittance converter with its output power controlled by varying switching frequency and tuning the resonant frequency of a resonant immittance network to track the switching frequency. Below 40% of the rated output power, the converter transforms to a tunable 3p-DAB series-resonant converter with its output power controlled by varying the impedance of a series-resonant network while keeping the switching frequency and phase shift constant. Reference [9] evaluates and proposes various compensation methods for three-level ZSIs under semiconductor-failure conditions. Unlike the fault-tolerant techniques used in traditional three-level inverters, where either an extra phase-leg or collective switching states are used, the proposed methods for three-level ZSIs simply reconfigure their relevant gating signals so as to ride through the failed semiconductor conditions smoothly without any significant decrease in their ac-output quality and amplitude. These features are partly attributed to the inherent boost characteristics of a ZSI, in addition to its usual voltage-buck operation. By focusing on specific types of three-level ZSIs, it can also be shown that, for the dual ZSIs, a unique feature accompanying it is its extra ability to force common-mode voltage to zero, even under semiconductor-failure conditions.

15.4 Redundancy

One of the reconfiguration methods is redundancy. Redundancy is the provision of functional capabilities that would be unnecessary in a fault-free environment. This can consist of backup components which automatically replace the failed components. Figure 15.16 shows a power supply with two converters as the redundant modules and isolating switches.

Providing a fault-tolerant design for every component is normally impossible. Associated redundancy brings a number of penalties: increase in weight, size, power consumption, cost, as well as time to design, verify, and test. Therefore, a number of choices have to be examined to determine which components should be fault-tolerant:

1) How critical is the component?
2) How likely is the component to fail?
3) How expensive is it to make the component fault tolerant?

The basic characteristics of fault tolerance require:

1) If a system experiences a failure, it must continue to operate without interruption during the repair process.
2) When a failure occurs, the system must be able to isolate the failure to the offending component. This requires the addition of dedicated failure detection mechanisms that exist only for the purpose of fault isolation. Recovery from a fault condition requires classifying the fault or failing component.
3) Some failure mechanisms can cause a system to fail by propagating the failure to the rest of the system.

Spare components address the first fundamental characteristic of fault tolerance in three ways:

1) Replication. Providing multiple identical instances of the same system or subsystem, directing tasks or requests to all of them in parallel, and choosing the correct result on the basis of a quorum.
2) Redundancy. Providing multiple identical instances of the same system and switching to one of the remaining instances in case of a failure (failover).

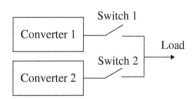

Figure 15.16 A converter and its redundant module with isolating switches.

3) Diversity. Providing multiple implementations of the same specification and using them like replicated systems to cope with errors in a specific implementation.

Fault-tolerant design advantages are obvious, while its disadvantages are:

1) Interference with fault detection in the same component.
2) Interference with fault detection in another component. Another variation of this problem is when fault tolerance in one component prevents fault detection in a different component.
3) Reduction of priority of a fault correction. Even if the operator is aware of the fault, having a fault-tolerant system is likely to reduce the importance of repairing the fault. If the faults are not corrected, this will eventually lead to system failure, when the fault-tolerant component fails completely or when all redundant components have also failed.
4) Test difficulty. For certain critical fault-tolerant systems, such as a nuclear reactor, there is no easy way to verify that the backup components are functional. The most infamous example of this is Chernobyl, where operators tested the emergency backup cooling by disabling primary and secondary cooling. The backup failed, resulting in a core meltdown and massive release of radiation.
5) Cost. Both fault-tolerant components and redundant components tend to increase cost. This can be a purely economic cost or can include other measures, such as weight. Manned spaceships, for example, have so many redundant and fault-tolerant components that their weight is increased dramatically over unmanned systems, which do not require the same level of safety.
6) Inferior components. A fault-tolerant design may allow for the use of inferior components, which would have otherwise made the system inoperable. While this practice has the potential to mitigate the cost increase, use of multiple inferior components may lower the reliability of the system to a level equal to, or even worse than, a comparable non-fault-tolerant system.

Active redundancy in active components requires reconfiguration when failure occurs. Computer programming must recognize the failure and automatically reconfigure to restore operation. Figure 15.17 shows two states of a power supply in hot redundancy. Isolator switches are used to bypass the damaged unit.

In this case there is a device that is performing something and a device that is idle and ready to take over the task of the other device in case it fails. This means that there must be some switching system that takes care of starting the idle device and of stopping the active device and switches both on the input side and on the output side. Figure 15.18 shows a power supply with a standby redundant device. One of the units operates at each moment.

Reference [10] presents design and control methods for fault-tolerant operations with redundant converter modules, one of the most prominent features in MMC topology. In fully implementing

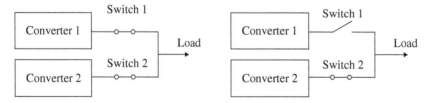

Figure 15.17 Two possible states for a converter with active redundancy: (left) both converters are normal, (right) converter 1 failed.

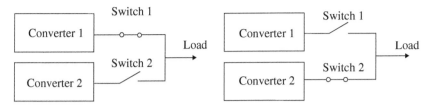

Figure 15.18 Two possible states for a converter with standby redundancy: (left) converter 2 operates and converter 1 is OFF, (right) converter 2 was failed and converter 1 operates.

MMC functionalities, a nearest-level control is applied as a low-switching modulation method. A dual sorting algorithm has been newly proposed for effectively reducing the switching commutations of each power module as well as for voltage balancing control. Built upon these primary MMC topological and control features, its redundant operation is comprehensively investigated for a fail-safe energy transfer. In particular, a novel spare process is proposed to handle an emergency situation when the number of faulty power modules exceeds the module redundancy. Since topological redundancy may cause the switching commutations of power modules in an arm to be unevenly distributed, a practical and effective mitigation measure is incorporated to keep the energy balance while avoiding the undesired switching stresses. The MMC is based on the cascaded connection of identical SMs, enabling additional redundancies. Reference [11] proposes the configuration of the MMC topology with redundant SMs and demonstrates the effects of active redundancies. The proposed configuration decreases the switching frequency per SM while reducing the SM capacitor voltage ripples. An analytical model for determining the SM capacitor voltage ripple and voltage dynamics is derived. The results from the analytical model are compared with the switching model for a 21-level MMC with five redundant (25 in total) SMs per arm. Reference [12] proposes a four-step open-circuit fault diagnosis and fault-tolerant scheme for isolated phase-shifted full-bridge (PSFB) dc–dc converters to improve the reliability. The fault diagnostic method utilizes the primary voltage of the transformer as the diagnostic criterion, which can be obtained easily by adding an auxiliary winding. When an open-circuit fault occurs in any switch of the PSFB converter, the proposed fault detection method can generate an indication of the abnormal state and trigger a fault in the control system. Under the APS state, it is very easy to locate the exact position of a faulty switch because the voltage waveform of the primary winding heavily depends on the location of the faulty switch. After locating the position of the faulty switch, the PSFB converter is reconfigured into an asymmetrical half-bridge (AHB) converter by turning ON the normal switch in the faulty leg and adding a redundant winding to the secondary side. Therefore, the rebuilt AHB converter can keep the output voltage constant under a reduced power rating after the open-circuit fault.

The important issues in both reconfiguration methods and redundancy are the system recovery time and the depth of drop of the system performance index. Figure 15.19a shows the performance of a power supply with redundancy but a non-resilient characteristic. It is seen that the voltage drop and the recovery time are considerable. Figure 15.19b shows the same converter with resilient operation.

To maintain a continuous power supply, a fault-tolerant control method based on the voltage single-loop control is proposed in reference [13], where the rectifier-side output square voltage is regulated. Nevertheless, it may excite the resonance between the resonant inductors and dc capacitors, leading to severe low-frequency oscillations (appearing as the envelope of the high-frequency current). This may trigger the overcurrent protection and the SRDAB fails to ride

(a)

(b)

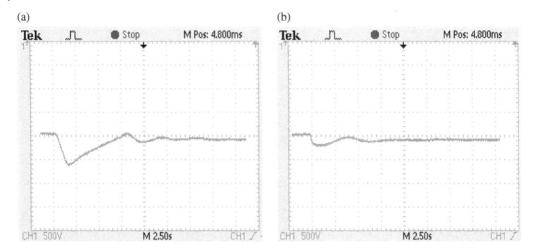

Figure 15.19 Non-resilient operation of a power supply (a) and resilient operation (b).

through the fault. To address this issue, low-frequency equivalent models are proposed first for the bidirectional power-flow of the SRDAB, enabling a frequency-domain analysis of the single-loop voltage control. The analysis reveals that the oscillation depends on the duty-cycle and control parameters, and it is more likely to occur when the converter operates in the boost mode. However, it is not possible to suppress the oscillations by the voltage single-loop control. Thus, a dual-loop fault-tolerant control method is developed. The proposed control strategy includes an outer-loop voltage control, an inner-loop current envelope control, and a nonlinear correction unit.

15.5 Working Under the Fault Threshold

One of the methods of resilience is keeping the operating point of the system under the threshold of the faults. The HVDC link contains a high-voltage high-frequency transformer (HVHFT) with high-voltage secondary winding and a rectifier stage. The important factor threating the HVHFT lifetime is the partial discharge (PD) phenomenon deteriorating the HVHFT insulators in a very short time. To avoid the PDs, an upcoming approach is to oversize the insulators, which will result in an extra volume. It also deteriorates the electrical characteristics of the HVHFT such as leakage inductance. This section concentrates on the design of the HVDC link to reach an extended HVHFT lifetime without oversizing and also presents a comparison between the HVHFT with a single-winding secondary and the HVHFT with a multi-winding secondary. In the second type, the secondary windings are rectified individually and then serially connected. Accordingly, the major part of the voltage applied to the insulators consists of the dc component. Thus, the ac voltage stresses over the insulators are reduced. HVDC links are very widespread in power electronics fields. They are mainly adopted for high-voltage dc power supplies and solid-state transformers, where a high-frequency ac voltage needs to be rectified and filtered. The schematic of the high-voltage dc links is provided in Figure 15.20. According to Figure 15.20, the key components of the HVDC link are an HVHFT, a high-voltage rectifier, and a filter stage.

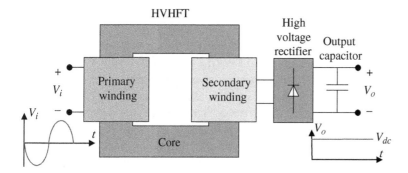

Figure 15.20 The schematic of an HVDC link and its key waveforms.

The reliable operation and long lifetime expectancy for HVDC links are absolutely necessary. The reason is related to the presence of the high voltage, which makes the insertion of the redundant modules quite expensive and bulky. Accordingly, reliability considerations in the design phase seem to be more practical. Through the components of the HVDC link, the extant concentrations are on the HV rectifier and HVDC capacitors. To this end, a reinforced structure of series-connected diodes with robustness against the diodes short-circuit fault is proposed. On the other hand, efforts have been made to replace the electrolytic capacitors with film capacitors to extend the lifetime of the capacitor bank. The HVHFT reliable design has not been specifically put into perspective by the researchers since it has a longer expected lifetime with respect to the semiconductor devices and capacitors based on the reliability standards such as MIL-HDBK-217. For the HVHFT, two critical issues threating the component reliability are the temperature of the windings and the insulator lifetime. The insulation design is usually carried out to avoid the insulation breakdown. To this end, the maximum value for the electrical field is calculated through the transformer and the insulation thickness is determined to avoid the direct breakdown. For high-frequency applications, the insulator deteriorations due to PD must be considered more intensely. The origin of PD is due to the ununiform voltage distribution over the insulator materials and the existing voids through the insulator. A PD occurs whenever the voltage exceeds the breakdown voltage of the air in the voids or on the insulator surface. One or several PDs occur in each switching period. In each PD occurrence, an amount of energy releases to the insulator, which weakens its voltage withstand characteristics. This deterioration eventually breaks the insulator over time. This deterioration is not very critical in dc or low-frequency applications. However, in high-frequency applications (e.g. more than 20 kHz), it may break the insulator in less than a few hours. Hence, the PD effect on the insulation's lifetime should be considered with a strong emphasis on high-frequency applications. Avoiding PD is carried out by increasing the distance between the winding layers and the winding distance to the core. To avoid the PDs, the voltage over the insulator should be less than the level at which the PDs occur and is known as the inception voltage. However, this approach leads to thick insulations. Irrespective to the cost increment, the leakage inductance of the HVHFTs increases, which is not desirable in many circumstances. To avoid PD in insulations, we present a comparison between the conventional HVHFT with a single-winding secondary and an HVHFT with a multiwinding secondary. In the HVHFT with the multiwinding secondary, the windings are serially connected after rectification. The number of separated windings equals the transformer layers. Using this approach, the ac voltage across the winding decreases. Hence, the voltage across the windings can be set to values less than the inception voltage more easily. The voltage difference between two adjacent winding layers also has only a dc component. Consequently, the PD occurrences decrease drastically in such insulators inserted between the layers. The voltage difference between the windings and the core, however, has a dc bias superimposed with an ac voltage. The deterioration of the insulators rises depending on the ac components and regardless of dc voltages. Hence, using the proposed design the chance of the PDs in both differential voltages, which are for the windings turns or winding layers, and the common voltages, which are for the windings and the core, is decreased. Figure 15.21a shows the winding diagram of the constructed HVHFT, which consists of a three-winding secondary. Figure 15.21b and c show the connections between the HVHFT and the rectifier in the case of single-stage and multistage rectification, respectively. The amplitude of the voltage of each secondary winding is lower than the inception voltage of the applied insulator. The HVHFT is supplied with a high-frequency inverter whose switching frequency is 50 kHz. The measured HVHFT core current contains considerable PD current, as shown in Figure 15.22. The PD current results from the high amplitude of the AC component of V_{com}. In the case of multistage rectification, each secondary winding is rectified separately. Therefore, V_{com} has a low-amplitude ac component, as shown in Figure 15.23. This prevents PD in the HVHFT insulator.

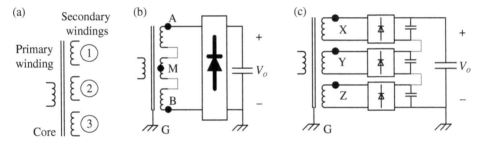

Figure 15.21 Two various connections of the HVHFT secondary windings: (a) the basic windings diagram of the studied HVHFT, (b) single-stage rectification, (c) multistage rectification.

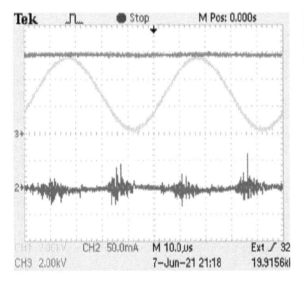

Figure 15.22 Operation of HVHFT with single-winding secondary: CH1: output voltage, CH2: PD current, CH3: ac component of the transformer voltage.

15.6 Inherently Resilient Elements

Some of the elements have the certain design to prevent the failure occurrence. One of the major reasons for ceramic element cracks in MLCCs (Multilayer Ceramic Chip Capacitors) is due to board flexure stress, which happens for a variety of reasons: handling during manufacture, stresses by suction nozzle, an inappropriate solder amount, different thermal expansion between the MLCC and board, board splitting, screw cramp, and excessive board bending, as shown in Figure 15.24. The crack may lead to a short-circuit failure, which can cause abnormal heat generation or ignition. Therefore, applications where reliability is important absolutely require suitable countermeasures. The designers offer a series of high-reliability MLCC products, which are designed to reduce the risk of flexure stress related to a short circuit in order to improve equipment reliability.

Ceramic elements are strong against compression stress, though weak against tensile stress. Thus, a crack is easily generated in the element when excessive stress is acting on a soldered MLCC

Figure 15.23 Operation of HVHFT with multiwinding secondary: CH1: output voltage, CH2: PD current, CH3: ac component of the transformer voltage.

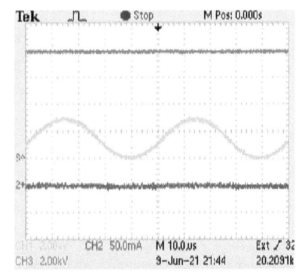

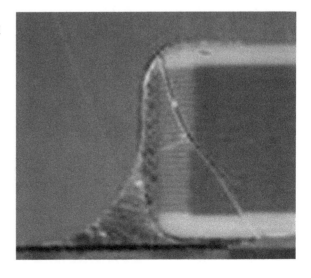

Figure 15.24 The crack growth in conventional ceramic capacitors. *Source:* TDK Co. (with permission).

from the board direction. Then, a short-circuit failure mode occurs if electrical conduction happens between opposite internal electrodes. Even if there is an open circuit mode at the beginning of the crack occurrence, it may progress to a short-circuit mode while the product is on the market. The short-circuit mode may cause problems such as abnormal heat generation or ignition of the MLCC, and so suitable countermeasures are required. There is a risk that microcracks occurring in the various processes from component mounting to set assembly will progress to ceramic element cracks while the end product is on the market. Special caution is required, especially in case of the following applications: equipment that is constantly exposed to vibrations and/or shocks and equipment that is exposed to frequent shock by dropping.

Furthermore, with equipment used in a humid environment, moisture by condensation may penetrate into the element crack, thereby increasing the risk of an open-circuit mode progressing to a short-circuit mode. The terminal structure of the soft termination type is different from the standard type, as shown in Figure 15.25. The standard type has a three-layered structure made of copper, nickel, and tin, whereas the soft termination type has a resin layer between the copper and nickel, and thus a four-layered structure is formed. The resin layer relieves mechanical stress and provides protection from cracks.

When excessive stress was continuously applied, cracks occurred in the ceramics element in a standard product. In contrast, with the soft termination product, no cracks were found in the element, though there was a peeling of the nickel layer and the resin layer, as shown in Figures 15.26 and 15.27. This shows that the resin layer has an excellent effect in suppressing element cracks.

15.7   Summary and Conclusions

In this chapter, fault-tolerant electric power converters were described. In these systems, the continuous operation with usually a nominal specification has no interruption, even with the occurrence of a failure. The main topics of this chapter are summarized as follows:

1) In inherently immune converters, the failure factor is applied to the converter but its impact is not sensed by the converter.
2) Elimination of weak elements such as electrolyte capacitors is another method of resilience in power electronic converters. Applications of active replacement methods and usage of high reliable elements are the methods for resilient operation of the converters.

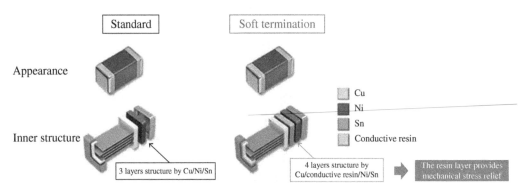

Figure 15.25   Soft terminals in ceramic capacitors. *Source:* TDK Co. (with permission).

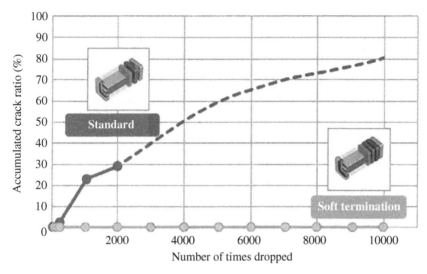

Figure 15.26 Comparison between the standard and modified designs of the ceramic capacitors. *Source:* TDK Co. (with permission).

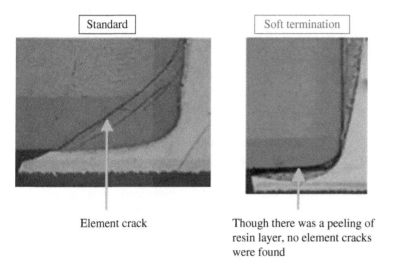

Standard	Soft termination
Element crack	Though there was a peeling of resin layer, no element cracks were found

Figure 15.27 Prevention of crack in the soft terminals of ceramic capacitors. *Source:* TDK Co. (with permission).

3) There are two main approaches in a fault tolerant system: using redundant systems and re-configuration. Reconfiguring can be used by normally operating converters to compensate the I/O variations of the converter (for example: shadowing in PV systems). The method of redundancy can consist of backup components that automatically "kick in" should one component fail. Redundancy can be in hot and standby forms. There is no delay in hot redundant systems. However, all of the redundant parts continue operation under stress in this type. Standby systems need a delay time for stable operation instead of faulty parts, but they are free from stress before failure occurs in the main parts.

References

1 Peng, F.Z. (2003). Z-source inverter. *IEEE Transactions on Industry Applications* 39 (2): 504–510.

2 Wang, H., Wang, H., and Blaabjerg, F. (2020). A cost-constrained active capacitor for a single-phase inverter. *IEEE Transactions on Power Electronics* 35 (7): 6746–6760.

3 Liu, P., Liao, C., and Kuo, M. (2018). A spur-reduction DC–DC converter with active ripple cancelation technique. *IEEE Journal of Emerging and Selected Topics in Power Electronics* 6 (4): 2206–2214.

4 Liu, H., Loh, P.C., and Blaabjerg, F. (2013). Review of fault diagnosis and fault-tolerant control for modular multilevel converter of HVDC. In: *IECON 2013 – 39th Annual Conference of the IEEE Industrial Electronics Society*, Vienna, Austria, 1242–1247.

5 Nguyen, T.H., Hosani, K.A., Moursi, M.S.E., and Blaabjerg, F. (2019). An overview of modular multilevel converters in HVDC transmission systems with STATCOM operation during pole-to-pole DC short circuits. *IEEE Transactions on Power Electronics* 34 (5): 4137–4160.

6 Vinnikov, D., Chub, A., Liivik, E. et al. (2018). Boost half-bridge DC–DC converter with reconfigurable rectifier for ultra-wide input voltage range applications. In: *2018 IEEE Applied Power Electronics Conference and Exposition (APEC)*, San Antonio, TX, USA, 1528–1532.

7 Shen, Y., Wang, H., Al-Durra, A. et al. (2019). A structure-reconfigurable series resonant DC–DC converter with wide-input and configurable-output voltages. *IEEE Transactions on Industry Applications* 55 (2): 1752–1764.

8 Khan, A.Z., Chan, Y.P., Yaqoob, M. et al. (2021). A multistructure multimode three-phase dual-active-bridge converter targeting wide-range high-efficiency performance. *IEEE Transactions on Power Electronics* 36 (3): 3078–3098.

9 Gao, F., Loh, P.C., Blaabjerg, F., and Vilathgamuwa, D.M. (2009). Performance evaluation of three-level Z-source inverters under semiconductor-failure conditions. *IEEE Transactions on Industry Applications* 45 (3): 971–981.

10 Son, G.T., Lee, H.J., Nam, T.S. et al. (2012). Design and control of a modular multilevel HVDC converter with redundant power modules for noninterruptible energy transfer. *IEEE Transactions on Power Delivery* 27 (3): 1611–1619.

11 Konstantinou, G., Pou, J., Ceballos, S., and Agelidis, V.G. (2013). Active redundant submodule configuration in modular multilevel converters. *IEEE Transactions on Power Delivery* 28 (4): 2333–2341.

12 Pei, X., Nie, S., Chen, Y., and Kang, Y. (2012). Open-circuit fault diagnosis and fault-tolerant strategies for full-bridge DC–DC converters. *IEEE Transactions on Power Electronics* 27 (5): 2550–2565.

13 Pan, Y., Yang, Y., He, J. et al. (2020). A dual-loop control to ensure fast and stable fault-tolerant operation of series resonant DAB converters. *IEEE Transactions on Power Electronics* 35 (10): 10994–11012.

Index

Resilient Power Electronic Systems, First Edition. Shahriyar Kaboli, Saeed Peyghami, and Frede Blaabjerg.
© 2022 John Wiley & Sons Ltd. Published 2022 by John Wiley & Sons Ltd.
Companion website: www.wiley.com/go/kaboli/resilientpower